W9-AUG-102

THE SECOND SENSE

Language, music & hearing

Robin Maconie

VISUAL & PERFORMING ARTS

THE SCARECROW PRESS

LANHAM, MARYLAND • LONDON

2002

SCARECROW PRESS, INC.

Published in the United States of America
by Scarecrow Press, Inc.
4720 Boston Way, Lanham, Maryland 20706
www.scarecrowpress.com

4 Pleydell Gardens, Folkestone
Kent CT20 2DN, England

British Library Cataloguing in Publication Information Available

Library of Congress Cataloging-in-Publication Data

Maconie, Robin.
  The second sense : language, music, hearing / Robin Maconie.
    p. cm.
  Includes bibliographical references and indexes.
  Discography: p.
  ISBN 0-8108-4242-4 (alk. paper)
    1. Music—Philosophy and aesthetics. 2. Music—Acoustics and physics.
  3. Music appreciation. I. Title.

  ML3800 .M2373 2002
    781.1'7—dc21   2002017572

FOR
John Mansfield Thomson

CONTENTS

	Introduction	vii
1	Sound and vision	1
2	Processing sound	17
3	Instruments	31
4	Signatures	51
5	Reflections	69
6	Directions	83
7	Space	101
8	Visible sound	119
9	In camera	141
10	Team players	155
11	Resonance	175
12	Leadership	189
13	Time and motion	203
14	Noise	221
15	Mechanical music	237
16	Mouth music	251
17	Relativity	265
18	Inspiration	283
19	Memory	303
20	Outer space	327
	Feedback	341
	Select bibliography	351
	Select discography	357
	Index	367
	About the author	373

Introduction

NITA is an interior designer. Her children had grown up and left home, so she thought she would give up estate management and train in something she really wanted to do. She came into my first class not ever having studied music before and not knowing quite what to expect. About halfway through the term something happened to make her stay back after class one evening. We had been talking about classical music that you listen to and film and background music that you don't necessarily listen to, and how each has its own purpose and is designed in a particular way.

She said, "It's just dawned on me that my line of work has to do with the things we have been talking about in class. In interior design we talk about line, pattern, form, rhythm, color, harmony—exactly the same as listening to music." I said, "That's wonderful; why not make a project of it?"

Toward the end of the course as the deadline for projects approached Nita stayed behind after class to bring me up to date. She had been working with clients on a project to redesign their livingroom. It had not been going well. The clients were a couple of her own age whose children no longer lived at home. They wanted to make their home more comfortable for entertaining, but were unable to agree on how the room should be redecorated. After a while Nita had a brainwave. "Do you like music?" she asked them. They did. He used to be in the military and liked trumpet music. She enjoyed classical music and was a regular concert-goer.

On her way home Nita stopped at a record store and bought a recording of a trumpet concerto, actually an arrangement of a sonata for flute, string orchestra, and harpsichord by the classical Italian composer Benedetto Marcello, who lived at the time of J.S. Bach.[1] At home she listened to the recording and made some notes. The people who liked this music, she reasoned, would like a design with a controlled sense of line, plain images, structural rather than applied contrast, a limited but harmonious range of tone, relatively neutral colors, regular rhythms, no fussy ornamentation, good finish, well-defined textures, and patterning in the weave rather than by color contrast in the print.[2] Using these notes as a guide, she looked through her pattern book again until she found a picture that corresponded to this image of design harmony. On her next visit she showed the image to her clients. They were very impressed, and Nita of course was delighted.

Let's replay that episode in slow motion. Nita is not a trained musician. Her clients are not musicians. They are people like you. The information she got from her clients was simple: one liked trumpet, the other liked classical. That is all. No history, no names, no preferences. When she got to the record store she did not know what music title or performer she was looking for. She went to the classical music section and chose a budget price disc from the trumpet music display. She had not heard this music before in her life. Nor had her clients. At home she played it on her hi-fi and asked herself, What kind of music is this? What distinctive features does it have? What would a person be liking for that person to appreciate this music? How is it organized? What preferences does it express? Is it loud or soft? Is it bold or discreet, colorful or monochrome, factual or emotional? Music tells you a lot when you ask the right questions. The information Nita got from the music guided her choice among a range of interior design pattern-book options. She then referred her choice, based on the music, back to her clients. What they recognized in Nita's suggestion was a style combining the functional discipline of military music with the restrained elegance of the classical style. And they got to hear the trumpet concerto of Marcello, too, which was an extra bonus.

This volume of ruminations about music is based on the same course in music appreciation that Nita attended. It is a bit different from my earlier titles *The Concept of Music* and *The Science of Music,* in two ways: first, because it is an actual learning program for real people who like listening to music but don't necessarily have any musical training or the time to acquire it; second, because this book talks about how the same rules of communication that apply in real life can be applied to any kind of music you choose. This is a course in *listening* skills that emphasizes acoustic and classical music simply because music designed for live performance without microphones or speakers is the best material for training people in listening skills.

Music has strong connections with language, feelings and memory. There has been a lot of publicity recently given to "The Mozart Effect": the idea that exposing babies to classical music encourages brain development and makes them better learners. Some of the benefit I am sure comes from parents rearranging their life to listen to classical music. When you perceive a value in acoustic music of any kind you also begin to see the value in hearing it clearly reproduced on good quality hi-fi equipment, and the value in setting aside time in your daily schedule just to listen to it. Classical music has value because it is a much richer listening experience, reflecting the range and depth of sounds in the real world. It's not just for babies. Through becoming more aware of the richness and variety in classical music, a listener is acquiring new powers of attention and observation that can be put to use in any learning task.

Critical listening is a skill whose time has come as a result of a revolution in the music industry. Media guru Marshall McLuhan observed "the medium is the message," and the message of the compact disc medium is a promise of greater listening satisfaction through exposure to a higher definition aural experience. For the first time in history the public has access to any and every kind of music, and the power to enjoy listening to it on an inexpensive audio system that is able to deliver the clarity and detail of an original master recording.

McLuhan also characterized technologies as extensions of the senses. It's not just that more classical music is available than ever before; it's because what you *want* to listen to is influenced by the degree of image resolution your audio system is able to provide. Compact discs in effect *improve your hearing.* With improved hearing you become more critical. The old rock and roll doesn't bring the same satisfaction in digital as it used to on vinyl or tape. You look for better listening satisfaction in a music that offers enhanced definition, structure, texture, and complexity. Classical music is the music of choice for a new generation of critical listeners.

It doesn't really matter what classical music you purchase, whether new release or vintage recording. I am a great fan of classical samplers, which are not only cheap and available but offer an informed range of unexpected items for a listener to explore: material that by and large is a lot healthier, of a higher quality, and available at much lower cost than maintaining a pulp fiction or video habit. With more and more music becoming available online, access is becoming even easier.

My class members come from vastly different backgrounds and cultures. They range in age from 18 to 48. English is not always their first language. They are architects, art historians, painters, graphic designers, computer animators, video artists and film-makers, fashion designers, and furniture makers. They come into the class thinking that classical music is an optional extra to their lives and discover that classical music is in fact directly relevant to their own special line of interest. For architects music is a way of learning about the acoustics of space. Art historians study the hidden messages of musical and acoustic imagery in great paintings. Painters compare techniques of representation of harmony, line, and color. Graphic designers study the evolution of music notations. Computer animators discover to their surprise that music is a form of data processing that uses a sophisticated software developed over five hundred years ago to model human movement. Video artists and film-makers investigate how music and sound effects address specific emotions and actions. Fashion designers explore the range of options available in music as a presentation feature for a show and as a means of managing and coordinating models on the catwalk. Furniture makers take a long hard look at the design and manufacture of musical instruments, and the structural and musical properties of wood, metal, varnish, skin, and gut.

For many students it is a tough assignment and their first sustained exposure to the other channel, the music of their parents or grandparents. But at the end of the course most students have a new understanding of what this music is about and can write with a degree of accuracy and insight under examination conditions about a movement of classical or folk music they are hearing *for the first time.* In my experience many young artists are wary of inquiring too closely into the meaning of art as it affects their own work. For them, the classical music experience is a useful exercise in critical self-evaluation that can be safely undertaken without any risk to their artistic instincts. On the other hand, many students bring a freshness of view to music that is often breathtaking to people like me who have been brought up in music from childhood and take more for granted than we care to admit. I am profoundly grateful to those individual artists whose imagination has been seized by classical music and who have found words of extraordinary poetry to express their insights. Some of their words are incorporated in this book as an inspiration to the musician in all of us.

Notes

1. Benedetto Marcello, Concerto for trumpet and strings Op. 2, No. 11. Miroslav Kejmar, Capella Istropolitana cond. Petr Skvor (Naxos 8.550243, n.d.).
2. To find out how Nita arrived at her conclusions, *see* 191-92.

Picture credits

Artworks reproduced courtesy of AT&T Archives (64); AKG London (123, 124, 126, 132, 134); Bridgeman Art Library International (136, 138).

CHAPTER ONE

Sound and vision

ALL movies are silent. True. All movies are silent because you can turn the sound down. Vision and sound are two different things. Vision comes out of the screen, sound out of a little speaker or pair of speakers underneath. The lips move, but the sound comes from somewhere else.

Movies take advantage of the fact that sound and vision are distinct realities and the fact that the brain is constantly working to check that the two senses are in agreement. But because of the way human beings are built, you can never be absolutely sure in the movies or even in real life that what you see and what you hear correspond to the same event or experience. Common sense says that the world is the same for all of our senses and that we don't in fact inhabit a series of parallel universes. But common sense is only a guess. We cannot know it for sure. Other cultures recognize invisible as well as visible realities and have evolved magic and ritual behaviors to deal with them. We have the movies, and we watch to see that the lips are moving in time with the sounds that issue from them in order to know if the actors are really speaking or only lip-synching. With speech you can tell. With sound and music it is never quite so clear. Not all of the sounds you hear in the movies are true to life. Whoever heard the sound of a punch on the jaw in real life? And why should it sound the same as hitting a ripe cantaloupe with a baseball bat?

Real life is much the same, but real life has nothing to prove. You still have to make the connection between seeing and hearing, but people trust life because it doesn't have an agenda and again because it's so messy. Unlike the movies, in real life you can decide what to look at and what to listen to. On top of that, reality is a lot more helpful. It provides a listener with a great deal more information, including unexpected information. And the sound quality is a lot better. But you still have to deal with the disconnect between the world you see and the world you hear. One world is visible and you need to see it in the light. The other world is invisible. You never see it.

Hearing is one of five senses that connect human beings and other living creatures to one another and to the real world. Each of the senses defines its own environment or reality. The most intimately circumscribed is taste, limited to the mouth and reach of the tongue. The boundaries of touch from foot to fingertip enclose a notional privacy zone or personal space. The sense of smell comes into play in defining a larger habitation or sphere of influence. The greater range of hearing and vision make these the two dominant senses in everyday social and intellectual life. Sounds in the environment can be detected from hundreds of yards, or in the case of aircraft, many miles away. The range of vision by contrast is for all practical purposes infinite: we can see distant stars.

An enigmatic drawing by Hieronymus Bosch depicts an owl in a hollow tree in a field of eyes, while in the background a small clump of trees gives shelter to two ears. "The field has eyes, the forest has ears, I want to see, be silent and hear" reads the legend on a contemporary woodcut of 1546 depicting the same image. To a present-day reader familiar with the language of advertising the message is simpler, less esoteric: "In the light, we use our eyes, and in the dark, our ears." The owl, of course, is the brain.[1]

While vision is generally considered the superior sense, hearing has distinct advantages. We can hear sounds in the dark, around corners, and behind our backs. Through hearing we are able to signal intention by means of speech and music, recognize a familiar voice, and evaluate the space and build quality of an unseen environment. Above all, hearing involves *time*. Vision, taste, smell, and touch deliver information virtually instant-aneously about conditions in the environment that are relatively stable. The mountain stays put. The perfume lingers. The hot sun continues to burn. Hearing is radically different. Sounds take time, and they do not last.

> In manipulation of form music can achieve results which are beyond the reach of painting. On the other hand, painting is ahead of music in several particulars. Music, for example, has at its disposal duration of time; while painting can present to the spectator the whole content of its message at one moment. Music, which is outwardly unfettered by nature, needs no definite form for its expression. Painting today is almost exclusively concerned with the reproduction of natural forms and phenomena.[2]

Kandinsky's admiration of music has to do not just with its representation of a reality beyond the visual, but as an art of time as well as space. It is precisely through learning to deal with the transitional nature of sound and hearing that human beings and other creatures acquire *and express* an understanding of the world and life itself as dynamic processes.

Among musicians, music is a medium of communication with rules and conventions that have to be learned, like a language. The world of expert knowledge is an exclusive domain of finely-tuned distinctions accessible only to initiates. The language analogy is attractive but limiting. It ignores the fundamental reality that the fine terminology of expertise is also a subset of speech, the vocal activity, and that voice signaling has its own parallel agenda of territorial definition and control (*see* 19ff.), a function of hearing behavior that connects regular people with other life forms, and academics with other academics.

For our purposes the definition of music is *any acoustic activity intended to influence the behavior of others.* It treats the roar of a lion, squeal of a dolphin, or chirping of a bird as the same kind of activity in principle as a concerto or symphony: each can be understood as an acoustic signaling process adapted to a specific organism, environment, and survival strategy. Music is more evolved and more complex, involves invented technologies and related skills, and has a lot more to say over a longer attention span. But the basic terms and conditions of audition are similar. In every case there is a need to get the attention of an audience and deliver a message.

Music is a subset of environmental sound. There are four main types of listening experience: noise, speech, music, and silence. The ear is a sensing device. It does not discriminate. All sound is evaluated initially in the same way, because you never know what a sound is until you have heard it. The world of sound is an unpredictable place, but music belongs to that category of activity designed to bring coherence and harmony to everyday experience.

For a majority of listeners music is a medium of escape or entertainment, subject to fashion and of no practical use. The widespread perception of music as a "black" art—literally, an art of the unseen—lifts it out of humdrum reality into a private world of the supernatural, of esoteric belief, mysticism, magic, and mind control. For many a college student

> music has the ability to calm, to excite, to make one sad or
> happy. Music is a lot like a drug: once you've had some
> music you like, you want more. It makes noise beautiful,
> and while silence is a need, music fills silence with more
> than just the loneliness or emptiness silence can bring.[3]

The powers attributed to music have as much to say about the real or implied incoherence of sensory information as about any attendant intellectual or developmental anxiety. To anyone struggling with the paradoxes of a fixed reality the invisible realm of music offers a vital, dynamic, and hedonistic alternative worldview—one reason why since the time of Plato the practice of music has repeatedly been linked to antisocial and taboo activities, in particular illicit sex and the consumption of mind-altering substances.

Whatever view we take of music, as private entertainment or public ritual, *the very fact that it is designed to be listened to* means that music is bound to conform to the structures of human hearing. Music works the way it does because it acknowledges a basic uncertainty in human experience that other excessively visual modalities of thought and expression have long been able conveniently to ignore.

Basic angst

The ongoing problem of reconciling conflicting realities of vision and sound not only affects the process by which the ear receives and the brain interprets information, but also underpins the popularity of the movie as a form of entertainment. Movies present an exciting alternative to humdrum reality. More to the point, success or failure at the box office is a matter of design, which involves the elimination of guesswork about audience response. Content is a matter of art, but success is a matter of knowing how audiences actually work, and while every movie is always a financial risk, designing the product is a standard procedure in perceptual orientation. A movie theater is in essence an isolation chamber in which the subject fixates on a screen and has little opportunity to move freely. An audience's grasp of movie reality is limited to only two of the five senses. Success in box-office terms is certainly a measure of the success of the perceptual *experiment* in conveying a coherent sensory *experience* (the two terms are related, after all) but, needless to say, it is an experience delivered under twilight zone conditions to a group of volunteers whose starkly impoverished sensory reality relegates the normally corroborative roles of smell, taste, and touch to an infantile compulsion to guzzle large quantities of warm popcorn and other comfort food.

Existential angst pervades not just the movies and the paperback industry but everyday life as well. Love, death, the insurance industry, Wall Street, and the American Dream alike thrive on uncertainty. Doubt not only is more real than any rational or moral imperative the collective human mind can concoct, it also provides a motivation for a great majority of palliative social behaviors. In their several ways art, fiction, religion, history, and stock car racing, as well as the movies, offer model strategies for coping with unknown and unexpected situations that arise not only from hurricanes, flying asteroids, and other circumstantial acts of which we are unwitting victims, but also from basic inequalities in the human sensorium itself.

Imputing angst to an ordinary videocassette sounds like a fun idea, but it is one that conceals a deeper truth. While it makes perfect sense to attach anxiety motives to scary movies and horror fiction—that is, to human *intention* (the script)—the idea that similar anxieties attach with equal force to technologies and delivery systems, including the humble video, may catch a reader off guard, even a reader concurring in principle with Marshall McLuhan that "the medium is the message." Focusing on design and technology rather than content brings significant advantages to the task of understanding music. It avoids unnecessary disputes over history or culture, and opens the meaning of classical and other music idioms directly to a non-specialist audience. If for example I make the point that present-day society is primarily visual and hearing is secondary on the basis of movie industry custom and practice, that is arguably a cultural determination of limited reference. But there are no such objections to observing the separate controls for sound and vision on your television. Their existence is not an issue of taste or tradition but of human design. The buttons on your remote are not cultural variables. They refer to a structural distinction in human sensory performance. That difference is crucial.

Any human interaction that involves design can be studied from a design perspective. Doing so avoids the danger of being sidetracked by issues of morality, judgment, or personal taste. For a movie, magic illusion, or song to succeed emotionally or artistically it has first to succeed technically. Through studying the latter it is in fact possible to arrive at a consensus about the former. Music and art as human activities do at least *conform* with the world in the fundamental sense that the materials, motivations, and skills involved are conditioned by what is possible in a putatively real world. For the audience, on the other hand, the music and art experience also has to *conform* to the world that each person inhabits.

Paradoxically this means that for an art experience to succeed as *illusion* it has to be grounded in a *reality* beyond the sampling error, a worldview more generous and more reliable than the single vision. In providing a temporary retreat from individual reality, the art experience is also affirming the existence of a plausible ulterior reality. If I like it, that is only my opinion, but if others like it as well, then the art work itself becomes *a term of agreement about the real world*. The consensus value of artistic activity may in the long run owe less to its capacity to entertain than to its usefulness as an antidote to personal doubt and social anarchy. Consensus, however, has no truth value. It has no cultural significance. It may emerge gradually as public opinion or erupt spontaneously as popular acclaim. Societies tend to legislate approval for the former and materially reward the latter.

The world of the movie soundtrack has three layers: speech,

sound effects, and music. Each has a particular function and caters to a particular realism. Dialogue engages the mind, tells the story, creates character, establishes motive, and incidentally provides the personalities on screen with something to say and thereby a reason for existing. Sound effects enhance the action. Music sets the mood. And there is always a subtext. Hearing is selective, involving the brain as well as the ear. The ear is a receptor: it misses little. The brain does the listening. It filters incoming sound for new information that requires attention, such as "a punch on the jaw":

> Deep in the Amazon rain forest, some 50 Yanomami scramble, scream, shout and threaten each other with poles, axes and machetes. The climactic scene of "The Ax Fight" ends with a sickening thud, as anthropologist Napoleon Chagnon narrates that a youngster has been "knocked unconscious" and "almost killed.". . .
> Digging further into the CD, you learn that the thud that seemed to mark the fatal ax blow was, in fact, added in postproduction at a sound lab. The film-maker created it by smacking a watermelon.[4]

Why *do* people listen? People listen because they have to, they can't help it. It's not why you listen but how you use the information. Listening is peculiar. You can't switch it off. You go to bed at night, switch off the light, close your eyes. But your ears stay awake. If somebody opens the refrigerator door in the middle of the night, you hear it. The alarm goes off, you wake up.

Having to resolve potentially conflicting evidence between what your eyes are telling you and what your ears are hearing is an ongoing task and reason enough, one might think, for so much of our lifetime having to be spent in sleep. Experience or logic might lead people to believe that we are dealing with different perspectives on the same environment, but the fact remains that often we find ourselves quite happily dealing simultaneously with completely unrelated sensory images, for example, watching baseball, listening to the crowd, and eating a hot dog. A coherent and unified world may make sense—but then what does "making sense" mean? That it is more reasonable to suppose that the world is the same for all of the senses than to imagine that we are dealing with a number of parallel universes at once? Wishful thinking. There is no proof. We can't know for sure.

Art and religion and science and music and poetry all play on the same basic uncertainty, and all of them share the same objective, which is to reduce or eliminate doubt. Existential angst is a personal rather than a social condition because society operates on the basis of collective assumptions—language, laws, morals, which side of the road you drive on, etc.—that entail acceptance of a higher reality even if they can't prove it exists. For a society to function agreement is necessary even though it may involve a compromise with personal truth, logic, or the available evidence. Religion develops the concept of different worlds, reality, heaven, and hell. Science plays with extrasensory dimensions. Art, music, and poetry create images of an elevated and coherent reality. Architecture designs human environments that are stable, secure, and sense-enhancing. If the separate realities of vision and hearing (not to mention touch, taste, and smell) were ever totally reconciled it would do away at a stroke with the inborn anxiety that drives people to be creative as well as to take aspirin or seek solace in esoteric beliefs. And that is not likely to happen. So there will always be a market for belief in other worlds, and there will always be a fear of the invisible, and the attraction of betting on the unknown. As also for works of art and fiction that invent connections, real or unreal.

Simultaneity

Music, along with life in the real world, begins with the voice. The first sign of life of a healthy newborn is a healthy cry. A healthy cry is a spontaneous expression of breath-taking, not a voluntary act. To those in attendance the crying act is an assertion of being alive: *Here I am*. But the newborn infant has no way of knowing of the existence of other people. It doesn't come into the world already equipped with communication skills. And yet everything flows from the crying reflex.

It's illogical to imagine a newborn infant behaving intentionally or having advance knowledge of a world of other people. It doesn't make sense. And yet somehow we have to explain how knowledge is acquired and where it comes from because within a relatively short time the infant is showing very clear signs of understanding and manipulating the world around it, including other people.

That one eventually grows up and acquires skills for negotiating with the world does not mean losing the structures of perception you are born with. Ignorance of music among philosophers can be related to the fact that they also don't understand how babies learn. Intellectuals deal in words and aspire to certainty, whereas lower forms of animate life such as small children and musicians have to deal with sensory processes. These processes are *structured into the body*. They are what you have and what you always use to extract information from the environment. Music works because we remember what it was like long ago when we were born and didn't know where we were. Deep down we remember because the basic structures and processes don't change. All that changes is how we interpret the information.

Primal scream

People scream out of excitement. It's a fundamental reflex. In the movies women scream in fear and men yell in anger, but it is not really about specific emotion but rather about the intensity of the adrenaline rush that people experience in a particular situation. Babies scream for no intellectual reason, and fans at a concert are neither frightened nor angry. There are *no words* in a scream. A scream is pure energy released in the sound of an individual voice at a constant and high pitch and at a constant and high amplitude. The length of a scream is the length of a breath. The primary satisfaction of a scream is in the release of energy.

If you scream it's always loud and steady, and always high-pitched—the highest your voice can reach. A scream is a particular kind of acoustic signal. It is physical, loud, musical, stable, and it lasts for a reasonable length of time. *Physical* means stress, and stress means body awareness; *loud* means acoustic energy, especially high frequency energy, going out into the environment; *musical* means tone rather than noise, and tone is a structured and efficient means of disposal of excess energy; *stable* means tone of constant pitch, which means a better signal for the ears to hear; and *duration* means your brain has time to figure out the environmental impact of the signal. For maximum effect when you scream *you stand still*, like the figure in Edvard Munch's painting *The Scream*.

I was thinking that it might be interesting to invent a situation where people are trying to communicate but all they can do is scream at one another. Then I realized the situation already exists. It's called opera. In grand opera everybody stands around *and they all scream.* They do. In a television interview Metropolitan Opera diva Renée Fleming described the technique

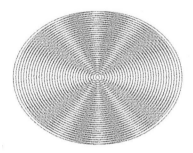

There is meaning in a scream

of opera singing as a *controlled scream.*[5] A scream is not about language, nor is it about action. People do not attend opera for the words: in these days of original language production you would need to understand Czech and Russian as well as French, German, and Italian, so an audience no longer asks nor expects to understand what is being said or rather sung. Nor is there much in the way of stage action, other than geared to targeting the person on stage at whom a character is screaming. Screaming involves very little body movement; in fact, any display of enormous amounts of energy, as expressed in opera in very pure and intense musical signals, will tend to inhibit other action. A top note is designed to bring emotion to a climax, the orchestra to a complete stop, and to last for what seems an eternity. Now *that* is what screaming is about. A scream is a moment of Zen.

Screaming and music work in the same way, only music is usually more intricately structured and calculated. Either way, you make the noise and you listen. There's a controlled version of screaming in everyday life when you say "Hey!" or "Hello!" to get a reaction from somebody else, or sound the horn in traffic. In the long run it doesn't matter who makes the noise; you can still listen to it and evaluate it in a positive way. However, if you are making the noise yourself the situation is a bit more complicated because the action of doing it is overlaid on, and interferes with, what you hear. In both cases you start with the raw data, which is the noise itself. Then you match the data to the structure of the ear in order to assess what information the noise can convey. If you put a noise into the environment you always get information back, whatever the noise.

Music is a special kind of outburst, structured and differentiated and easy to hear compared with the noises of everyday life.

The meaning of hum

Communication is making the connection between what you do and what you observe, learning what to expect. A newborn baby has the sense of sight but cannot see properly, first because the eyes are not in focus and second because all it knows of the visible world is the warm, comforting reddish glow of closed eyes. On the other hand a healthy baby can hear perfectly at birth and has already been listening and physically responding to sound for some time. Whether it learns to sing in the womb is unclear.

Maybe it hums; we don't know. (Well, maybe not—just try humming and pinching your nose at the same time. A hum expels air through the nose. Unborn children don't have that luxury.) People hum *to themselves*. Humming is a comforting sort of noise that resembles the kind of sound a baby would hear in the womb. Humming has very particular associations. It's the complete opposite of a scream. If you wanted to attract somebody's attention you wouldn't hum at them. That would be silly. When you hum your lips are closed, so the sound can't get out. Humming is an internal sensation. You can *feel* it.

If you say *Mm-mm-mm!* it means "That looks tasty!" If you say *Yumm!* it means "That *was* tasty!" Say *Hmm-mm...* and you are saying "I'll think about it." Say *Umm-mm...* and you are saying "I'm still thinking." If you are Buddhist, you might meditate on the sacred syllable *Omm*, which is a sound by which an initiate strives to attain harmony with the divine. Sacred or profane, every hum refers to an internal process. *That* is the meaning of hum.

Whatever sound world is already imprinted on the newborn is bound to be limited in very particular ways. The musical

world of the unborn includes the constant rhythmic sound of the
mother's heart and breathing and the occasional rhythms of
music that the mother may sing or listen to from time to time
during the hours of waking. The sound world of the not yet born
will also include noises of irregular occurrence, such as digestive
processes, the mother's speaking voice, or the television; the

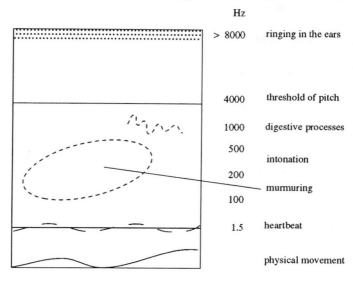

The sound world of the unborn child

baby may also be disturbed by occasional insistent mechanical
sounds of loud processes in the environment, for example the
sound of a car engine, coffee grinder, or vacuum cleaner.

    For the unborn listener, however, all of these sounds are
critically limited in frequency range. Inside the womb only low
and medium frequencies are audible. High frequencies cannot
penetrate from outside, and so the infant does not learn to deal
with them even though it would be perfectly able to hear them.
There is all the same spontaneous activity in the auditory nerves
that produces a sensation of ringing in the ears as a byproduct;
this ringing is the only high-frequency signal of which an unborn
baby would conceivably be aware. The presence of a constant
signal in the high frequency band might sensitize the developing infant to that region of sensation, but as the involuntary

byproduct of an internal process the ringing is unrelated to any external acoustic event. It is interesting that in situations of extreme danger people become aware of this internal ringing in the ears as a symptom of excitement or anticipation. In radio terms one would describe it as more like a carrier frequency than a genuine signal.

If the stress of being born is enough to sensitize the infant to this high frequency region, then the energy expended in that first cry on being born unleashes a huge storm of high frequency information precisely targeted on this completely new region of sensation. By an extraordinary coincidence it is the exact same high frequency zone of hearing that is uniquely adapted for learning about space, the final frontier. And the vital difference between ringing in the ears and high frequency sound in the real world is that in the real world high frequency sounds are *subject to change*. High frequencies are very directional; they cast shadows, and they reveal objects and movements. First of all there is the shock of an expanded sound world, a world of harsh sibilants and brutal clatter. High frequency sounds signal a new world of space and separation and direction and distance, of up and down, here and there, back and front. High frequency consonants break up the comforting flow of a mother's voice into random segments accompanied by unwelcome pulses of air on the face.

High frequency cries are the radar with which the struggling infant scans the world beyond its immediate reach. Crying is the audible byproduct of a spontaneous response to a catastrophic change of environment. The noise is not a deliberate act. Taking that first breath is a basic reflex action. Air is sucked into newly-exercised lungs and then forcibly exhaled, in the process agitating a valve structure in the throat that vibrates rapidly back and forth to create a siren-like airflow of 500 or more pulses a second.

A baby cannot *intend* to make a sound, and for all we know it isn't even aware of making a sound. After all, it's the first time. But because a baby can hear perfectly well, it can hear that sound. *The whole world is making a sound*. There is one important difference. A newborn's first cry is *louder* than anything it has ever heard before and probably louder than anything it will ever hear again. Really loud. And loud in a zone of hearing, the high frequency zone, that has never ever been exercised before.

You hardly ever hear the sound of your own voice as the rest of the world hears it. When you do, it's not pleasant. Why is everybody put off by the sound of his or her own voice? There's no rational basis. It's a powerful reflex that signifies that something about the sound is recognized as a danger signal. In real life the sound of your own voice is always accompanied by the physical sensation of speaking. Hearing your recorded voice signifies a major disconnection in time and space: a dislocation in time because you are not aware of speaking those words at the time you are hearing them, and a dislocation in space because you can hear that the sound is coming from *some other place* than where you happen to be, because all of your voice in a recording is perceived externally, whereas in normal speech the voice is mostly located inside your head. A person in the act of speaking hears his or her voice in large measure from *inside the head*, not as an effect of the environment.

The question then is, how do you recognize that it's *your* voice if you never hear it at a distance? In fact there is part of the voice sound that you do hear from a distance because you can't hear it inside your head. High frequency sounds do not transmit very well through tissue and bone. When the dentist is drilling out a tooth cavity with a high-speed drill, the sound that transmits through your jaw is a low-frequency whine, while the sound in the room is a high-pitched whistle. Both sounds stem from the same physical process, and yet you hear two distinct sounds, a low-frequency sound associated with the actual vibration of the drill on the tooth, and a high-frequency sound associated with the environment. The high-frequency sound doesn't seem connected in any way to the drilling sensation. It belongs to the zone of acoustic stimulation on which everybody relies for information about the outside world, because it is the *only* source of reliable acoustic evidence that is manifestly unconnected with internal body processes.

Above the limit

Acoustics science treats high-frequency sound as a simple extension of middle and low frequencies. But that's not the way we hear it. To the newborn high-frequency sound is both a new sensation, and also information that has to be processed in quite a different way. Lack of exposure to high frequency sound in the womb might seem to account for the inability of the inner ear mechanism to track frequencies higher than 4,000 hertz (cycles per second) with the same accuracy as lower range frequencies. But despite a lack of practice there are also physical limitations to how fast the inner ear structures can move and the auditory nerves switch on and off. These limitations are felt starting at the 4K frequency threshold, with the outcome that for higher frequencies the sense of distinct pitch is lost. Images in this uppermost range of hearing become shadowy and statistical in nature. This is the region of percussion, of spoken consonants, of perceptions of tonal color, brightness, and texture. *And space.*

A crying newborn is getting two kinds of acoustic information from a single source. It makes the effort; the physical act distorts the physical body and sends a vibration through the body corresponding to the low- to mid-frequency component of the vocal signal. When a person speaks, or even hums, low to mid-frequency sound is detected within the body, and because it reaches the inner ear *first* and is a *stronger* signal it takes precedence over the same low- to mid-frequencies in the airborne signal. For high frequency information there is not the same match of inner and outer sound images. High frequencies do not travel inside the head but instead escape into the atmosphere where their residue is monitored externally as evidence of the larger environment "out there." And there is invariably *a delay* between high-frequency signal emission and response. To the individual simultaneously generating and monitoring an acoustic signal the high-frequency information is perceived as *perpetually out of synch* with the physical act of speaking or singing. In addition it comes back to the ears with added resonance: reflected images from walls and ceiling, some of which have ricocheted back and forth from surface to surface before heading back in the direction of the singer or speaker. The word for the accumulation of delayed reflection is *reverberation.* It is an acoustic effect determined by exterior structures and providing the listener with coded information about the size of a space and other shape and material clues. This information is *different* for left and right ears.

Honey, I'm home

Body tissue is not designed to conduct high frequencies. To be heard properly the higher components of any vocalization have to be expelled into the environment, there to encounter hard surfaces from which an accumulation of much weakened wave-fronts is reflected as an airborne pressure surge reaching the eardrum after a delay measurable in milliseconds. It follows that when an infant cries it is primed to become aware of, or at least attuned to, the low and midrange part of the signal as an effect coinciding with the physical act, whereas the high frequency disturbance is *always delayed*, and not simply in respect to the vocal action but *differently for each ear*. Whereas the low and midrange frequencies of the speaking voice are monitored inside the head in mono, the high frequency band is perceived uniquely in stereo.

Late medieval philosophers argued whether the world was real or whether it was just a figment of the imagination. From the evidence available to a newborn child it is possible to conclude that the world is real because modulated high frequency *binaural* sound cannot be internal in origin.

Notes

1. Hieronymus Bosch, "The Hearing Forest and the Seeing Field." Charles van Beuningen, *The Complete Drawings of Hieronymus Bosch* (London: Academy Editions, 1973), 26-7.

2. Wassily Kandinsky, *Concerning the Spiritual in Art*, tr. M.T.H. Sadler (New York: Dover Publications, 1977), 20.

3. Jamie Bowerman, student, in answer to the examination question "If silence is a human need and noise is a necessary evil, where and how does music fit in?"

4. Sharon Begley, "Into the Heart of Darkness," *Newsweek*, 27 November 2000, 70.

5. Interview with Renée Fleming on *Sixty Minutes* (CBS, December 1999).

CHAPTER TWO

Processing sound

OVER millions of years of evolution human beings and other forms of animal life have learned to use the body's sensitivity to atmospheric pressure change to develop ways of communication in sound through speech and music and other forms of acoustic signaling. Using sound pressure waves to relay signals is a way for individuals to communicate with others who are physically out of reach. Along with birdsong and the alarm calls of different animal species music and speech are acoustic behaviors *designed* to influence the actions of members of the same or other species from a position of safety. In a natural environment where direct physical contact is often a life and death affair the evolutionary advantages of being able to manage or inhibit the actions of others by remote control are obvious, especially as sound signals function equally well if not better in situations of poor visibility such as fog, night-time, or a dark forest or cave where unseen dangers may lurk.

Survival issues are not the first to come to mind where music is concerned, but they are there all the same. The cathedral or concert hall is a protected area. There are behavioral constraints on an audience. To a younger generation the protocols associated with classical music functions are nakedly obvious and pose a very real threat to its independence of action. Youth culture responds to music in a decidedly more visceral fashion. Adults in turn respond with alarm to the intensity and energy of youth

17

music, which they rightly see as intimidating, and make futile efforts to restrict it on health grounds. As I said, survival.

The adult paradigm, if we can call it that, is of classical music as a communication process. Received wisdom, carefully scripted, faithfully delivered, gratefully received: a rite of passage for the young, an act of conservation for the old. The trouble with this model of music-making is that it doesn't explain a thing. It reduces the process, in effect, to a smoking gun and a bleeding heart: targeted action in a straight line from here to there. This will not do.

It won't do because it involves too many assumptions for which there is not enough evidence. What information is being communicated? How much does the performer know? Is the information mental or physical? If mental, how does it translate into the physical and acoustical domain at the point of transmission, and from physical sensation back to mental process at the point of reception? What role does the instrument play in the process, or the human ear? And aren't we forgetting that the sound we hear is completely dependent on the air we breathe?

Enough questions. Whatever meanings may be attached to classical music or any other acoustic behavior, none have any chance of taking effect in the absence of an atmosphere. Now the bigger step. The meaning of a musical performance is subject to natural processes in materials, in the air, and in the auditory mechanism. Only after these natural processes are acknowledged is it possible to infer any message the composer may have intended and whether or not it has been faithfully delivered. An even more radical conclusion would be that the *only* information of any consequence music has to offer comes down to the quality of the acoustic signal, the sound of the instrument in a particular location, and how much of it a person can hear. We do not intend to suggest that a knowledge of the history of music and performance cannot add to the enjoyment of a finely executed concert. *Well, yes, actually we do.* What matters is the experience. The experience is acoustical. Anybody with functioning ears is in a position to get as much out of a performance of classical music as the sound itself has to say, in addition to which the innocent non-musician is in the distinctly advantageous position of having a mind uncontaminated with aesthetic theories or expectations to bring to the auditing process. Let us see if this all works out.

It's a dog

You go out. You hear a dog bark, just once. You cannot *see* any dog. How much do you *know* as a result of hearing a dog bark once? At last count there were over ten items of factual information derivable from that momentary disturbance. They include some—no, wait, *all*—of the big issues of what it is to be a human being.

One, it's a dog. You know it's a dog because you have heard the sound before. Sounds come and go. We survive by storing sounds in memory. If you know it's a dog, you have a long-term memory of the sound, so you have a past existence as well.

Two, you hear it. You have functioning hearing. You can hear it and continue to evaluate it even after the sound has died away. Most incidental sounds in nature are of short duration. We only recognize what they are after they have already happened. For that you need a short-term memory.

Three, you exist; you are awake. A dog barking is a wake-up call. Maybe you were day-dreaming. Now you are aware of a real world beyond the realm of fantasy. How do you arrive at the idea of a real world? Because the dog barking was not in the script. You did not intend it. It caught you by surprise. Hence

Four, the world exists. We seem to be going backwards here. The distinction is between an ideal world that is a dream or mental projection, and a real world that has its own agenda and is therefore less predictable and potentially less safe.

Five, you can tell how far away the dog is, from the level of energy of the bark. This is a subtle assessment involving the relative amplitude of the signal (the farther away, the weaker the sound pressure level), and also the balance of high and low frequencies (the farther away, the less bass and top).

Six, you know where the dog is, and what direction the sound is coming from. This knowledge comes from having two ears receiving two slightly different signals and the brain making a calculation from comparing them. These differences are in the time of detection of the sound, since unless the dog is directly in front of you, its bark will reach one ear before it reaches the other; and also a phase difference, whether the two sets of wave fronts are in or out of synch. The binaural signals from a moving dog will vary subtly in both amplitude and phase.

Seven, if the dog is large or small. A smaller body produces a higher-pitched sound. That's axiomatic.

Eight, you can tell what sort of mood the dog is in. That's a musical issue, because it refers to pitch, tone, melody, and rhythm. A howl of pain is different from a joyous bark or a menacing growl. Irregular barking is a sign of anxiety, and a regular rhythm, of high spirits.

Nine, the dog is responding to something. A bark, like a lion's roar or a baby's cry, is a territorial signal. "If you can hear this, you are in my manor," to borrow a gangland term. Watch it. Even a benign bark carries a territorial message. The dog is not necessarily talking to anyone in particular. Barking is first and foremost a gesture of self-assertion and only incidentally a warning to others who may be listening. A barking dog may mean a foreign presence in the neighborhood. After all, that's one of the reasons people have dogs.

Ten, the environment. You are not only hearing the direct sound, but also multiple reflections from other structures in the vicinity. If you are in a built-up area you can hear it.

Shall we go on?

Eleven, where the dog is, in relation to the built environment. You can tell if it is in the front garden, around the back, on the street, or inside the apartment behind a window.

Twelve, the breed of dog. Not everyone will get that one, but on the evidence of size and temperament, most people should be able to distinguish a lap-dog from something closer to a pit bull terrier.

Thirteen, the weather. If the air is still, or if there is a breeze, either will affect the clarity of the sound you hear. A humid atmosphere is denser than normal, and droplets of moisture suspended in the air are light enough to blanket the higher frequencies and muffle the overall sound quality. On the other hand, if it is a sunny day with snow on the ground the sound of a dog barking may seem a lot more reverberant than normal, since a blanket of snow transforms absorbent features like hedges and lawn into sculpted, sound-reflective surfaces.

Fourteen, the dog is domesticated. A wild dog only howls, we are told: a barking dog is actually imitating human speech. Oh, the irony that in teaching a dog to sing the proud owner is actually awakening the animal's primitive survival instincts. Or is it ironic after all? Think, think opera.

Fifteen and last, what to do next. The dog is large, it is close by, it has a vicious temper. You go the other way.

Traffic

Understanding the environment and being able to make life decisions on the basis of the sounds we hear is not a trivial skill. And you don't have a choice, because you don't know what kind of sound it is until you have heard it. Every sound a person hears is potentially life-threatening. It's not very likely because people create communities and structures that offer relative security, but a sense of the possibility of danger is ever-present, as we know from the popularity of life-threatening subjects, both real and imaginary, in the media. Classical music is simply a sub-category of normal listening. In that respect the sounds it makes come into the environmental assessment process as a matter of course. The more intriguing question is whether that process can have anything to contribute to a valid musical response—which for most people is emotional rather than rational in any case.

One of the more dangerous environments of modern-day life is the road. Drivers on the road and pedestrians crossing the road respond to a simple vocabulary of alarm signals ranging from a bicycle bell to the blare of an interstate truck's horn. The tone and complexity of signal a vehicle makes identifies the vehicle, signals its presence, and provides a coded indication of what action should be taken. These alarm signals operate in exactly the same way as animal cries in the wild; indeed they have to be simple enough for a stray elk, let alone a fourth-grader, to understand. The dialects of car horns and ambulance sirens may be mechanical, but the sounds themselves are musical in origin and function, and the gestural language is universal. A crowded road is a dynamic environment where reflexes are keyed up and drivers on permanent alert.

The bicycle bell generates a pleasant tremolo designed to be heard by another cyclist or a pedestrian, but not by the driver of a larger vehicle to whom it does not represent a threat of physical injury. Its sound has brightness, texture, and rhythm. Motor vehicles such as cars, trucks, vans, and trains are equipped with louder horns to produce signals that can be heard at greater distances and by other drivers with the windows

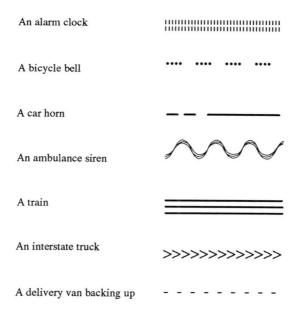

An alarm clock

A bicycle bell

A car horn

An ambulance siren

A train

An interstate truck

A delivery van backing up

A selection of alarm signals

raised and the radio switched on. Emergency vehicles are equipped with a combination of flashing lights and sirens of oscillating pitch that make their presence even more conspicuous, added to which most have regular horns as well.

Motor vehicles operate in a crowded environment and their alarm signals tend to cluster in a relatively narrow bandwidth optimized for the attention of other drivers. That is why larger road vehicles do not have horns set at a pitch corresponding to size as ships or dinosaurs do, for example. Low-pitched sounds carry well over water, and the open sea is a relatively noise-free environment compared to the high street. Larger land vehicles, however, tend to signal their size and authority by producing more powerful signals. Lower-pitched horns on land also make the vehicle less easy to locate, so the higher pitch level is a safety feature.

We normally make a distinction between a driver-operated signal and a continuous alarm such as a traffic siren. A regular car or van horn is designed to produce a short warning beep to alert a driver or pedestrian ahead of possible danger. The horn itself is a mechanical version of a regular bugle or trumpet. It produces a monotone signal at a constant loudness level, but despite these limitations it can still convey an emotional message if the regular beep is converted into a steady unbroken blare (in imitation of a truck) or sounded repeatedly in a rhythm, which is a way of saying that there is more to the situation than merely wanting to draw attention to a minor traffic problem.

Larger vehicles such as trucks and trains operate clusters of horns producing a close harmony that softens the aggressiveness of a more powerful alarm without lessening its impact. These horn signals tend to be longer in duration, which is a classic indication of magnitude: the longer the breath, the bigger the beast. Bigness also means dinosaur reflexes, so a long blast is a friendly way of saying "Get out of my way, because I cannot stop." A delivery van or school bus backing up sounds a gentler "beep-beep-beep," pitched at pedestrians within a relatively small backup area, and of course the speed of the vehicle is very much slower driving in reverse than full steam ahead.

Emergency vehicles produce signals that oscillate or warble back and forth between pitches. The warble is a primitive and dissonant melody set at a relatively high pitch to indicate a combination of stress, determination, and urgency. It is switched on for the duration and the sound is designed to be easily located and tracked by drivers as the vehicle approaches or recedes. When an emergency vehicle passes by, the Doppler shift on a warble tone is very easy to hear. A continuous warble at high pitch combines the intensity of a primal scream with the presence of mind of a modulated voice.

These vehicle signals are musical in nature and function. In tone, pitch, and harmony they embody the original signaling function of the horn instruments of the orchestra on which they are modeled. In addition, their musical language of short, long, intermittent, pulsed, modulated, or harmonious gestures is both apt to convey character and conventional in message.

Water Music

Handel's *Water Music* suite is a collection of pieces assembled from various sources to provide musical entertainment for King George I and his retinue in procession up the river Thames from

Lambeth Palace to Chelsea and back on a mid-July evening in 1717. This is about as close to a traffic situation as classical music gets. (A view of the sumptuously ornate Royal Barge is preserved in a Canaletto painting of London and the Thames dating from the 1730s.) Handel's orchestra of harpsichord, strings, trumpets, and horns occupied a second barge following at a respectful distance. "Many other barges with persons of quality attended, and so great a number of boats, that the whole river in a manner was covered," reported the London *Daily Courant*. This cheerful music is designed for playing out of doors, and presents the king to his subjects as a powerful and kindly ruler. Our example is the Andante in D of just under three minutes' duration.[1]

What does the music say? First of all, it signifies a special occasion. In 1717 there is no music radio, no television, no personal CD player. Music is a rare event for most citizens and hiring an orchestra is expensive. Music in public is rarer still. This music is powerful enough to be heard at a distance. It has to be, because the performance is taking place on the river with only the water surface as an acoustic reflector.

It gives out a typically strong signal designed to be heard over the background noise of everyday life. To the king's audience the orchestra is the royal public address system and the music itself is the voice of the king, delivering a structured, disciplined, and positive message saying right away that:

1. King George is keen to communicate with his subjects;
2. he has a powerful voice;
3. he represents order and harmony;
4. he is in charge.

Compared to the sound of a dog barking, the more complicated patterns of Handel's music signal that the royal personality is both civilized and has something to say. The king controls the air-waves. He asserts his royal right to he heard, and his sphere of influence extends as far as the music can reach on either side (farther still, if the echo effect between trumpets and horns is to be believed). The power and range of the royal voice proclaim the king to be a larger-than-life being of superhuman powers.

The royal authority of Handel's music has its present-day counterpart in the familiar Eddie Murphy character driving a beat-up Cadillac pulsating with hip-hop music at 120 dB that draws alongside at a red light. (I have a fantasy that one day the car will shake itself into a pile of scrap, leaving the driver at the wheel with nowhere to go.) Both are using the power of music to assert their presence (and their taste in music) on the public from a position of safety in a moving vehicle. But the king's music is not hip-hop with a powerful beat and jazzy lyric, but rather an image of a team effort, a society working in harmony. Harmony is about teamwork and good relations between a leader and his people.

There are functional resemblances here to the sound of an emergency vehicle. Trumpets and horns are designed for outdoor signaling and deliver their uniformly loud and bright messages within the same narrow waveband as an ambulance or approaching train. However, from making such a comparison we realize straight away that despite being loud and penetrating this music is not intended to alarm or intimidate an audience or make people get out of the way. Instead it delivers an upbeat and welcoming message that says, "Come and see me, I have good news for you." The musical message is continuous for maximum attention like the horn of a truck or train, and therefore hard to avoid, but unlike a blaring horn or the incessant warble of an ambulance or fire engine this musical message is interestingly varied, a music of action that says, "All systems go."

The shape of Handel's music is clear and bold. Its melodies are not soft curves but straight lines and exact repeats. The echo dialogue between high, penetrating trumpets and deeper, more mellow horns tells the audience that this is a listening king, and the exact repeats signify an agreement that is both natural (the echo) and civil (between the ruler and his subjects). The royal trumpets reach out, and the answering horns give assent and add body and weight.

The king's vital signs are good: the music is upbeat, vigorous, displaying youthful energy and remarkable stamina. This is an intelligent king who shows interest in his people, is articulate and outgoing, and not dull, boring, or monotone. The quick, measured pace of the music expresses the virtues of confidence, determination, and self-control, while the festive character of the music depicts a king in high spirits and good humor. If you were to work out to this music, you would be in great shape.

Finally, the combination of trumpets and horns expresses leadership, since both are instruments of command. Trumpets lead the charge on the field of battle, and horns give direction in the hunt, a corresponding peacetime recreation. Such a king can be relied on to lead his people to victory in war and prosperity in times of peace.

Big Bang

A sensation of sound arises from a spontaneous release of excess energy into the atmosphere. We will confine our attention to atmospheric vibration since that is where most human hearing activity takes place. The benchmark example of sound in nature is a clap of thunder, detectable over the entire range of audible frequencies. When architects want to test the acoustic of a newly constructed concert hall they fire a starting pistol, which is the next best thing. A sharp clap of the hands in an empty room can tell you a lot about how a building is made.

What ears or microphones detect is a pattern of vibrations or rapid changes of pressure in the atmosphere arising from inequalities in the distribution of energy within an air space. An impression of sound is caused when minute fluctuations of positive and negative pressure are registered at the surface of a person's eardrum or the surface of a microphone ribbon or diaphragm, making it vibrate.

Some materials in the environment simply cave in to the pressure of a traveling sound wave. If that happens the excess energy is absorbed. Soft materials, like cloth or external body tissue that soak up excess pressure, are sound-absorbent. For audible vibration to occur, as in a musical instrument, the material has to show *resistance* to a sudden change in pressure. Where there is resistance energy is *stored* and later *released*, and the process of storing and releasing energy creates a pattern of vibration that can register in the brain as environmental information.

We can think of a source of sound in terms of a floating gas-filled balloon bursting in a room. Energy is stored in a balloon as compressed air or gas. It takes energy to make the balloon inflate, and that energy is contained in the higher pressure that is trapped inside the balloon compared to that of the surrounding atmosphere. The gas inside has to be at a higher pressure, otherwise the balloon would not inflate. It can also be lighter than air, to allow the balloon to float upwards, but in order to float the resulting added buoyancy has to be enough to overcome the weight of the balloon and its gas and moisture content.

An inflated balloon does not make any sound because the energy that has been converted into excess pressure is safely contained inside the membrane. The higher pressure inside the balloon is isolated from the remaining air in the room by an impervious barrier.

A burst balloon

When a balloon bursts the membrane drops away, exposing a zone of higher pressure in a lower-pressure atmosphere. This is an unstable situation. Most people understand the phrase "Nature abhors a vacuum," but the converse is also true, which we may formulate in the phrase "Nature isn't too partial to localized excess pressure either." Instead of being at a constant pressure throughout, the ambient atmosphere now consists of a zone of higher pressure where the balloon used to be, in the middle of a larger volume of air at normal atmospheric pressure. The only way for the excess energy to be redistributed is by expansion into the adjacent lower pressure zone, which occurs equally in every direction until the initially localized excess energy is redistributed throughout the air space. The bang we hear is this process of energy and pressure redistribution taking place.

Aftershock

A burst balloon releases a fixed amount of energy, the excess it contained in the first place. The farther a given energy front expands, the more that quantum of available energy is attenuated and consequently the less there is to hear at any one point. It follows that the farther a listener is from a source of sound, the less of the original energy front impacts directly on the ear. That is also a reason why sounds die away rapidly in an open-air environment. In an enclosed space, however, the energy front is contained. When a pressure wave encounters an obstacle such as a wall the more powerful low frequency component may interact with the structural mass while the remaining middle and high frequency energies change direction. High frequencies that make a room sound crisp and sharp are absorbed by soft furnishings, curtains, carpets, and people, but are reflected by metal, glass, wood, and tile. The combination of partial absorption and partial reflection retains a sound's energy in circulation for a period of time, which is why a person listening inside a closed room is able to intercept more of the original energy and hear a louder signal from a given source than a person listening out of doors where fewer reflective surfaces are available to reinforce the direct signal.

Since the average room in a house or apartment is not very large in relation to the 2-10 feet waveband of speech or melody there is a lot of energy in reflected sound and it comes back pretty quickly. People like to sing in the bathroom because it is the smallest and most acoustically reflective room in the house. Early reflections interact with the voice to strengthen its attractive features and conceal its weak features. The qualities we admire in a musical voice are strength and steadiness of tone. When an ordinary singing voice is combined with one or more strong early reflections the consistent features interact positively and are reinforced in relation to random fluctuations and tonal inconsistencies. The same effect happens in a choir of multiple voices: what you hear is a consensus, a consistent average sound.

Sound travels in different materials at different speeds. It depends on the density of the material, which expresses how closely packed the molecules are and how they are arranged. Sound in air at sea level travels at around 1,100 feet per second, but along the grain of wood at 13,000-14,000 feet per second —slower through the spruce of a violin and more rapidly through the mahogany of expensive floors. Iron and steel are even denser than wood, and through these and other metals a pressure front travels even faster, at over 18,000 feet per second.

In old cowboy movies you sometimes come across a scene where a grizzled old bandit is able to tell the robber gang when the mail train is approaching by putting his ear to the railtrack. The vibrations of an approaching train can be detected in the iron railroad a lot earlier than the sound of a whistle traveling

through the air. In rigid materials such as iron sound tends to
flow with the material and not leak into the surrounding atmosphere.

Monitoring structural vibrations dates at least from the early
nineteenth century. In old London streets until fairly recent times
leaks in mains water pipes were being detected by means of a

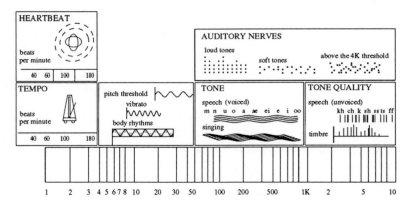

The frequency continuum

simple tube and earpiece applied to the pavement picking up
vibrations transmitted to the surface. The doctor's stethoscope is
another contact listening device, in this case for monitoring the
low-frequency sound of a beating heart at the surface of the
body.

Saying that sound travels at different speeds in different
materials does not mean that the sound of a violin and the sound
of a gong travel at different speeds in a musical context. What a
listener in the audience hears at a concert performance is airborne sound produced at the interface between the body of the
instrument and the surrounding atmosphere. Most musical
instruments are designed to transmit a range of pitches from low
to high, but in the same way that all the colors in the spectrum
are of different frequencies yet all travel at the speed of light, so
all the pitches or frequencies of sound travel at the speed of
sound, and not at different speeds. High notes do not get there
first; if they did, speech and music would be impossible. The
speed of sound varies with the medium conducting the sound,
and not with the relative frequency or density of sound waves.

In order to change the speed of sound in a medium such as air one has to change the density of the air. At high altitude sound travels at a slower rate. Sound pressure waves travel faster in the lighter helium-oxygen mix of gases deep-sea divers breathe as a protection against decompression sickness, raising the baseline pitch to give an impression when the diver speaks of the high-pitched tones of a cartoon character or baby doll. This effect cannot be replicated by sampling a voice into your computer and varying the tone, because a voice sample that is compressed or expanded in time from a recording or synthesizer changes the speed of talking along with the pitch of the voice.

Paradigm shift

According to the orthodox representation music and speech are information transactions in which a message is transferred from a person or a body to an audience. That view, if not mistaken, is no longer adequate. It does not account for the information that is communicated except by the disingenuous argument that what it means is what an expert or authority figure other than the performer normally decides. Furthermore, the "smoking-gun, bleeding-heart" theory of musical communication ignores the acoustic process. Restoring the experience of hearing to center stage makes the meaning of music accessible to every listener and not just a few experts.

The conventional view of music and speech as sending and receiving messages separates the message from the process of sending it, and identifies the sender as the dominant active partner and the receiver as passive. The new paradigm identifies music and speech as real-time signaling processes to be understood in terms of basic survival issues. For the performer, music is a territorial claim; for the listener, an index of environmental safety. Both experiences partake in the most fundamental human apprehensions: danger, security, integrity, self-control. The real interaction is not so much between performer and listener as between the individual and the acoustic environment.

Notes

1. Georg Frideric Handel, *The Water Music* (original 1717 version). English Chamber Orchestra, cond. Johannes Somary (Vanguard Classics SVC-47, 1996).

CHAPTER THREE

Instruments

SOUND is often represented as an external force and the ear as an organ offering only passive resistance. In reality the human ear is a robust and extraordinarily sensitive device capable of registering the slightest changes affecting the *balance* of pressure on either side of the eardrum. In addition to filling the lungs, mouth, and nasal cavities, air conducted by the eustachian tubes presses on the inner surface of the eardrum with a force equal to normal atmospheric pressure. At ground level the effect is barely noticeable, but at 30,000 feet travelers use the swallowing reflex to relieve a discomforting excess pressure in the inner ear when the external cabin pressure drops below normal. The experience is a reminder that opposing pressures as well as natural stiffness maintain the eardrum in tension, and that it is the opposing air pressure in the inner ear that causes the eardrum to flex when the external air pressure changes abruptly.

The eardrum is maintained in a normal rest position by the air pressure on outer and inner surfaces being equal:

+ | +

An incoming sound is detected as an increase in air pressure on the exposed surface, causing the eardrum to bulge inward:

- (+

This is followed by a drop in outside pressure relative to the eustachian tube, making the eardrum flex back outward:

31

+) -

The back and forth motion continues as long as a difference in relative pressure can be detected. To be perceived as sound these back and forth vibrations can be as low as 20 per second and as rapid as 16-20,000 per second.

For sound to be transmitted two conditions have to be met. First, the medium has to be elastic since if there is no elasticity there is no resistance. We may not think of the air as an elastic medium and indeed at slow speeds it moves like a fluid, but for rates of pressure change of 20 in a second and higher air behaves effectively as a solid.

To reach a listener sound may travel through a succession of different materials, but at every point along the way there has to be an elastic medium of some kind to carry the energy forward. Gaps in the flow can arise either from an absence of conducting material, or from the intervention of a barrier of sound-absorbing material. Double glazing works as sound-proofing by virtue of sandwiching a vacuum layer between panes of glass. Sound cannot travel in a vacuum. Heavy curtains reduce the penetration of sound to a lesser degree, but that is because curtain material has insufficient stiffness to vibrate at audible frequencies. A source of vibration can be detected at the same time in a number of ways. A windowpane set in motion by the noise of a passing train excites a corresponding pattern of pressure waves in the air inside the room. At the same time vibrations from the wheels of the train travel underground to the foundations of a building and are conducted through the structure to set the floor and walls of the same room in sympathetic vibration.

Cats and dogs have swiveling antennae, but the human ear is a fixed dish structure mounted on the side of the head, designed to collect incoming sound. The distinctive folds and creases of individual ears are not simply for ornament. They help the owner to identify where a sound is coming from by imposing distinct reflective profiles on sounds according to the angle and direction of arrival: front or back, left or right, up or down. People have differentlyshaped ears, so it follows that in subtle ways everybody hears a slightly different image of the world.

Just as objects are seen in perspective by the brain comparing images from left and right eyes, so sources of sound in the environment are located by the brain comparing the acoustic images received by the ears. The eyes are designed to swivel independently so that the human observer can cover a wider field of view, or focus on a particular object. Ears on the other hand do not need to be pointed directly at a signal in order to hear it, and can hear sounds from all around. Perspective vision

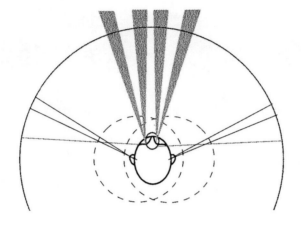

Vision and hearing

is an acquired skill involving the memory as well as the eyes. In practice it is impossible for an observer to see an entire field of vision completely in perspective or in complete focus, because only a small region can be in focus at any one time. Hearing is arguably more efficient than vision because the field of hearing extends through a full 360 degrees and the sounds you hear are always in focus.

Sideways and down

Sound enters the ear canals from the front at an angle of about minus 60 degrees on the left and plus 60 degrees on the right in relation to your nose, unless of course your nose is bent out of shape. Carefully insert a Q-tip in each ear, look at yourself in the mirror, and you will see that not only do they indicate angles of approach of plus and minus 60 degrees in relation to straight ahead, but that the ear canals are also angled *downward* at 30 to 45 degrees. It follows that while vision is aimed straight ahead and horizontal, hearing is oriented sideways and downward. The auditory sense is therefore naturally *inclined* to monitor local and lateral disturbances in the near field environment, allowing vision to *focus* on more distant regions, in particular objects out

20 feet = 55 Hz

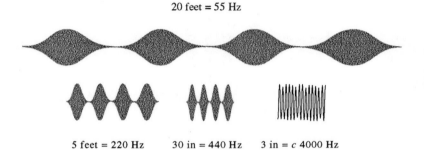

5 feet = 220 Hz   30 in = 440 Hz   3 in = c 4000 Hz

Frequency corresponding to wavelength

of earshot.

The *frequency* of a tone—whether it is heard as high or low in pitch—depends on the number of regular pressure oscillations detected in a given measure of time. For frequency in hertz (Hz) the unit of measure is one second, so that for instance a 55 Hz tone corresponds to a source vibrating at a rate of 55 cycles every second. Since sound propagates through the air at the same speed whatever the frequency, the term *frequency* ultimately refers to the *distance* or *wavelength* between successive pressure peaks. A tone of 55 Hz corresponds to a wavelength of about 20 feet and the pitch A1, an octave above the lowest note on the piano keyboard. Five feet or 220 Hz corresponds to A3 below middle C, 30 inches to A4, violin A, and a wavelength of 3 inches, approximately to the top C of a piano keyboard.

For sounds in the low to middle range of frequencies the size of the human body does not represent a significant barrier, but for frequencies greater than 2,000 cycles a second (equivalent to a wavelength of seven inches or less) the head acts as a baffle and casts an acoustic shadow, with the effect that a signal containing a significant high frequency content (such as speech) will appear louder and clearer, as well as register fractionally earlier, at the nearside ear than at the ear farther away.

The ear is a remarkable detector and transducer or converter of audio information. Acoustical energy received in the form of pressure waves is converted to mechanical vibration in the bony

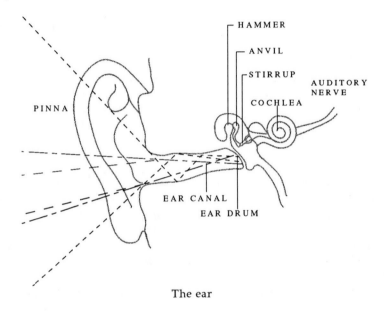

The ear

structure of the inner ear, and then to electrical energy in the form of nerve impulses that go to the brain. The eardrum vibrating in response to pressure changes in the ear canal activates the ossicles, a miniature bone structure in the inner ear comprising the hammer (*malleus*), anvil (*incus*) and stirrup (*stapes*). Working in a way like the hammer action of a piano, the ossicles transmit percussive vibrations to a self-damping fluid sealed within a folded tube rigidly enclosed within a spiral bony formation called the *cochlea*. Pressure pulses introduced at one end of the spiral tube set up oscillations in the fluid along either side of an intervening ribbon called the basilar membrane, a conductive band of tapering thickness around 35mm in length. Since the separation in time and space between successive pressure fronts is a function of wavelength, and wavelength is a function of frequency, the cochlea behaves effectively as a filter. Local perpendicular motions of the basilar membrane cause attached hairlike structures called *cilia* to bend back and forth; the bending of the cilia in turn triggers a reaction in adjacent hair cells to fire off nerve impulses that travel along associated neural pathways leading to the auditory centers of the brain. These patterns of nerve impulses are the information the brain identifies as sound.

Energy conversion

The ear takes in complex, ever-changing sounds and sorts out their different frequencies, passing on the information to the brain as a pattern of low and high bursts of electrical energy. A musical instrument by comparison takes in energy in a relatively raw or disorganized state and converts it into sound pressure waves that are relatively well-defined. In their roles as energy transformers musical instruments have a great deal to say about the development of engineering and materials science, structural mechanics, geometry, ballistics, architectural acoustics, fluid dynamics, metallurgy, and related areas of scientific inquiry.

Musical instruments incorporate structures that vibrate in different ways. The human larynx operates like a simple buzzer, the string of a guitar or harp like a bowstring, while the player's lips in a horn and the reed in the mouthpiece of a woodwind instrument act as valves admitting pulses of pressurized air into a tube. Every musical instrument can be understood as a combination of driving and resonating structures, and its special sound as the solution to a particular equation of forces.

The musician is the primary source of acoustic energy: bang the drum, blow into the tube, press the key, draw the bow across the string, pluck the string, and so on. What the instrument does first is offer *resistance*: the drumskin resists the blow of the stick, the mouthpiece resists the pressure of air, the violin string pulls back against the bow, the piano string bends under the hammer. What happens as a result is that the energy transferred by the musician through either continuous or intermittent action is *stored* temporarily by the instrument and then released in a controlled manner. The excess energy is either released or absorbed. If energy is absorbed, the instrument heats up; in fact a tiny amount of energy is always absorbed when an instrument is played. Musical instruments, however, are intentionally designed to radiate energy into the atmosphere in the form of sound for others to hear, because music is about sending signals from one person to another. So a performer operating a musical instrument is in effect charging up a transformer for it to emit a particular kind of signal.

There is a myth about telephone switchboard operators in the early days having to wear a primitive version of lip microphone, essentially a telephone mouthpiece with a button of fine granular carbon particles as a diaphragm. A carbon button microphone is not particularly efficient, so a proportion of the incoming sound energy is absorbed in the movement of the carbon granules. Over a period of hours, so the story goes, these mouthpieces used to heat up, sometimes to a point where lips began to blister. Operators learned to keep a replacement mouthpiece on hand and to change mouthpieces regularly to prevent them from overheating and their lips from being burned.

Incidentally, the ritual of tapping a microphone at a social event to make sure it is working also dates from the era of the carbon microphone. From the constant agitation of the voice over time the carbon granules, in addition to heating up, tended to consolidate into a compact mass with a resulting loss of sensitivity. The problem was easily put right by a sharp tap to loosen the carbon and restore the microphone to normal function. Today nobody uses carbon microphones at social events, but the tapping ritual and accompanying "Hello, hello!" still go on.

The art of music is an art of combining sounds of different materials and densities to produce an intended aesthetic effect. Classical music, like classical art, is valued for the clarity and complexity of information it is able to transmit. Classical music is not always as easy to follow as folk or popular music because it often employs more instruments and is designed to carry a higher density of acoustic information. Its special significance in modern culture resides in the fact that classical music represents a higher level of information management and enables listeners with appropriate training to process complex information more effectively.

Complex vibration

A body in motion can vibrate in a simple or complex fashion. A simple vibration is one where the energy distribution follows a simple curve. In general, a simple vibration expresses a simple structure, for example, the pulsation of air in an organ pipe. A complex vibration is one where the energy distribution is complicated by internal delays, as in the case of a drum or the body of a violin. Where a surface is in complex vibration the different frequencies interact in the air with the result that certain components are enhanced and others are canceled in a pattern that varies relative to the location of the individual listener. If a pattern of fluctuation is complex but cyclical (periodically repeating), the effect is a rich and complex musical tone; if it is complex and disorganized the effect is noisy.

Noise is sound without discernible structure, but tone is *always* the product of a periodic oscillating process. A regular oscillation of pressure maintained for a minimum period of time and within the range of twenty to four thousand oscillations per second is identifiable by a human listener as a *tone* of *constant pitch*. When a tone is heard you know that the structure responsible is vibrating coherently and efficiently.

Music is sound created for effective transmission and easy reception involving a process that is energy efficient. A musical instrument is structured to discharge excess energy from the performer into the atmosphere in an efficient manner. The most energy-efficient sounds are also the most musical sounds, because in musical sounds the component vibrations oscillate in regular sequence, which in turn makes them the sounds the ear is best able to process efficiently.

Resistance

All sounding bodies in a stable environment have a natural tendency to maintain their shape. Nature resists change. The air we breathe is an environment within which objects are subject to a constant pressure of 14.7 pounds per square inch. We don't notice atmospheric pressure because our bodies are used to it and because the pressure *inside* our bodies conforms with the external pressure. What bodies are sensitive at monitoring is not absolute pressure so much as sudden or subtle *change* in pressure. Being able to sense changes in air pressure has survival implications. Such fluctuations can be extremely gradual, occupying hours or even days, or they can be extremely rapid, oscillating thousands of times in a second. On the macro scale changes in the weather involve changes in atmospheric pressure affecting the swim bladders of surface-feeding fish, making them reluctant to rise if a storm is on the way. Changes in atmospheric pressure on this timescale are generally irregular and hard to predict. If one were to take the continuous barometric record of a given location over a number of years and somehow speed it up until the fluctuation in atmospheric pressure became audible, it would sound like a noise. The only periodic elements in such a data stream might be changes in temperature as night follows day or through the seasonal cycle. Converting any sort of numerical information into music is actually a useful activity since the ear is a filtering device designed to analyse a continuous stream of data and detect any regularities. In *Earth's Magnetic Field*, realized in 1970 by the American composer Charles Dodge, changes in terrestial magnetism resulting from the action of the solar wind over the course of a year are interpreted as melodic and rhythmic cadences.[1]

Resonators

The source vibration of a musical instrument is a stretched string in the case of a violin, piano, guitar, or harp, a flexible reed in the case of a clarinet, oboe or saxophone, a membrane for a drum, and so on. In order for the actions of a musician to communicate as sound, the source vibration has to be transferred to the surrounding air. A drum-skin presents a relatively large surface area, likewise a cymbal or gong, but a stretched string has very little effective surface, so in order for its vibration to be heard its energy is transferred to a passive resonating structure that provides the necessary surface area. The hollow box structure has to be *strong*, to maintain the considerable tension of the strings, and also *flexible*, for efficiency in converting string vibration to surface vibration. The same engineering principles are involved in the design of a violin or piano as in the design of a suspension bridge.

String instruments consist of a number of strings as sources of vibration, maintained at high tension over a hollow box acting as a resonator. A resonator has two components: a container and the volume of air inside it. The container influences the sound by its *material* (wood, skin, metal) and by its *shape* (flat, curved,

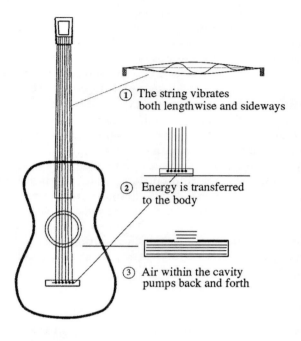

① The string vibrates
   both lengthwise and sideways

② Energy is transferred
   to the body

③ Air within the cavity
   pumps back and forth

Transfer of energy from string to body in a string instrument

shallow, deep). The volume of air excites a cavity resonance that
influences the tone quality and transfer of energy to the
container in the same way as the design of a loudspeaker cabinet
influences the vibration of a speaker unit. In a sealed enclosure
the air inside is trapped and the cabinet cannot vibrate freely.
This has the effect of tuning a resonator so that it vibrates in a
less random manner. The copper bowl of an orchestral kettle-
drum or Indian *tabla* is a sealed enclosure that forces the otherwise random motion of a drum-skin to vibrate more coherently
and generate a more musical tone than an open or two-ended
drum. A violin or guitar string does not need to be forced into
coherent vibration, but its pitch is subject to change from one
moment to the next, so the sound box of a string instrument is
required to vibrate in a manner equally serviceable to a wide
range of different frequencies. Unlike the sealed bowl of a kettle-
drum, the sound box of a string instrument is a *vented resonator*,
one that allows excess air pressure inside to escape. In a guitar or

lute the vent is a circular opening in the center of the front plate, sometimes elaborately worked, in the middle of the upper surface, while in instruments of the violin type fluctuations in air pressure are equalized through the two f holes on either side of the bridge.

Energy from a vibrating string is transferred to the resonator via a projecting rib or cantilever located near the vent that intercepts the lengthwise string vibration and, in the case of the violin bridge, converts it into an up and down squat thrust motion on the front plate that, in piston fashion, compresses the airspace within. It is useful to imagine the resonant body cavity of a string instrument as a scale model of a room in order to understand how the internal shape of the air cavity influences tone quality and direction. The motion of flat front and back plates of a guitar, for example, follows a relatively simple pump action tending to favor notes of certain pitch more than others, roughly corresponding to the clap tones of a rectangular room. A practical consequence of flexing back and forth in relatively simple invariant motion year after year is that the wood of a guitar front plate eventually loses its flexibility and becomes brittle or "crystallizes" and has to be replaced.

By contrast instruments of the violin family have very complex sculpted boxes, vented resonators with clearly defined and reinforced right-angle corners defining front and back plates and sides, but everywhere else a deliberate absence of parallel surfaces, like the curvaceous walls of Baroque architecture. Four strings of similar length but differing thickness are stretched over a high wedge-shaped bridge with cutouts in it that direct the pathway of vibrations from the strings to the plates. The bridge has two feet, one foot resting directly on the front plate and the other foot transferring vibrations via a wooden sound post to the back plate. This has the effect of causing the front and back surfaces to vibrate in synch and in parallel at the more powerful lower frequencies, improving output and helping to make the violin a much more expressive instrumental voice than its older relative, the viol. Curved plates are also more resistant to vibratory wear and tear; in fact some of the earliest violins made in Cremona, Italy, by Amati, Stradivarius, Guarneri and others are said to sound even better today than when they were first played many centuries ago. The same curves create complex interacting resonances that give the violin family of instruments a ringing, volatile tone reminiscent of the human voice.

The headlamp effect

The teardrop shape of a lute behaves in some ways like a dome in a church, collecting sound and bringing it to a focus, and in other ways like the headlight of a modern automobile. As a headlamp takes light energy from the light bulb and radiates it as a directional beam, so the lute takes acoustic energy from a line source, a plucked string, and radiates it as a narrow beam of sound brought to focus by the curved interior and transmitted as parallel energy waves by the flat front plate. In theory the young squire serenading his lady love should be able to aim the soft-spoken sound of his lute directly at her balcony window without attracting the attention of her husband in the next room.

Trapped lips and valves

Wind instruments fall into two groups. The brass are the trumpet, trombone, cornet, tuba, and horn families (though in an orchestra, for some reason, the horns are considered as honorary woodwinds). All of these instruments operate by compressed air using the player's lips as the vibrating source. The lips are compressed into a mouthpiece that forms a seal with the face allowing the trapped lips to vibrate. For instruments of the modern trumpet, trombone, cornet, and tuba families the mouthpiece is a shallow cup and the lips are retained in place under pressure; for instruments of the French horn family the mouthpiece is a tapering tube that draws in and elongates the player's lips allowing the extremities to vibrate more freely. The orchestral or transverse flute also uses the player's lips as an energy source. In this case, however, the lips are not vibrating under compression inside a mouthpiece but under muscular tension controlled by the player.

Air released under pressure from the lungs is forced through the lips to create an initial buzz. The action of forcing the air through a narrow aperture converts a continuous movement of air into an audible periodic sequence of pulses of compressed air. At this stage the sound is in a raw state; the function of the tube is to convert a relatively noisy buzz into musical tones of variable pitch. This is achieved in two principal ways: by the air contained within the tube acting as a tuned resonator, and by

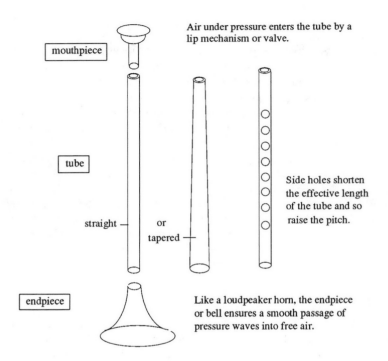

Air under pressure enters the tube by a lip mechanism or valve.

mouthpiece

tube

straight — or tapered —

Side holes shorten the effective length of the tube and so raise the pitch.

endpiece

Like a loudpeaker horn, the endpiece or bell ensures a smooth passage of pressure waves into free air.

Components of a wind instrument.

feedback from the pumping action of pressure waves within the column, forcing the player's lips to vibrate in synch.

Edge resonances play a significant role in trumpet, trombone and French horn sound because in these brass instruments every note comes out of the flared bell, the metal surround of which vibrates to give a brilliant finish to the instrumental tone. All wind instruments change the note by changing the length of the tube, but in brass instruments the added length is produced within a sealed air column, in a trombone by manual operation of a slide changing the length of the tube, and in other brass instruments such as trumpet and French horn by the use of piston or rotary valves to divert the air flow into auxiliary lengths of tubing.

In a brass instrument the lips contained in a cup mouthpiece form a soft valve of thick tissue, material that is relatively inert; the lips forced apart emit compressed air in a series of bursts at low frequency with a high noise content. Woodwind instruments employ a vibrating reed instead of using the lips as a valve. A reed consists of a single or double layer of split cane, carefully tapered to provide a stiff but flexible airfoil capable of responding extremely smoothly to changes in air pressure from the player's lungs. In single-reed winds, such as instruments of the clarinet and saxophone families, the thumbnail-sized reed valve allows the player to fade notes in and out almost imperceptibly. Among double-reeds such as oboe, bassoon, and the bagpipes, the valve mechanism consists of two fingernail-sized tapered reeds bound together at the base. The player's lips clamped around the reeds form a seal, and air pressure in the mouth builds to a point where it is released by the two reeds opening and closing like miniature swing doors. The smaller size of double reeds, and the fact that there are two of them vibrating in opposite directions, makes the tone quality of double-reed woodwinds particularly rich in high frequencies.

Air columns

Not all woodwinds are in fact made of wood. Among those that are, the tube is a heavily constructed resonator of dense ebony or similar close-grained wood that unlike the lightweight box of a violin or guitar is designed not to vibrate freely but to stabilize the air column within. Modern orchestral flutes together with the saxophone family by contrast employ tube resonators of relatively thin metal that is allowed to vibrate to a certain extent and add color to the tone. For woodwinds just as for brass instruments, changing the note involves changing the length of the tube or air column, but for woodwinds this is effected by the player opening finger holes at different locations along the tube. When a traveling pressure wave reaches an open finger hole its pressure escapes into the surrounding air and is prevented from traveling any farther. In this way the location of the open hole determines the effective length of the tube and thus the pitch of the tone. Because woodwinds do not employ the whole of the length of the tube all of the time, their sound does not all emerge from the endpiece or bell as it does in a brass instrument; in consequence woodwinds lack the brilliance and directional efficiency that in olden times made the trumpet, trombone, and French horn such reliable signaling instruments on the battlefield and in the hunt.

The interior *shape* of an enclosed air column also plays a role in determining the timbre or tone signature of a wind instrument. If the tube is cylindrical or parallel bore in shape, as for a flute, clarinet, or trumpet, pulses within the air column remain at a constant pressure from end to end and every note has considerable stability. Tapered-bore instruments such as the oboe, bassoon, or saxophone, are designed with an air column wider at the base than at the mouthpiece, the effect of which is to draw pressure pulses away from the player's mouth and down the tube, making intonation less critical and allowing the player a degree of control over the exact intonation. This ability to "bend" notes gives tapered-bore instruments a decided advantage in tonal expression, from the subtle singing quality of an oboe to the almost speaking inflection of an alto saxophone.

The voice

The human voice, unlike other musical instruments, is a product of flexible components made of soft tissue. This has significant consequences. A regular musical instrument made of *hard stuff* is a *passive* resonating structure that responds *consistently* to a *variable* input. That consistency is a function of material and cavity resonances, and is perceived as an instrument's characteristic tone color or timbre. The voice by contrast is an *active* resonating structure made of *soft stuff* that is *changing* its shape and resonances all the time. We value consistency of tone in a regular musical instrument, that it can be relied upon to produce virtually the same sound from one performance to the next. A voice, on the other hand, is valued for versatility of tone on which the richness and variety of spoken language relies.

Normally human body tissue would be classified as energy- or sound-*absorbent* and for that reason unsuitable material for sound generation. Voice production involves a combination of muscular tension and forced vibration of bone and tissue structures, together with the lungs, mouth, and nasal airways as resonators. Air stored under pressure in the lungs is released into the vocal passages through the *larynx*, a valve consisting of flexible tissue folds on either side of the airway to the lungs. These curtain-like folds or *vocal cords* can be made to vary in tension and aperture at will. When compressed air is forced through the larynx, a stream of pulses is produced in much the same manner as the lips buzz in a trumpet mouthpiece. The pulsed sound is transmitted simultaneously back into the lungs and forward into the resonant chambers of the nose and also the mouth, a semi-rigid structure containing a variety of control surfaces. The tone quality of a person's voice is influenced both by the length of the vocal cords and by the volume and shape of individual mouth and nasal cavities.

To vary the pitch of the voice a speaker alters the *tension* of the vocal cords; different *vowel* sounds [a] [e] [i] [o] [u] are created by varying the *cavity resonance* in the mouth by changing the shape and volume of the tongue; while to produce a variety of *consonantal* sounds [t] [p] [f] [g] [k] a speaker employs various modes of *frictional obstruction* of the passage of air through the throat, palate, and teeth. The richness of speech sounds testifies to the sophistication and complexity of the speech process. The pitch range of a voice depends on the length of the vocal cords. This is why children have higher-pitched voices than adults, and why women in general have higher voices than men. The *timbre* or personal tone quality of an individual voice is affected by *permanent* features of larynx, mouth, and nasal structures, for example, a speaker's jaw size and shape. The nasal cavity acts as an important passive resonator, which is why the voice of a person suffering a heavy cold and blocked nose is noticeably muffled in tone. In addition to providing an air space allowing the roof of the mouth and palate to resonate, the nasal airway contributes its own distinctive cavity resonance to a person's voice.

Tongue

Elements in the mouth directly influencing speech production are the palate, teeth, throat, and most of all the tongue, a remarkable organ that never seems to get tired and has the ability to change its shape and volume at great speed. The tongue acts as a bi-directional organ, monitoring *taste* in food and drink entering the mouth on the way to the stomach, and modulating *information* in the outward passage of pulsed airwaves from the lungs.

How do speech sounds differ from one another? In general, vowels are sustained harmonious speech tones of identifiable resonance determined by the volume of air in the mouth cavity.

A cavity resonance is the sound produced for example by blowing across the top of a soda bottle. The tone produced is related
to the volume of air: if the volume is reduced, for example by
adding water, the cavity resonance rises in pitch. In the voice the

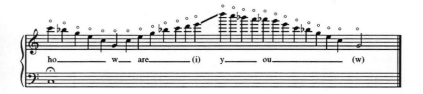

Formant resonances of the phrase "How are you?"

nasal cavity is a *fixed* resonator and the mouth cavity is a *variable*
resonator. Since nasal resonance is normally invariant, it is also a
means of identifying an individual voice.

Vowel sounds are the varying harmonic resonances produced by the tongue changing shape and thereby altering the
volume of air in the mouth cavity. Pitch and timbre involve
separate and distinct voice mechanisms. *Pitch* is the buzz tone
generated in the vocal cords and is under the muscular control of
the larynx at the throat. *Timbre* or tone color, on the other hand,
is an effect of resonance overlaid on the buzz tone by mouth and
nasal cavities acting together: it is under the muscular controls of
the jaw, lips, and tongue.

These control mechanisms operate completely independently. For example, a person singing up the scale on the same *vowel*
[a] is controlling the pitch by varying the muscular tension of the
vocal cords at the larynx without having to change the positions
of the mouth or tongue. Alternatively, a person singing successive vowels [o] [a] [æ] [ø] [i] [u] etc. on the same *pitch* will be
exercising control over the tongue and mouth while leaving the
vocal cords to vibrate unaffected.

In everyday speech activity elements of both pitch and
timbre are constantly in motion and seemingly interlocked. Understanding speech involves making a clear distinction between
the two. Pitch is a *dynamic* effect of muscular tension and air
pressure associated with the expression of *emotion*. Cavity resonances or *formants* express the distribution of energy within the
tone, and articulate *meaningful distinctions*. Everybody, whatever

their age or sex, uses the same words to express similar ideas. That is a fundamental reason for language being able to function effectively, and why it does so despite enormous differences in human body size and structure. The words *Hey! Yo! Here! Wow!* sound alike whatever the pitch of the voice because the resonances are the same for a small child as for an adult, and it is the

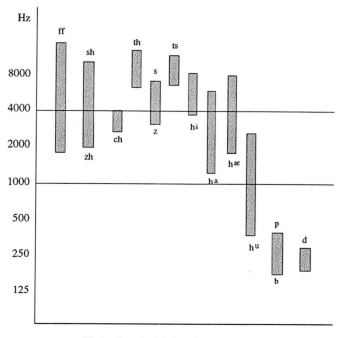

Noise bandwidths of consonants

resonance that gives a vowel its distinctive identity. An interesting if overlooked feature of medieval chant is its aptness for inquiring into the distinctive features of vowel resonances in the absence of present-day microphones and oscilloscopes. Often in an "Alleluia" voices are heard to sing a range of different notes to the same vowel sound, indicating that the vowel sound or cavity resonance remains audibly the same even while the accompanying melody varies continuously in pitch. At other times the choir will sing a succession of different vowels on the same note, and in those instances a careful listener can hear the change of vowel as a change of timbre.

Consonants are like formants in that they occupy specific frequency zones, but unlike vowels they are indeterminate or noisy in structure. Consonants are in fact colored noises that occupy distinct regions or bands of frequency. Some are continuous or *diffuse*, ranging from [ff] and [ss] at the highest extremes of vocal range down through [sh] and [ch] to voiced consonants like [x] and [zz], while others are *abrupt*, like [t] [k] [p] [b] [d] or [g]. Consonants are a rich source of vocal mimicry. Many words in everyday use actually sound what they mean: for example the words *click*, *thud*, and *bang* correspond to the sound effects they describe.

Notes

1. Charles Dodge, *Earth's Magnetic Field: Realizations in Computed Electronic Sound*. Bruce R. Boller, Carl Frederick, Stephen G. Ungar, scientific associates (vinyl, Nonesuch Records, H-71250, 1970).

CHAPTER FOUR

Signatures

MUSICAL instruments are made of materials such as wood and metal, materials denser than air. Before a violin or a gong begins to radiate pressure waves into the atmosphere, the energy input first propagates throughout the entire body structure. Excess energy travels many times more rapidly through solid wood or metal than through the atmosphere. If you are listening to a violin or gong, what you are hearing is the entire object in vibration and not just a localized disturbance at the point of contact. On the other hand, if you are listening to a person speaking or singing, the energy source is confined to the region of the open mouth, because the human body is largely made up of absorbent tissue and protective clothing and does not resonate audibly over its entire surface area. The rigid character masks of ancient Greek drama and African tribal ritual serve not only to conceal the face but also to amplify and direct the energy of a speaking or singing voice.

Rhythm and melody

There is sound you cannot hear but to which the body still responds. *Infrasound* is a regular oscillation of atmospheric pressure on a scale of single figures per second, fast enough to qualify as sound but still too low in frequency for the human ear to detect as pitch. In Japan, where earthquakes are a regular occurrence, scientists watch out for changes in the behavior of farm animals. In the build-up to an earthquake the ground starts to oscillate at very low frequencies that are felt rather than heard. Cattle feel nauseous, stop feeding, and settle down, while horses and sheep tend to panic and run about. Infrasound at high energy levels can be dangerous and interfere with the body's natural rhythms. The

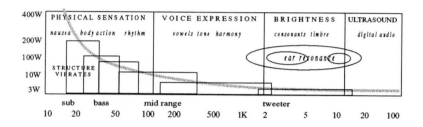

Frequency in relation to power and sensation

use of low frequency sound was briefly considered during the seventies as an alternative to batons or riot gas for crowd control. Infrasound affects a person's bladder and bowel retention without producing any long-term ill effects. However, fortunately for the public it can't have any effect on a protest group without first having the same effect on the operators as well. Low frequencies are like that, powerful and omnipresent; that's why the subwoofer of a hi-fi system can be hidden anywhere in the room. In music low frequency sounds and pressure fluctuations within the range of one to ten pulsations per second are associated with *rhythm*, in particular the bass drum and bass-line.

Low frequency sounds are the most powerful sources of energy in music. The concentration of energy in the bass influences the body directly and makes a listener want to move in time with the beat.

Voice communication and musical expression rely on mid-range frequencies. When we are listening to a song we are paying attention to a *dynamic pattern of change*. A melody is a signal expressing movement in pitch and time. Melody in speech or song communicates a sense of *line*. For all practical purposes melody occupies a region from a low of around sixty per second (the pitch of a low bass voice or instrument such as a bassoon or cello) to a high of 1,200 per second or more (associated with a soprano voice, a flute, violin, or the lead guitar of a heavy metal band). Melody range is equivalent to the active *pitch* domain.

The human ear is especially sensitive to changes of frequency in the melody range. These changes correspond to fluctuations of pitch in the speaking voice.

Scales and modes

Different cultures and styles of music are also identified with different scales or modes representing degrees of sensitivity to pitch change. Although they associate with different nationalities and traditions these scales or modes are not related to any biological differences in ear structure or brain function, but rather to the social and cultural circumstances in which music is created and enjoyed, including instruments, performance environments, and local behavioral conventions such as speech patterns. Most scales relate in some way, however, to the range of pitch of a normal speaking voice—that is, around an octave. The extended vocal range of much classical music has to do, interestingly enough, with the larger capacity of opera houses and concert halls in the West, and a Western desire for the voice to compete in expressive range with a violin or trumpet.

Western music has established a unit of pitch change called the semitone, representing a difference in frequency of around 6 percent between adjacent pitches. The semitonal or twelve-note chromatic scale corresponds to a stepwise progression of black and white notes on the piano keyboard. Finer gradations of pitch are easily detectable by the human ear, but the 6 percent threshold for musical performance has evolved over many centuries of experiment and evolution as the optimum definition for melody perception and recall under the closed conditions in which classical music is ordinarily performed. Tonal and popular music that is composed in a major or minor key uses a seven-note diatonic scale corresponding to the white notes of the piano in sequence. Here the unit of difference between adjacent notes is closer to twelve percent. This classical scale is easier to remember and more robust as well. It allows you to pick out your favorite country ballad on the car radio over the noise of the engine, and you are still able to recognize the melody even when the singing is a little—or even a lot—out of tune.

Among oral cultures worldwide folk songs are traditionally created in a five-note pentatonic scale corresponding to the black notes in sequence on a piano. Here the stepwise movement note

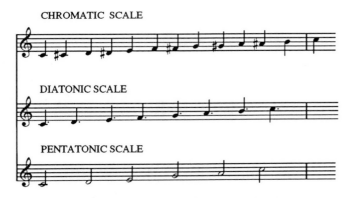

Chromatic, diatonic, and pentatonic scales

to note is even larger, between 13 and 20 percent. This reflects the fact that much of this music is performed out of doors, where finer distinctions of pitch and tone are not so easy to hear.

Classical music, especially music of the twentieth century, has evolved to exploit relatively fine gradations of pitch for a number of reasons, partly cultural evolution, partly because the acoustic conditions for classical music-making and listening are strictly controlled. Classical 6 percent music takes place in acoustically isolated halls in front of audiences that are socially and financially obligated to keep still, stay quiet, and pay attention. The 12 percent scales of popular music are optimized for listening in noisier environments where audiences enjoy the freedom to respond spontaneously either vocally or in a physical manner, for example, by dancing. The 13 to 20 percent resolution of pentatonic folk melody is associated with a strong, in-your-face style of performance, well adapted for signaling out of doors and over long distances where reception can often be problematic.

Tone color

The upper extreme range of hearing, from 4,000 to 16,000 hertz and beyond, is the region of *timbre* or tone coloration. Timbre is a response that goes beyond melody information to register tonal quality associated with a particular voice or instru-ment. It's a different kind of perception. Whereas rhythm and melody are *dynamic*, exciting an awareness of degrees of change, timbre is *stable*, a perception of overriding consistencies in a signal, reflecting aspects of sound production that are relatively constant. Timbre perception allows a listener to identify a person's tone of voice, or to distinguish the sound of a guitar from that of a trumpet or violin. To explain consistencies in tone quality one looks for consistencies in tone production determined by permanent features of the voice or instrument, the materials of which it is made, the mechanism by which sound is activated, and the shape of the airspace within which a sound is produced. When a person says "I know that voice," the remark expresses an *aural* perception of the consistent features of voice timbre. On the other hand, when a person says "I understand that argument," it expresses an *intellectual* perception of the dynamics of voice modulation and change.

Stress patterns

When a person sings or speaks, the voice *inflects*. It fluctuates in pitch. It moves up and down. This up and down motion is not random but directed. It has meaning in the sense that it helps the listener to break down the flow of speech into individual words, but it has additional meaning in that the pattern of fluctuation is also a *melody* or line that has its own expressive dynamic. One is for sense; the other is for meaning. Pitch is related to the rate or density of pressure changes, otherwise known as the *frequency* of a sound. The movement of pitch is called a melody. For the voice and for many musical instruments such as winds (flute, trumpet) and strings (guitar, violin) a movement up or down in pitch is associated with a change of energy level or stress. It takes more effort to sing a higher note, or the string is at a higher tension for a higher than a lower note; in addition the higher note in the pitch range will tend to sound stronger and more brilliant.

The ancient Greeks recognized the relationship of pitch and stress and assigned a higher moral value to music of higher pitch. The image survives in language when we speak of an individual being *highly-strung* or brilliant on the one hand, or dull or *slack* on the other. Today pitch is more abstract. The piano

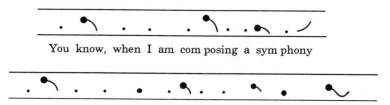

You know, when I am com posing a sym phony

I want some thing more ex citing than two poached eggs.

Example of intonation in English (after Daniel Jones)[1]

keyboard expresses a Renaissance conception of pitch as a spatial continuum of levers or keys, each assigned to a different note arranged in succession from low at the extreme left to high at the extreme right. This convenient but highly artificial system of pitch selection conforms to the universal standard of Western musical notation on which the symphonic tradition is based.

Movement in pitch is useful and necessary in speech because a change in pitch is able to indicate a change of word or syllable. If each word in a sentence is given a distinct pitch, the listener then has a way of knowing where one word ends and another begins. This can help to make the individual *words* clear, but it does not always make their *meaning* clear. The meaning of a statement, poem, or song has to do with the pattern of thought that words express in sequence, and to convey meaning involves a speaker in communicating a sense of direction in the voice line.

Expression is conveyed in three ways: 1. *line* (melody), 2. *timing* (rhythm), and 3. *emphasis* (accentuation). In the notation above, devised in the twentieth century by the phonetician Daniel Jones to record intonation in the *speaking* voice, there are only approximate indications of pitch and timing within a simple two-line grid indicating high and low extremes of range. Although it bears some resemblance to standard musical notation and even more to late medieval plainchant, this notation is self-evidently not designed for exact reproduction, but rather as a visual guide to the interpretation of speech patterns in an existing sound recording. Between the lines the *position* of a dot indicates the relative *pitch* of a syllable, higher or lower relative to the individual voice. The *size* of a dot indicates loudness or emphasis. Three degrees of emphasis are indicated. Finally, the curved lines attached to some of the dots indicate a movement of pitch within a syllable, as, for example, when someone says *A-ha!* or *Wow!* with a rise and fall of pitch.

Exact timing is not a significant factor in speech transcription because there is only one voice involved and it has already spoken. What the *lack* of precision of this notation tells us is that exact pitch and timing are not essential for natural speech communication—in fact, they indicate a degree of seriousness in intention, formalism in manner, and intelligence in execution. Exact timing and intonation only come into play where special significance attaches to their contribution to meaning. Nowadays of course we have audio recording to preserve the minutest inflection of a word or phrase. Long before the invention of sound recording people already recognized the role of intonation and timing in the expression of meaning, and one of the reasons we can adduce from the development of a more precise musical notation is to preserve an original reading or intonation of significant written texts.

Reliving disaster

The "Hindenburg Disaster" radio broadcast by Herbert Morrison is one of the most anthologized news items of the twentieth century—not for its factual content, which is minimal, but as a human reaction to an unfolding catastrophe.[2] At the moment of explosion the detached newscaster is transformed into a terrified human being. Listeners witness the horror of the burning airship through the speaker's struggle to regain composure. There is something of Greek tragedy in the spectacle of a witness reduced to near incoherence by a powerful and unexpected event. Before the explosion Morrison's voice is steady, relaxed, and evenly paced; after it his voice is raised, shrill, his words unscripted, repetitive, in bursts. Losing control, he loses grip of the language as well: "It's crashing terrible!"

It burst into flames! it burst into flames, and it's falling, it's
crashing! Watch it! Watch it! . . . Get this, Charlie, get this,
Charlie! It's afire, and it's crashing, it's crashing terrible!
Oh, my, get out of the way, please! It's burning, bursting
into flames and is falling on the mooring mast and all the
folks agree that this is terrible, this is one of the worst cat-
astrophes in the world! . . .[3]

The point of a memorable event is *to be remembered*. Facts are
facts, and can be written down in books, but a memorable event
is an experience, and for the experience to be remembered words
alone are not enough, it is how they are said. The Hindenburg
Disaster newscast is memorable for the very reason that the witness could not control his emotion. The death of Oedipus in
Sophocles' tragedy is rendered more moving and memorable as
witnessed through the tears of the messenger, than in full view.
Written down, the words of any witness to a disaster convey
only part of the experience. To convey the full impact requires
the sound of a human voice modulated in accordance with the
emotion of the event. Language (*la langue*) is the system and has
dictionaries; speech (*la parole*) is the individual act. In language
meaning resides in the words, but in speech meaning lies in the
expression. That expression is *music*.

Voice recording

The most basic of all musical instruments is the voice. Singing is
the oldest form of music. People use the voice in various ways:
- for making noises that express sensation:
  "Ouch!" "Wow!" "Aaargh!" "Mmmm!" etc.
- for communicating in factual language:
  "That will be two dollars and fifty cents."
- for expressing thoughts and ideas with emphasis:
  "That looks grreat!"
- for conveying meaning through rhythm and timing:
  "I wandered lonely as a cloud . . ."
- for conveying meaning though the rise and fall of melody:
  "My country 'tis of thee . . ."
- for imitating noises or musical sounds:
  "A wop bop a lumop, a lom bam boom. . ."

Virtually everyone has a voice, and we listen to it and use it
every day. We talk and listen to ourselves speaking, and we talk

and on occasion listen to other people. As children we are taught that the purpose of language is to communicate useful information to others. Of course that is not the case—most of what people say is designed either to remind ourselves and other people that we in fact exist, or to blank out the sound of other people, or to influence the pecking order in the group to which we belong. Meaningful communication has very little to do with any of this activity; what it is mainly about is self-assertion. Speech is vocal signaling disguised as language, and both casual and serious communication alike are blended into the ongoing self-awareness process of making and monitoring environmental noises.

Florence Nightingale's message

After a distinguished career in health care reform Florence Nightingale, the "lady of the lamp" of the Crimean War, was invited to make a voice recording on the new phonograph. The year was 1890 and Florence Nightingale was one of many celebrities to record a message as part of a thinly-disguised advertising venture by the young industry. The Edison cylinder recording has survived. This is her message:

> *Female announcer*:
> At Florence Nightingale's house, London:
> July the thirtieth, eighteen hundred, and ninety.
>
> *Florence Nightingale*:
> When I am no longer
> Even a mem'ry, Just a name,
> I hope my voice may perpetuate
> The great work of my life.
> God bless my dear old comrades of Balaclava
> And bring them safe to shore.
> FLORENCE NIGHTINGALE.

Hers is a short statement, of only two sentences, but it shows a very shrewd understanding of what recording is all about.[4] To an observer reading the words on the page, the speech contains very little information of any consequence. A reader who does not know who Florence Nightingale is cannot know anything of her "great work," or who her "dear old comrades of Balaclava" may

When I am no longer E-ven a mem'ry, Just a name,

I hope my voice may— per pet u ate— The great— work— of my

life.— God— bless my dear— old com-rades of Ba la cla va,—

— and bring them— safe to shore.

The voice of Florence Nightingale transcribed in
musical notation

have been. This statement is not about factual information. You
can find facts in books. Rather, it is a record of her *identity*, of the
person she is. A reader cannot find that in any book. It is in the
sound of her voice.

Her statement is in two parts. In the first part she is
addressing the future:

When I am no longer *even a memory*
　　　—no longer remembered as a person by anyone living—
just *a name*
　　　—a name in the history books—
I hope *my voice*
　　　—she says "my voice," meaning *the sound of her voice*—
may *perpetuate*
　　　—may inspire listeners to continue—
the *great work* of my life
　　　—her health care reforms which a reader can find in the
　　　history books.

And to whom is she speaking? She is speaking to *you. Now.* At
this very moment.

In the second part of her speech she is looking back to the
past, addressing the wounded soldiers for whom she cared in the
Crimean War. "My dear old comrades" she calls them, and her

voice is full of emotion. "May they rest in peace." And she signs off with a flourish: "Florence . . . Nightingale."

There is a symmetry here between the words she addresses with a sense of urgency and strength to future generations and the words with which she evokes the sufferings of generations now dead. Her message is designed to awaken in the listener a sense of the reality of past events in order that the lesson of history may be remembered in the future. This is a statement about memory, a time capsule connecting past and future. Her physical presence has vanished, but her voice remains fixed in a present moment that is 1890 and for ever.

Print and sign your name

It happens all the time. PLEASE PRINT YOUR NAME it says at the top of the form, and at the bottom it says PLEASE SIGN YOUR NAME. Why should it be necessary to do both? Printing and signing your name are just two ways of telling the reader who you are. But of course they are very different. Anybody can print your name on a form. It has to be legible, but it doesn't have to be in your handwriting. The information of your name in print is your *social identity*, the name you go by in society. It identifies you as a member of society.

Your signature has quite a different meaning. Nobody else is allowed to copy it. A signature is a person's distinctive mark. It is often completely illegible. It identifies you as an individual. And when you sign something you are making a promise.

It is the same for a person's voice. Presidents don't always write their own speeches, but they *authenticate* a text through the sound of their own voice. The voice gives added value to a statement that is otherwise only words on paper. It is a doubly added value, the first being *personal endorsement* of the content, the second being a *precise and unassailable interpretation* of the content.

Florence Nightingale's recorded speech has practically no content. The meaning is all in the voice. We recognise the difference between factual and expressive meanings in those situations where a head of state visits a foreign country and makes a speech. The president or prime minister is visiting Japan and gives a press conference. He speaks in English or his own language (his own *tongue*) and an interpreter translates into Japanese. A Japanese listener who does not understand English listens to the speaker to understand the expressive intention of what is being said, and to the interpreter to understand the information content of the statement. The role of the interpreter is strictly limited to translating the factual content. The voice of an interpreter is flat and robotic in character. An interpreter cannot attempt to convey the expressive intention of what is being said. That has to come from the original speaker. It is the authentic signature of the speaker's voice, along with the speaker's identifiable voice timbre.

Signature tune

Florence Nightingale's voice is pure added value. What are the features of her voice that a listener can identify? First, there is the individual *sound* of her voice. Since she is no longer here in person you have to take her voice on trust, just as you have to believe the face in the photograph is indeed Abraham Lincoln as the caption alleges, and not an imposter. But we know that.

The sound of her voice is her signature and endorsement of what she says. A listener who believes the voice is authentic is open to believe that she means what she says. To be listening to the voice of somebody long dead is a weird experience. It confounds normal perceptions of time and space. But this is also true of memory in general.

Then there is the way she delivers the message. Voice expression employs three modalities: 1. intonation, 2. timing, and 3. emphasis. Intonation (tone) is expressed in the line or melody that contributes a sense of direction to voice pitch. Another word for intonation is inflection, the way the voice bends up and down. Timing (tempo, rhythm) covers speed of articulation and use of pauses. Emphasis is the use of greater force to accentuate important words.

Although Florence Nightingale is not singing, her recorded voice has a solidity and steadiness of pitch that is quite musical, and her voice rhythm is also formal and regular, so much so, that it is possible to transcribe her speech in musical notation. The musical quality is partly a style or mannerism, a way of speaking adopted instinctively by public speakers. In the case of a cylinder recording, however, there is the added incentive to speak out because of the nature of the recording process itself. In those early days of acoustic recording it was the unaided power of the voice speaking into the horn that drove the recording stylus cutting into the wax. So the voice had to speak clearly, and speak up, in order to be heard at all.

In normal conversation a speaking voice is a flexible instrument responding directly to emotional impulse. A speech, on the other hand, is a prepared statement. You want people clearly to hear what you are saying, and remember and pass on the message to others not present. For an audience to understand a prepared statement simple emotional impulses are not enough—you need structure. Structure implies a degree of steadiness in pitch and timing to make the message easier to hear, and the same regularities also make the message easier to remember. It is the rule of the sound-bite.

Abraham Lincoln's picture

Does anybody *not* recognize this image? Salvador Dalí liked it so much he incorporated it twice into his 1976 painting *Gala Looking at the Mediterranean Sea which from a Distance of 20 Meters is Transformed into a Portrait of Abraham Lincoln*, which he also subtitles "Homage to Rothko." *Computer Cubism?* by Leon D. Harmon is a 1971 image produced at Bell Laboratories after an original photoportrait of the American president.

The first question, as Stravinsky was fond of saying, is "Who needs it?" or what is the point of the exercise. The second question is "How is the image recognized?" or what can we learn from the process. This famous picture is an early exercise in digital imaging, aimed at finding out whether computers could be made to store pictures as well as words. Pictures are complex images. In today's computers images are stored in extremely high definition as hundreds of thousands of data points, but in 1971 computers had very limited memory. More than ten years after this image was produced the desktop computer I used to work with boasted just 64K of memory and you can't store complex images with only 64K.

In order to store a complex image in a computer memory of very limited capacity, some way had to be found of reducing the amount of information to be stored. Since the original image in this case is already reduced to black and white, it can be stored as a grid formation of a minimum number of data points (pixels), together with the brightness value (on the gray scale) of each.

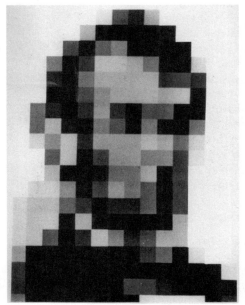

Leon D. Harmon: *Computer Cubism?* (1971)

The number of data points required to record a complex visual image is theoretically without limit; likewise degrees of brightness. The more pixels, the more detailed a resolution; the more brightness values, the greater smoothness of contour.

What the image reveals is *how little* definition is required for an already familiar image to be recalled. By the way, the title *Computer Cubism?* is really a misnomer. The cubist movement in painting dealt with arrested movement and representing objects in more than three dimensions, whereas this computer image is more in the tradition of the Impressionists. They were the ones interested in breaking through the surface of the visible to reveal underlying structures and relationships that give images their real identity. Impressionism is like looking through a ripple glass window at the world outside, and this image is simply using an electronic form of ripple glass; either way you know it's the world outside but it has been reduced to a patchwork of colors —or in this example, shades of gray. The questions for the viewer in both cases are what it is and how you know what it is.

An image reduced to a grid of 200 pixels and a scale of sixteen brightness values starts to look like a random pattern of linoleum squares on a kitchen floor. The issue is how far the process of data reduction can be taken before the image is completely lost. The same process is employed in television to conceal parts of an image from view, for instance, the faces of interviewees who wish to remain anonymous, or parts of the body in an X-rated movie. There is a fine line between exposure and concealment. In this case recognition is the criterion, and the exercise is to find the minimum resolution possible for the portrait of Lincoln still to be recognized.

Now comes the interesting bit. When all you can see is an arrangement of squares of different shades of gray, how is it possible to identify it as a picture of Lincoln, and by what mechanism is his identity conveyed? It cannot be outline, because the distinction between figure and background has been eliminated. It cannot be modeling. It cannot be expression. All that remains is *a schema of essential relationships*: of height and width in the face, and of light and dark (forehead, eyes, jawline, beard). It is a blurry image disguised as a crossword puzzle, and what the viewer must do is disregard the grid formation and allow the pattern to register as an out-of-focus image. Perhaps we should not be unduly surprised that deep-down relationships can be recognized even when the surface features are eliminated, since that is how the very young learn to recognize images in the first place.

Images like this are familiar as software icons on your computer screen—tiny, low-resolution pictures indicating different software functions. But the lessons of data reduction are more complex and fascinating when transferred from the computer to the human domain. We can sum them up in the following way:

Data reduction enables complex information to be stored in limited memory and later retrieved. For an image to be retrieved, essential relationships have to be preserved.

This was all very new in 1971. New, that is, for the science of visual imaging. Not at all new, however, for the science of *acoustic* imaging. The science of acoustic imaging is as old as civilization.

Western music is based on *acoustic imaging procedures*.

Where it all began

Imagine being an ancient Greek, or Chinese, or Mesopotamian scholar, or priest, or lawgiver. There are no computers. There are few books. Only a few software specialists can write. They are called scribes. People use scribes to write messages that can carry information to relatives far away. Otherwise civilization operates by word of mouth, not just to get along, but to govern.

In any society there are rules. In any society there are great events and achievements. In any society there is awareness of change. Old people complain of the passing of the good old days. And since old people run society, they like to preserve the rules that govern how people interact, and they are also keen to preserve the memory of their achievements and those of earlier generations.

You have writing; you can write the rules down. But if most of the population can't read, that means control is given to those who can read. Those who can read still have to communicate the rules to those who cannot read. And they do it by converting a written text into audible speech.

The scholar is concerned that the written record does not preserve all of the meaning of the rules. The written text has no sound, no tone of voice, no melody, no rhythm, nothing directly to convey an original sense of purpose. It is a real concern because the information stored in a written text is open to misinterpretation by ignorant or perhaps unscrupulous people eager to seize power by claiming that an interpretation favorable to their interests is the truth.

The *only way* of preserving the original meaning of an official decree is by finding some way of preserving the essential acoustic features of the original statement along with the written record. These essential acoustic features are tone of voice, inflection, timing, and emphasis. They can be preserved in musical form, but to do so the infinite flexibility of the voice has to be organized and greatly simplified. Realize that this is exactly the same problem facing Leon D. Harmon and his colleagues at Bell Labs in 1971, but this time translated into the acoustic domain. Instead of computer memory, it is a problem of limited *human* memory. Instead of seeking out the limits of data reduction for storage and retrieval of complex visual images, it is finding the limits of data reduction for storage and retrieval of complex *acoustic* images. Instead of having to accommodate the reduced data structure to the mechanism of a digital computer, it is having to design a system of data reduction suitable for implementation on existing acoustic structures: *musical instruments.*

There are no tape recorders in ancient Greece, China, or Mesopotamia. There is no way of preserving an exact copy of the human voice. So what you do is *reduce* the pattern of inflection of the human voice *to a limited scale of discrete pitches,* and the rhythmic cadence of the voice to a regular pulsation. Fortunately, as we can hear in the recording of Florence Nightingale, a person wishing to make a significant statement already adopts a simplified and more structured pattern of inflection, one that can even be written down in musical notation. Words and syllables are expressed in note values corresponding to distinct pitches and regular time values. Only a little help is required to firm up those voice pitches into a musical scale, one that can then be *stored* in the tuning of a stringed instrument such as a lyre or koto or sitar or harp, to be later *reproduced* by a poet declaiming the original words using the instrument to keep the voice in tune and thus preserve the actual meaning of the words.

Modes or moods

Every voice finds its own level. Some voices are naturally high, others are low. It has to do, as was said before, with the physical particulars of the individual, the length of the vocal cords and the shape and volume of the mouth and nasal cavities. But there is also an emotional level in the voice that relates more to the degree of motivation that a voice conveys. Motivation can be defined as an indicator of seriousness, and seriousness as a measure of the elimination of ambiguity. If you want your words to be understood exactly as you mean them, the pitch of every syllable is important, and when all of the permitted variations in pitch are sequenced together a musical scale or mode is the result. By preserving a knowledge of the scale or mode of a person's speech as a musical formula of notes and intervals, it is possible to reconstruct a pattern of intonation, and thus the implicit meaning of a written text, long after the original speaker has passed from the scene.

Notes

1. Original text from *Dodo* by E.F. Benson, illus. after Daniel Jones, *The Pronunciation of English* (Cambridge: Cambridge University Press, 1956), 195.

2. Herbert Morrison, "Hindenberg (*sic*) Disaster." In Glenn Korman ed., *20th Century Time Capsule* (Buddha Records 7446599633 2, 1999), track 8. Date of original recording 6 May, 1937. The hard-edge, newsreel quality of this audio transcription contrasts with the softer tone and wider bandwidth of "1937 Hindenburg Air Disaster" in Richard Fairman ed., *The Century in Sound* (The British Library NSA CD8, 1999), track 19, which loses in dramatic urgency what it gains in presence.

3. <http://www.hindenburg.net/disaster.htm.> (accessed 29 October 2000).

4. Kevin Daly ed., *The Wonder of the Age: Mister Edison's New Talking Phonograph* (vinyl, Argo ZPR 122-3, 1977), disc 1, track 1.4.

CHAPTER FIVE

Reflections

ARCHAEOLOGY used to be about discovering the past by digging it out of the ground, a forensic science stressing the visual and tactile senses. In recent years, however, interest has begun to focus on the acoustics of the location as well as the objects and artifacts discovered within it. In a 1988 paper two French researchers claimed to have established that ancient cave sites in the Pyrenees resonate at specific locations to monotone singing on specific pitches, and that cave paintings tend to congregate at such locations.[1] An American scientist has recorded the acoustics of rock art sites in Europe, North America, and Australia and found that the subject matter of rock art sites is consistently related to the acoustics of the site, and that prehistoric rock formations in Australia function as parabolic reflectors to focus echoes.[2] In a 1998 paper delivered to the Acoustical Society of America, a Californian acoustics consultant claimed to have computer modeled the chirp-like echo inside a Mayan pyramid and discovered a striking resemblance to the song of the quetzal, a bird sacred to the Mayan people.[3]

Children share with ancient oral cultures and the animal kingdom a fascination with sound signals and their powers to influence others. The instinct of a small child, introduced into a cave, a cathedral, or a vast mausoleum, is to shout long and loud. Adults are embarrassed by such behavior, believing it to show a lack of discipline, but the four-year-old who climbs under the rope and into the pulpit and cries "I'm the king of the castle!" is acting spontaneously with the same purpose as the priest in his official robes, a purpose with far more serious implications than mere self-expression. We have already observed that the cry of an agitated baby is more than just a spontaneous discharge of stressful sensations, acquiring over time (in association with hearing) additional territorial and spatial functions as well, by virtue of the usefulness of a protracted cry both as an acoustic marker and as a means of exerting a powerful leverage on the environment of other people. For a child in any new environment to shout and run about is not only normal but essential exploratory behavior. Adult societies make a distinction between the improvisatory actions of a child and the formal chant and choreography of a religious service or pop concert, but the underlying processes are exactly the same: 1. extension of the self through enhancing the voice; 2. monitoring the environment; and 3. controlling the actions of others.

A long song

Environmental consciousness, digital audio, and a growing interest in world music among the younger population coincide as expressions of a welcome revival of oral sensibilities and cultures among industrial societies. Among the many delights of a recent recording of music of central Siberian Tuva and Sakha traditions are musical imitations of animals and birds, the sounds of children playing among domestic animals, instrumental imitations of speech sounds, and techniques for modulating the voice in the rhythm of a galloping horse, the tremolo of a rippling stream, or reverberation of a cave.

A chant or formal song is a deliberate act. You do it in order to hear its effects. You know what you are doing, and because of that you are very aware of the relationship between what you are hearing within your head and what comes back as reverberation from the environment.

What reverberation tells a listener is true of the environment. Nature does not lie. Because the signal is known to the person singing, and because a chant is a complex and manipulated signal, the environmental information that can be gathered from monitoring a chant is more detailed and more reliable than from a random acoustic signal. If there is any discrepancy between the

sound you know you are making and the sound the environment
is returning to you, it signifies something true and meaningful
about the environment. In this example of traditional song the
performer sings across a river and his voice is reflected from a
cliff opposite.

> Kaigal-ool performs this "long song," so called because
> of its long reverberation time, in the traditional fashion,
> with the reverb provided by a cliff. Kaigal-ool recalled
> that herders often used to sing to one another across the
> banks of a river because song carried better than speech.
> "It wasn't easy to cross rivers then," he said. The cliff in
> this recording forms the north bank of the Kaa-Xem, a
> river which joins the great Yenisei at Kyzyl. Kaigal-ool
> stood on the south bank and sang across the river, his
> voice reflected by both the cliff and the water.[4]

Ritual chants across the world share a number of distinctive features related to their primary signaling and monitoring functions. Bursts of singing alternate with periods of silence as the
reverberation is assessed. A ritual chant is often repetitive in
nature, to reinforce a basic acoustic image, but also varied in
subtle detail of text, rhythm, and inflection, to examine the effect
of particular variations in signal strength. What is being listened
for are consistencies of reflected sound over an extended sample
period that indicate permanent rather than transient features of
an environment that may in fact be invisible, for instance, in the
dark of a cave (this particular recording took place at dead of
night).

What is remarkable about this recording is that a listener can
clearly hear that the voice reflected from the cliff beyond the
river *is an octave higher*. It is not the same male voice in the tenor
register produced by the singer, but a reflected voice in the soprano range. There is an acoustic explanation for the difference
in pitch. For the speaker or singer low and mid-range frequencies
are monitored internally and only noticed if the wall surface is
nearby and the time delay therefore so short as to cause the
reflected sound to merge with and amplify the original voice.
The deep tones that add authority to a radio or television newscaster are easy for a listener at home to hear because they are
picked up by microphones attached to the speaker's body, but
the same low frequencies do not carry well in the open air over

great distances. You would not recognize a famous voice so easily if the person were calling to you across an open field. When the reflecting surface is at a considerable distance the reflected sound will consist only of that part of the original voice signal that survives the journey and is easiest to hear, which for a human being will be in the region of 2,000-4,000 Hz where the

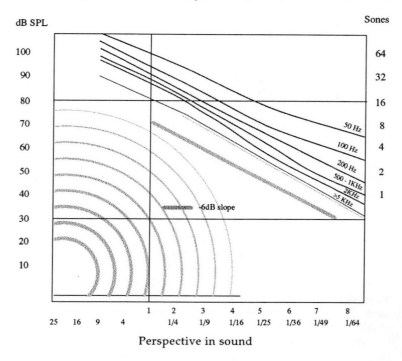

Perspective in sound

ear is most sensitive.

In the diagram above, lines descending from left to right indicate relative sound pressure levels (SPL) at which a listener perceives equivalent declines in loudness (sones) at increasing distances for low through high frequencies. Higher energies are required for low frequencies to remain as audible as high frequencies. The standard calculation, which is a 6dB attenuation in a signal for every doubling of distance, is a fair approximation and is based on a calculation of the thinning-out of energy within a wave-front as it expands away from a source.[5]

A prehistoric vocalist, however, has no technical knowledge

of acoustics, only the experience of sound as it happens, and a firm belief in the truth of sound evidence. To such a person the evidence of ritual chant is that in close environments the voice you hear is your own, more or less enhanced, but in a more expansive acoustic what comes back is *a different voice* at a different pitch: a higher, disembodied voice answering back. The logical explanation for the difference in voice is that the reflection is somebody else, an invisible spirit making its presence known in response to the person chanting. And that is powerful magic.

Robotic speech

Chanting is a form of meditation designed to ease the mind. Inward meditation combines with outward expression. The primordial act of singing resonating within the head radiates coherent tone into the world at large, connecting an awareness of self with acknowledgement of an external reality.

The simplest chant is monotone and without rhythm, that is to say, singing at the same pitch and to a constant and unvarying pulsation. The prayers recited in the courtyard of the shrine of Fushimi Inari Taisha by Japanese Shinto priests during the ritual Fire-Burning Festival (O-hitaki-matsuri) in Kyoto are sung in a monotone with only the merest inflection to indicate a new verse or phrase. It is not a quiet event. There is a crowd of worshipers milling about, and the listener's attention to the chanting is interrupted by the intermittent swish of burning torches. It is music with no melody, no rhythm, no expression, a mode of delivery similar to the style of robotic speech adopted by power-hungry computers in Cold War science fiction movies.[6]

Eliminating all of the essential ingredients of musical expression is a sure recipe for total and absolute boredom, or so one might think. And yet this style of singing is perfectly adapted to its meditative and social role.

Put yourself in the actual situation, in the midst of a crowd out of doors. There is crowd noise and a bonfire crackling away in the center of the courtyard. People are moving about. Somewhere in the midst of the crowd there is a priest. You cannot see him because he is a little old man hidden among normal-sized people. As a worshiper you want to find the priest and get closer to the action and join in the chanting. So what do you do? You listen for the sound of his voice. Monotone chant, in a steady even tone, at a constant loudness, and evenly pulsed, is the *ideal* acoustic signal for locating somebody in a noisy crowd. Noise taken as a whole is random sound without regular features. Campaigning politicians rallying in town squares hire microphones and powerful public address systems in order to make their voices heard above the crowd, but in a world without public address systems you have to find other ways. Vendors at street markets and newspaper boys on busy street corners shout their wares in distinctively musical phrases. Politicians with public address systems rely on their superior power to attract attention. Monotone chanting is a way of making yourself heard without having to shout. Since shouting distorts the signal and consumes energy it is clearly not an efficient mode of communication. It would not be right for a priest to lose control or lose his voice before the prayer is completed. A prayer has to be faithful to the text, and consistent to the end. The Shinto chant is a navigation signal showing a sophisticated understanding of the relationship of tone to noise. A signal that exhibits regular features is going to stand out in noisy surroundings *even though it is not loud.* In addition to constancy of *tone, pulsation,* and *loudness,* the signal also has relative consistency of *location,* which for the listener means a fixed angle of incidence. The ear notices these consistencies in an otherwise chaotic acoustic environment. That is what is meant by the "cocktail effect" of being able to zero in on a particular voice in a noisy crowd. As soon as the ears detect a constant feature in the acoustic environment it is a simple matter to continue to focus on that signal. It is like tuning in to a satellite transmission among all the static.

Try a different analogy. Suppose you are out sailing and it gets dark and the sea gets choppy. You head for port. The sky is clear and the stars are visible, but although being able to see the constellations can tell you in which direction you are heading, that information is useful only up to a point. At a certain angle the points of light change from white starlight to yellowish electric lights of houses along the coast, but the up and down swell makes it difficult to tell sometimes where starlight ends and the shoreline begins. At the harbor entrance there is a light with a difference. It blinks on and off. It doesn't have to change color. It isn't a neon sign. It doesn't flash a message in Morse code. It just blinks on and off, on and off, at always the same pace and intensity. Pure information.

A blinking light stands out. You can see it at once because it is the only light that blinks, and you can keep it in focus because the blinking never varies. That's important because your own position is changing all the time, as the boat rises and falls on the waves and as you head toward the harbor. The Shinto priest's monotone chant works in exactly the same way as a pulsating musical signal in an unstable noisy environment. If there were any variation in pitch or intensity or timing, contact could easily be lost.

Being impersonal

Absence of expression has other meanings as well as demonstrating the values of steadiness and coherence in individual action. Once you have located the priest and moved to a closer position you are able to join in the prayer. A monotone chant is a very easy chant to join. There is no melody to remember. You don't need to locate a particular melody line or chorus. All that you need to remember is the text of the prayer and the point in the text that the singer has reached. Once you have found your way close enough to recognize the words, then joining in with the prayer is no problem. And you sing the same note as everybody else. Singing in unison is the simplest and most basic form of vocal harmony, as well as being a symbolic gesture of spiritual unity.

Robotic speech is speech without melody or rhythm. Compared to normal speech it seems wooden and meaningless, but the absence of personality has a positive side. A monotone robot is saying "I have nothing to express, only information." The stereotype of a robot is of a powerful and single-minded intelligence that acts in consistency with an over-riding mission that has no place for human emotion or sympathy. It is all in the monotone. It is the same for a priest chanting a monotone in a sacred service. The message of monotone chant is clear. It says: "This is not about individual expression. *It is about the truth.* The voice is only a medium. The truth is not to be interpreted. The meaning is all in the words."

Monotone chant can also be viewed as a form of inquiry into the nature of language and the communication of meaning, in particular, how words as sounds are able to convey meaningful information. We know the answer—as sequences of consonant and vowel combinations—but asking the question acknowledges that the process remains a mystery. In order for meaning to be conveyed, distinctions have to be articulated. Monotone chant is speech rendered transparent to hearing by having all emotional or impulsive variables stripped away. When words are spoken or sung in a precise uninflected monotone the differences between syllables become absolutely clear. It is possible to hear different vowels as different resonances. This is the same point of departure as for developers of synthesized speech in the late twentieth century, and the movie robotic style of speech that dates from their early experiments.

The artificial voice of Professor Stephen Hawking has a modest degree of intonation and expression, and listeners accept it as a real voice characteristic of a powerful mind. In the evolution of speech synthesizers, intonation and expression are not issues of primary concern. They come later. First one has to isolate, define, and reproduce the different harmonic spectra of vowels [a] [e] [i] [o] [u]. In order to isolate and measure the differences in harmonic profile that distinguish one vowel from another, you compare how they all sound *at the same pitch*. In a monotone voice or chant the cavity resonances of each vowel are audible as variations in harmonic profile. The ancient Greeks and Egyptians were just as interested in the overtone structures of vowels as the robot creators of the twentieth century, but since they did not possess electronic harmonic analyzers, they experimented with a type of ritual incantation on a monotone in association with a sensing device tuned to the same pitch that could reveal by sympathetic vibration how the harmonic energy associated with each vowel is distributed. In George Crumb's song cycle *Ancient Voices of Children* a moment occurs where the soprano soloist sings into the open piano while the sustaining pedal is depressed, and her words are picked up and resonated by the strings that are in tune with her voice.[7] Ancient civilizations did not have pianos but they did have the harp, an instrument of multiple strings tuned over the entire frequency range of the human voice. If you sing different vowels in the presence of a musical instrument with a range of strings each sensitive to a specific frequency, only those strings in tune with the voice will vibrate in response, and the remainder will be undisturbed. That way, by singing and observing which of the harp strings are set in motion by the voice, it is possible to arrive at an understanding of the frequency distribution of partials in the different vowels.

In the fairy tale *Jack and the Beanstalk,* Jack encounters an ogre whose catch-phrase is *Fee, fie, foe, fum!.* The ogre also has a talking harp. What the story refers to is the popular super-stition concerning early studies of the acoustics of the voice. The ogre character is an object of fear. The land in the clouds where the ogre lives is the invisible world of sounds, sounds that are weightless and that inhabit the air. The talking harp is the ogre's sensing device. The ogre has gold because research of this kind is costly but also extremely valuable, a key to human understanding. Monotone chant is thus an expression of science as well as a statement of values. It deconstructs speech into syllables, and transforms syllables into harmony.

The one and the many

Plainchant is another expression of science preserved as musical art. It takes place in a resonant interior space, a cathedral or church. It is music for voices, solo and group. The voices fill the church space so that a listener can hear not only the singers but the space as well, in the way the singing rings out and dies away. The music is adapted to the acoustic of the space. It moves carefully and with precision from note to note. The singers take their time and do not hurry. There is no driving beat or rhythm to which they have to conform.

In plainchant the manifest purpose of singing is to articulate the meaning of a text. The words are ceremonial and are in the Latin language. Since Latin is not the language of everyday speech, it follows that this singing activity is not about self-expression or communication but rather about saying the right thing at the right time and in the right way. If the words are not exactly right and are not intoned in exactly the right way, the magic will not work because the meaning expressed will not be consistent with its original purpose. Even if the words are authentic, a variation in intonation and timing can change their meaning. Any statement can be changed into a question: "So, you are *certain* about this?" (upward inflection) does not convey the same meaning as "So, you are *certain* about this" (downward in-flection). For a form of words to be preserved in all of its original meaning, it is necessary to adhere to a particular melodic shape as well as to a particular text. Plainchant is about preserving a

Plainchant "Antiphona ad Offertorium"

formalized inflection and thereby the meaning of a spoken text in a form in which it can be written down, memorized, read, retrieved from memory, and passed on from one generation to the next. With the aid of a written notation, an exact pattern of melody can be preserved.[8]

How can one be sure that this form of singing is about preserving a text and its meaning? If only one person were singing it would be difficult to tell. But plainchant is also about a group of people singing the same words to the same melody. That situation can only arise if those who are singing already know what they are supposed to be doing, and it can only mean that the music is composed and not improvised. There is a script. Composed music is music where the melodic shape and the timing are laid down in advance. It tells the listener that the singing conforms to a pattern, and that the pattern has been agreed and accepted. To the ordinary listener who doesn't understand Latin the fact that everyone in the choir is singing to the same script is impressive in itself.

Notice the dialogue in plainchant. Sometimes the words are sung by a soloist, at other times by a chorus. There is only one melody line. This is a very old musical form, when you have verses sung by a lead solo voice alternating with choruses sung by multiple voices in unison (singing at the same pitch). Not much difference between the two, you might think. But while they may be singing the same words, and following the same melody, the messages the solo and chorus are giving out to the listener are essentially and profoundly different.

Charisma

The pay and celebrity attached to being a solo performer have always been a source of friction in the music profession. The lead singer, the conductor, the soloist in a concerto are the star performers. They get top billing, and they get the lion's share of the proceeds while the unfortunate backing vocals, rank and file players, and chorus line barely make a decent living. What is it that makes a star out of a mere soloist? It could be superior technical skill. It could be personal charisma. But mostly it could be that when the lead musician is performing there is always the possibility he or she *might be making it up*.

Audiences are excited by uncertainty. A performance is a dramatic confrontation of memory and actuality, of experience remembered and the same experience in real life. The lead performer intercedes with the world of memory. It is the soloist's responsibility to restore a collective memory to real life in the present. That is what the fans are waiting to hear. But the solo performer can also influence audience recollection in various ways. A change of pace, of emphasis, of melody can transform the meaning of a song. When there is only one person in control, it is possible for that person to rewrite history, or at least give it a new spin. And that is powerful magic. Audiences are attracted to power that can create collective uncertainty and replace it with a new certainty in one and the same instant. As long as only one player is in charge, the audience is going to be in a state of suspense. They have expectations, and want to find out if those expectations are confirmed. The lead singer has that power, and that is why the lead singer earns star billing. The person who can alter the given meaning of a text can change an audience's perception of reality.

It's the truth

A chorus can never do that. When two or more performers sing exactly the same melody, an audience knows immediately that they are singing from the same script. There is no way a group of people working independently could arrive at the same melody by chance, so they could not by any stretch of the imagination be making it up. There has to be a script. A soloist holds an audience in suspense because you can never know exactly what to expect. The experience is one of transcendental doubt. With a chorus the audience knows the meaning is already secure, confirmed, established, written down. A chorus of unison singers doesn't have the clarity or brilliance of a solo voice. When you add half a dozen individual voices together the subtle differences between them cancel out and you are left with an average inflection with fuzzy edges. *But it is the truth.* With a solo player there is always the possibility of a breakthrough, of a new meaning being revealed. People are excited by the idea of revelation or epiphany. But with the chorus you have the security of the truth, and people are also comforted by their belief in the truth and tradition.

In Greek drama the protagonists are understood as heroic or hapless individuals at the mercy of fate but also the ones ultimately responsible for their actions and motives. They are the soloists. Greek drama also has its chorus commenting on the actions of the protagonists. The chorus is still found in opera today, but has been an anachronism in drama from the time of Shakespeare. A chorus is unreal. To modern audiences, having a chorus commenting every so often interrupts the action and breaks the suspense. To Greek audiences, however, the chorus is a necessary reminder that the lead actors are working to a script, not the author's script as much as the script of the gods. The main characters may act as though they have freedom of choice. It may look as though the events depicted have happened by chance and that the players are inventing their own future. But of course we know they don't. Not because the gods are in control, but because the moral laws under seige cannot be overthrown.

Why not just talk?

So why sing at all? Would it not be sufficient if everybody spoke the text in unison? There are two ways of answering. First, if you speak the text you are speaking in *your own voice*, and there is not the same degree of control over the way you say the words; in turn that means less certainty over what you or anyone else will understand by the words, since the meaning of the words is dependent on the rise and fall of the voice. By singing to a given melody the soloist is, in effect, accepting a meaning that is given, not one that you make up for yourself. This is all part of belonging to a society and accepting common rules of behavior. And in sharing a particular melody the chorus singers are meditating on that collective meaning rather than each one trying to work it out individually. Chorus singing is a social act, and the symbolic expression of agreement on common goals is an important aspect of all ritualized behavior. These remarks apply to all music, because the simplifications of melody and timing that distinguish music from speech serve a similar purpose in every culture where particular forms of words of social significance are to be remembered and reproduced on special occasions.

Notes

1. Iegor Reznikoff and Michel Dauvois, *Bulletin de la Société Préhistorique Française* 85, 1988, 238-46.

2. Steven Wailer of the American Rock Art Research Association, as reported by Leigh Dayton, "Rock Art Evokes Beastly Echoes of the Past." *New Scientist*, 28 November 1992, 14.

3. David Lubman, in a paper presented to the Acoustical Society of America. Natasha Loder, "Findings in Brief," *The Times Higher Education Supplement*, 16 October 1998.

4. Kaigal-ool, singer, performing "Kyzyl Taiga" (Red Forest). In *Tuva, among the Spirits: Sound, Music and Nature in Sakha and Tuva,* prod. Ted Levin and Joel Gordon (Smithsonian Folkways SFW 40452, 1999), track 16. Liner note by Ted Levin.

5. After Deborah Ballantyne, *Handbook of Audiological Techniques* (London: Butterworth-Heinemann, 1990), 18.

6. "O-hitaki-matsuri" (Fire-Burning Festival), chant by unidentified celebrants. In *Kagura: Japanese Shinto Ritual Music* rec. János Kárpáti (Hungaroton SPLX 18193, 1988), track 7.

7. George Crumb, *Ancient Voices of Children* (Elektra Nonesuch

79149 2, 1970).

8. Offertorium "Domine Deus, in Simplicitate," Gregorian chant from the Proprium Missae in Epiphania Domini. Choralschola der Benediktinerabtei Münsterschwarzach, dir. Godehard Joppich, *Music of the Middle Ages: Gregorian chant, music of the Gothic era* (DG Klassikon 439424 2, 1982), track 6.

CHAPTER SIX

Directions

THE original purpose of notation is to preserve voice intonation. For that to happen we need to organize the slipping and
sliding fluctuations of everyday speech into a fixed, stepwise
scale of pitches from low to high. Preserving voice intonation is
important because written words alone do not convey the full
sense of a text, and if the text has moral or legal implications for
society, it is especially necessary to preserve an appropriate
mode of utterance if its meaning is not to become corrupted.
Because the human vocal apparatus is made of soft stuff and is
therefore unreliable, the simplest way of organizing a fixed scale
of pitch values is by constructing a musical instrument that can
produce a set scale of pitches.

The first Western music notations were not written at all.
They survive in actual musical instruments such as the panpipes,
the lyre, and the harp. Such instruments embody a complete
scale of pitches: a graduated series of tubes in the case of the
panpipes, a set of parallel strings graduated in *tension* in the case
of the lyre, and a sequence of strings graduated in *length* in the
case of the harp. The normal speaking or singing voice has an
expressive range of about an octave, the distance between a
fundamental tone and its first harmonic, or between the tone of a
stretched string and the tone of the same string at the half-way
point. An octave in the diatonic scale is eight notes in sequence
with some 12 percent separation in frequency between successive

pitches. Musical instruments with a separate string or pipe for each note of the scale are tunable in advance to a desired scale. Since many cultures regard the choice of scale or mode as expressing a particular mood, such instruments can be *pre-programmed to a given emotional state* from which the correct intonation then follows automatically. Western music recognizes the emotional implications of major and minor keys, and non-Western cultures such as India make a point of identifying the chosen mode at the beginning of a piece, and matching the mode of a performance to the appro-priate season and time of day.

The association of mode and mood comes from observation of the way people actually speak. If a person is excited and elated, the voice rises in tone and brightness, and differences in pitch are strongly defined; if someone is depressed, the voice drops in tone, sounds duller and flaccid, and less effort is made to distinguish levels of intonation. The Greeks valued excitement, and the tension and clarity that go with it. For this reason serious philosophers considered the lyre and its scale of tension values a more valid model for the singing voice than the even tone values of the harp and panpipes. The art of correct intonation consists of matching a particular inflection to a precise set of pitches. Since the general meaning of a text is in the words, its corresponding inflection curve is largely a matter of common sense.

The basic outline of a melodic *shape* is usually not difficult to recall, especially if the text is available in written form. The accompanying instrument is available in that case to ensure that the familiar shape is intoned using exactly the right set of pitches to convey an original intention. As a researcher and cataloger for the British Library confronted with the practical difficulty of identifying music titles for library customers unable to read music, and for whom conventional dictionaries of musical themes were no help, Denys Parsons devised a thematic catalog based on contour alone for which a customer only had to remember the up and down motion of a melody. Remarkably, he found that it is possible to identify the start or *incipit* of a famous tune or theme (the part people remember best) by the changes of direction encoded in a relatively short sequence, with an asterisk for the first note. For example, the opening of Burt Bacharach's song "Raindrops Keep Falling on My Head" is expressed as the sequence:

*RRUDD DUDUR RUDDD

and "God Save the Queen" (or "My Country 'Tis of Thee") as:
*RUDUU URUDD DUDDU [1]
There is a considerable body of evidence from research in musical memory that people recall a melody as a shape or contour rather than as a succession of pitches. What that tells us is that in order to store and reproduce an exact form of intonation, early civilizations would have had no option but to develop instruments in which essential pitch relationships were preserved.[2]

Steps and arenas

Mode is mood, but scale is measure. Words are clues. Someone who constructs or tunes a musical instrument can tune it to a set of pitches that suits an individual voice. Some people have lower and some have higher-pitched voices, but in general one expects a person's voice to conform to outward stature, a small person having a higher-pitched voice and a giant a deep booming voice. It doesn't always work out that way. The popular image of actor John Barrymore, a stereotypical hero of the silent movie era, was damaged beyond repair with the arrival of sound film, which revealed his voice as unheroically lightweight and high-pitched, not to mention refined.

Like a suit, a musical instrument can be tailored to the pitch of an individual performer. However, designing structures to serve a community of speakers or musicians involves finding a compromise tuning acceptable to a range of voices of different pitch. Among early Western civilizations the School of Pythagoras paid particular attention to the derivation of scales or standard measures based on the average male or female stature. Its efforts were directed less at corporate music-making, as we understand it today, than to the acoustical requirements of venues for public speaking and dramatic presentations. In their search for standard measures of pitch corresponding to wavelength, the Pythagoreans observed and took into account the acoustic character of intuitively designed public spaces and arenas.

Surviving examples of ancient Greek and Roman architecture, such as the Parthenon, the Theaters of Dionysos and Herodes Atticus in Athens, the Temple of Apollo at Didyma, and the Roman Capitolium at Nîmes, all embody structural features that can be interpreted as imposing uniform standards of intonation on speakers, actors, and singers alike. The Dutch astronomer Christiaan Huygens, on a visit to France in the late seventeenth century, noticed that the sound of a fountain reflected from a nearby set of steps was tuned to a specific pitch. A sensation of pitch is not normally associated with a fountain because its sound is among those classified as "white noise," with its energy

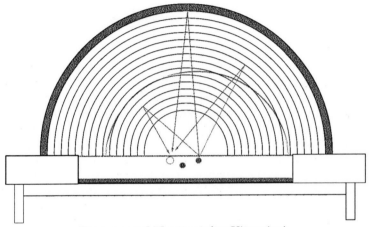

Roman amphïtheater (after Vitruvius)

being distributed across the entire audible spectrum. Huygens deduced that the tone he heard was an effect arising from the regular spacing between the risers of the stairway producing a cascade of reflections marching in step, as it were. We can be quite sure that he was not the first person to notice that staircases can impose a tuning or tonality on environmental sound. If such an effect is audible for white noise, it is likely to influence a listener's perception of sounds of human origin as well.

Stairways leading up to the entrance of temples and offices of government are a prominent feature of Greek and Roman architecture and are still a significant element of the architecture of law, religion, and government today. It is still possible for a listener to stand as Huygens did, at the foot of the steps leading up to the main entrance to a major public building, and hear the roar of traffic reflecting from the stairway as a shadowy tone of constant pitch.

Unlike a public stairway where a musical effect is created

spontaneously, the structure of a Greek or Roman amphitheater is designed specifically for acoustic performance. Its semi-circular shape shows a clear understanding of the spherical propagation of sound, and allows members of the audience at the extremes on either side to hear as clearly as those in the more desirable seats in the center.

For actors or vocalists on stage the semi-circular shape and regularly stepped seating of a classic Greek or Roman theater create a combination of acoustic conditions that are bound to influence not only the style but also the form of a dramatic or musical performance. Like the stairways leading up to a temple, the seating reflects sound back to the actors or singers on stage at a fixed repetition pitch. The seating, in effect, *imposes a tonality* on the theater, and the interesting thought is whether the acting or singing on stage are influenced by it. It has been suggested by some specialists in architectural acoustics that such a feature amounts to a serious error in design:

> The effect can easily be demonstrated by speaking, singing, or clapping hands on the stage of a typical Greek or Roman open-air theater. The sound produced from the tiers of benches produces a sustained echo whose characteristic pitch is determined by the distance separating adjacent risers. As a result, when speech or music is heard in an open-air theater, the reflected sound may suffer a serious distortion in frequency. Fortunately in an open-air theater these frequency-dependent reflections generally pass over the heads of the audience, but since the reflections come to a focus on the stage they can be extremely disturbing to performers rehearsing in an empty theater.[3]

However, it has to be said that the requirements of twentieth-century architectural acoustics are very different from those of ancient Greece or Rome. Today we have to cater for orchestras of seventy to one hundred players, then it was a few singers and instruments. Today music is complex and unpredictable in key, whereas in Greek and Roman times the guiding principles were uniformity and consistency of tone. It does not make sense to impose present-day assumptions on the artistic practices of two millennia ago. A more profitable line is to take the acoustics at face value and consider their impact on musical structure and performance.

Musical considerations figure significantly in the ten-volume *De Architectura* by Roman architect and builder Marcus Vitruvius Pollio, evidence perhaps of a greater awareness of architectural acoustics than we may have been led to believe. In laying out recommendations for the design of a Roman theater, he summarizes Aristoxenus on the subject of the Greek modes in their original sense of "moods" or *modes of speech*: normal, angry, subdued, and so on. Aristoxenus declares that there are certain fixed pitches, corresponding to A, B, D, and E in modern notation, that are common to all modes, their intervening pitches varying according to the chosen modal character.[4]

Elsewhere, in specifying the dimensions of seating, Vitruvius remarks "The steps for the spectators' places, where the seats are arranged, should be not less than a foot and a palm in height, nor more than a foot and six fingers; their depth should be fixed at not more than two and a half feet, nor less than two feet." Such measurements, of course, have to be interpreted according to the Roman system, by which a foot is one-sixth of the ideal height of a male; a foot consisting of four palms, and a palm of four fingers. The exact length of a Roman foot is still a matter of scholarly controversy, but the *proportions* of the ideal Roman figure are recorded in sculpture and art, most notably in the Vitruvian figure of Leonardo. Since the average height in Roman terms was closer to 5 feet 6 inches, a foot of one-sixth average height is 11 inches, and a palm 2¾ inches. These measurements turn out to agree with my own average Roman dimensions and give values for Vitruvius's seating of 13¾ - 15¼ inches for rise, and 22 - 27½ inches for depth. In calculating the repetition pitch, depth of seating is the significant dimension, and in this case the pitch corresponding to a wavelength of 22 inches is E, and for 27½ inches, B, in general agreement with Aristoxenus.[5]

What Knudsen means by "distortion of pitch" is an effect that arises from a mismatch of the voice with the repetition tone. When the two coincide, the voice is strengthened; when they do not, the voice is weakened and may waver uncontrollably. In performance terms this is likely to lead to a style of intonation centered on pitches relating to the repetition pitch of the seating. In drama by tuning the voice to the prevailing acoustic a hero's voice can sound strong and mighty, a sinister character unsteady, and a transformation of character such as the fall of a hero can be indicated by a change of mode. Furthermore, the inverted dome of a Greek or Roman amphitheater lends itself to symmetries of action and dialogue. For a person standing exactly at center stage the echo is heard at the same point, but for a character left or right of center the reflection from the auditorium will come to focus at an equidistant point beyond the center. It follows that if actors or musicians in such an arena are to conduct dialogue and hear one another clearly, the most efficient way of doing so is by arranging themselves in symmetrical formations.

Over a period of centuries one can imagine amphitheater acoustics to have influenced the evolution of staged drama and music in certain key ways: first, by reinforcing a style of intonation consistent with a fixed tonality, and second, by encouraging symmetrical formations. Evidence of the former may be found in the spread throughout Europe of drone instruments such as the bagpipes, and of drone-based ritual singing, for example, the archaic Beneventan liturgy known as Ambrosian chant.[6] Symmetrical formations, on the other hand, are a commonplace in most religious art, for example, the triptych, as well as classical ballet.

Notations

Once a scale of pitches is agreed a musician can begin to think about expressing it in written form. In the example of "Antiphona ad Offertorium" no less than *three* notations are competing for attention. It is not necessary to read musical notation in order to appreciate the different levels of precision and the different priorities expressed in each notational style. Above the four-line stave a reader can see an angular notation resembling a form of modern shorthand. This early form of musical sign language gives direction, but not pitch. It is like traffic signs that read "Turn Right" or "Straight Ahead" or "Stop." Such signs assume correctly that you know where you are coming from, so all you are asking is where to go from here. Each sign corresponds to one syllable. Plain ticks and dashes are plain notes; the more complicated hooks, loops, and curlicues stand for inflections of different kinds that correspond to units of expression in real speech, such as the up and down inflection of "Wow!" or the downward slide of "Oh no!"

Notation of the shorthand variety gives direction but nothing else. The singer needs an instrument to establish the mode and pitch of the voice. In prescribing inflection but not pitch the

shorthand style notation is allowing for the possibility of adjust-
ing the pitch level to suit the voice range of the individual singer.
Timing is also left to the individual. Note too that a choir reading
from this form of shorthand notation would have to follow the
intonation set by the lead voice, since it is unlikely to be

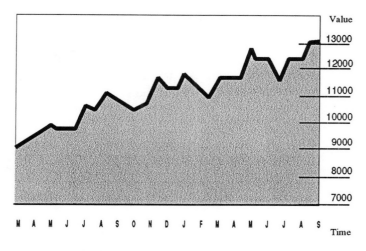

A typical graph derived from musical notation

interpreted spontaneously at the same pitch by every singer. So
this is not by any means a universal notation.

Stocks and shares

The four-line stave is a later development and a considerable
intellectual advance. If your average day begins with hot choco-
late, croissants, and the financial pages, take a good look at the
graphs showing movements in share performance during the
past week or year. A typical movement is a zig-zag line reading
from left to right, superimposed on a graph of horizontal lines
corresponding to a scale of values, and vertical, to units of time,
usually days. By following the curve and direction of the wavy
line, a reader can see how the market has fluctuated and estimate
where it seems to be heading. The concept of *movement* is embodied in the image of the wavy line, which a reader is intended
to see as *a dynamic progression*. In reality the wavy line is derived

from a series (usually regular) of *sample readings* taken at widely separated time intervals, and joined together by straight-line connections after all the samples have been recorded.

Your stocks and shares graph actually owes its existence to the generations of working musicians who began developing a universal standard music notation a thousand years ago. The

Plainchant melody as a line graph

four line notation of plainchant incorporates exactly the same kind of information as the graph in the financial pages of your daily paper charting the movement of oil prices or pork belly futures. It describes the movement of a voice part over time on a horizontal four-line graph or stave between arbitrarily selected low and high values of pitch with reference to a standard key signature or reference pitch (the angle bracket or clef heading the top line of every stave). Unlike the stock market graph the progression of a plainchant melody is laid out as a discontinuous sequence of samples. In one or two places data points (notes) are run together into ideographs representing tremors or movements of the voice within a single syllable. The horizontal coordinate indicates time order but not chronological time—for the historical reason that mechanical clocks did not yet exist, and for the practical reason that the timing of a sacred text is subject to local acoustics and the requirements of clarity and intelligibility.

Global positioning

This notation is not limited to giving directions. It embodies a universal code of pitch that anyone may read. Compared to the shorthand form of arrow signs above the stave, four-line plainchant tells you where you are in an absolute sense; not just where you have to go, but, like a satellite navigation system, exactly where you are in relation both to your starting point and final destination. A music notated in this way is not designed just for the lead singer but can safely be read by a number of voices or instruments simultaneously. This is a largely non-hierarchical notation, each basic note shape representing the same value. That value is positional information, meaning where a note is located in the scale and its position in the overall sequence. The graphics are uniform and easy to follow. Square and diamond-shaped notes, as well as the narrower vertical lines that connect them, are all derived from the thick horizontal and thin vertical marks a regular square-cut quill pen makes when held in a fixed position and manipulated efficiently. This is a notation designed for speed, clarity, and a high degree of standardization. It is a notation consistent with the concept of low and high pitches as degrees in a uniform and continuous scale corresponding to numerical quantity or absolute pitch. Nothing in this style of notation would suggest a tension gradient from low to high, which in turn aligns the thinking behind plainchant notation with the same Pythagorean science that valued the uniform tuning of the harp or pipe organ over the tension gradient of the lyre.

Punctuation marks

In his pioneering study of Western oral culture *From Memory to Written Record*, Michael Clanchy draws attention to the fact that prior to the Renaissance books were *always read aloud*.

> Traditional monastic reading (*lectio*) was more a process of rumination than reading, directed toward savoring the divine wisdom within a book rather than finding new ideas or novel information. As medieval Latin was an artificial language without any native speakers, an agreed correct punctuation—and hence correct pronunciation —helped to make sense of it, for both reader and listener. Roger Bacon in 1267 discussed the correct pronunciation of words in the Bible, particularly those with Hebrew or Greek roots. He places accentuation and punctuation under the heading of music: "because all things consist in the raising and lowering of the voice and are therefore like some kind of chant, it is obvious that the explanation of all these matters pertains to music. . . . John of Salisbury remarked that the ancients had used certain 'notations' (*notae*) to distinguish the 'modes' of writing, 'so it may be grasped which bit in them is clear and which obscure, which certain and which dubious, and many other things of that sort.'"[7]

Reading aloud and memorizing a text are still major learning
techniques among many of today's surviving religious and oral
cultures. In today's world libraries are places for silent study,

. ; : , " " ... ! ? ~ ´ ^ ` —

Punctuation marks derived from medieval plainchant notation

but in medieval Europe they were literally hives of activity, as
readers of rare and precious volumes focused on the sound of the
words as they read them aloud to themselves. Children at infant
school today still learn to read by reading aloud, in order to
grasp how it is that the words they read have meaning. For the

pá- ru- it   e-   i Dó-   mi- nus. •

Line of plainchant terminating in a *quilisma*

adult population silent reading may be more acceptable socially,
but it is not the same activity and is not designed to commit information to memory or to discover exact meaning.

The *modes* to which the medieval scholar John of Salisbury
refers in passing are like musical modes or scales, corresponding
also to *moods* or emotions that are embodied in the range and
manner of fluctuation of the voice. John acknowledges the persuasive force embodied in these "tiny notations" (*notulis*) "by
which singer musicians are able to indicate many variations in
the highness or lowness of the voices by a few marks."

These "few marks" correspond to the earlier shorthand notation visible above the plainchant four-line stave. The same power
to influence meaning has come down to the modern reader in the
form of punctuation marks and accents on the printed page. The
same dots, dashes, and curls that indicated voice direction to the
medieval singer of a holy text and eventually evolved into four-
line stave notation, also became formalized into the punctuation
marks of modern print. Since reading is a largely silent procedure in modern times, the musical implications of punctuation
are often forgotten. But whenever a text is read aloud, punctuation marks and accents are correctly interpreted as indicators of
relative pitch, timing, and inflection. The sign for a question

mark [?] corresponds exactly to the rising inflection of a *quilisma* (shown in the "Antiphona" as the rising inflection of "dicentes" at line 6 before the double bar, and in the second verse at "Dominus," the last syllable of the last line). The rising inflection of a question mark in modern speech means an answer is expected; in medieval plainchant a quilisma signifies that the verse is ended; it is a cue for the chorus to begin, "Your turn."

A modern reader interprets quotation marks as signs for a change in pitch, commas and periods as signs for breaks in the flow, and question and exclamation marks as signs indicating emphasis and inflection. They are musical signs in every respect. All that they lack in comparison to standard four-line or five-line music notation is a sense of absolute pitch and timing. In his study *Radio Drama*, Ian Rodger remarks:

> The emulation of natural speech in dramatic dialogue required much careful notation. This response to the revelation of real speech and the silences within it was not consciously realised by writers at the time [the mid-1950s]. Playwrights like [Harold] Pinter and [Samuel] Beckett began to look upon the business of constructing a play more in the manner of a composer of music. . . . The use of the word "pause" was soon found by the writers to be too vague and Beckett was one of the first to start using dots instead of pauses to indicate the length of hesitations. There was as a result an apocryphal story about Beckett once rebuking an actor for pausing for only three dots instead of four.[8]

Rodger omits to mention that Beckett's use of dots to indicate timing reverts to a medieval notation, the *distropha* (two dots) and *tristropha* (three dots) of plainchant—though, to be sure, these notations in a plainchant context signify not silence but prolongation of a syllable. For examples of a two-dot prolongation look at the syllables "lae-tus" in line 1 of "Antiphona," and for three-dot prolongations look at "Dominus" in lines 3, 5, and 8 (78, 93).

I said earlier that the purpose of singing is to articulate a text. When we listen to plainchant it is obvious that the text is not being articulated in a natural manner. So if the original purpose of notation was to preserve a particular inflection, that has been superseded in religious ritual by a more decorative and formalized style of delivery, akin perhaps to the elaborately decorated letterforms to be seen in manuscripts such as the Book of Kells. Plainchant self-consciously lacks the dynamic features of natural speech, namely, its flexibility in pitch and timing. In their place are features expressing *constancy* and *standardization*: a scale of pitches endlessly recycled so that the mode or mood is firmly imprinted on the memory, expressed in a natural pulsation that divides the text into sound units that the ear and mind can evaluate. Note the trivial but important feature that it is the *vowels* that carry the burden of the melody line.

Because plainchant is sung at an optimum loudness for the available space and does not raise the voice to emphasize a point, other ways have to be found to draw attention to significant features of a text. In its place plainchant has evolved textual enlargement techniques to focus attention on particular words. For example, in "Antiphona" the words "Dominus" and "Deus," references to the deity, are stretched out like product names on an advertising hoarding or on the side of an interstate freighter to show that the subject referred to is something really rather big.

Form

Form expresses those features of an image that allow it to be perceived as a whole, and as something that is designed for a purpose and not by accident. In speech that is formalized into poetry or music, form is expressed in structural patterns and sequences that are normally avoided in natural language but which serve to aid the process of memory storage and correct recall of information. We can call them the three Rs: repetition, rhythm, and rhyme.

Advances in software design allow for more complex and detailed programming. The enhanced definition of four-line notation allows a plainchant programmer to visualize musical compositions that are more complex in their inner structure in ways beyond the conception of a non-reader. One has to be able to read this music in order to perform it, and indeed music beyond the capacity of oral memory is now feasible. The solo singer is responsible for guiding verses and choruses alike, whereas the choir is heard only in the less demanding choruses. The verses include display features that are not only more demanding technically and mentally, but also involve exact repetition, for example, the extended flourishes on the words "Dominum" (78, line 4) and "Dominus" (line 8). Elaborate melodic repetitions such as these declare their origins as software-related inventions: they would not have been invented had the means of writing them not already existed.

Artistry and aesthetics in the modern sense of artistic self-consciousness and "star quality" have to do with demonstrations of special skills and techniques. Despite its religious function, plainchant is not immune from artistic self-advertisement. It requires exceptional breath and tone control to manage extended phrases and sequences of syllables with consistent purity and intensity of tone. In addition the soloist commands a greater range by three or four notes than the more modest vocal demands of the chorus, and he (it is always *he* in this politically incorrect era) needs unusual stamina to remain for long stretches in the more physically demanding but vocally more brilliant upper voice range, as in the "Dominus" flourishes already mentioned, and throughout the second verse. of the "Antiphona."

The qualities accentuated in the solo voice part express the virtues of self-control, charisma, physical endurance, memory, focus, dedication, and quickness of response, qualities associated from ancient times with courage and moral leadership.

Love song

In a delightful arrangement of possibly the oldest secular love song surviving in manuscript, the fourteenth-century "Mandad' ei Comigo" by Spanish composer Martin Codax, formal and expres-sive features of early music are drawn together in direct and simple imagery.[9] The poem has its own shape and form reflected in a strong melodic line and asymmetrical three-line verse structure of 3 + 4 + 4 measures. The vocal range is a six-note diatonic scale in the key of C. The voice is accompanied and assisted by a band of players, each with a distinctive function like a modern rock and roll band: folk harps to keep the singer's voice in tune, a drone played by a hurdy-gurdy to anchor the mode, small hand drums to maintain the rhythm, and a lower-sounding tambour or shallow drum to maintain the beat. The singer is female; the subject is love—"My love is coming home" —but this is not a private or sentimental expression of emotion but a proud and public declaration of marital fidelity expressed

"Mandad' ei Comigo" as interpreted by Sinfonye

in a repetitive verse structure and strong beat that invite the listener to dance. Such features distinguish secular music, with its sense of emotion and physical movement, from the religious music of the period.

Pérotin: Viderunt Omnes

Pérotin (we don't know his first name) was a French composer based in Paris, possibly at Nôtre-Dame cathedral, in the late twelfth and early thirteenth century. This work is an early example of multi-voice writing in four parts of which one, the *tenor* or *holding* part, has the function of keeping the other more decorative voices in tune. This is *art* music because it is about more than just preserving and reproducing a sacred text. The name of this kind of multi-voice music is *organum*. The music works simultaneously on two distinct timescales: in slow motion on a superhuman scale, and also in real time.[10]

A Gothic cathedral is a structure on a superhuman scale. The visitor's first impression of a cathedral is a sense of awe at its enormous size. It says, "this is big." The size of the structure is designed to inspire awe at the magnitude of divinity. A visitor approaching and entering the cathedral however becomes aware that this grandiose shell is encrusted with creatures and figures on a human scale. Some of them are solemn bishops and saints in prominent view standing in rigid and upright attitudes (the good guys). Others, however, are grimacing and smirking figures in humorous attitudes, tucked into corners and unexpected places.

In Pérotin's setting magnitude is identified with the opening words of the sacred text. The five syllables of "Viderunt Omnes" are stretched almost to breaking point, magnified right out of the timescale of human perception and rendered incomprehensible to a lay audience except as abstract speech sounds (in the recording the two words are prolonged for more than three minutes). Naturally these extended tones are sung to vowel sounds, since consonants, being transitional noises, cannot be sustained indefinitely in the same way. It is a way of saying that God can only be praised in vocal sounds that are intrinsically *harmonious,* and not in noisy consonants that belong in the same category as the sounds of indelicate bodily processes. Although the sacred words are stretched beyond normal human understanding, all four voices are nevertheless exactly synchronized and no single voice has priority. At the superhuman, sacred level, voices and text are consistent, therefore, with an over-riding image of harmonious union, solemn and uniform in tone, and attentive to correct textual and ritual protocols.

The second or secular level, however, expresses a human scale of a kind that can be seen in the sidebar decoration of a medieval illuminated text. In "Viderunt Omnes" human interest is revealed in jaunty dance-like voice rhythms that interweave like a maze. On a human timescale the tone is lively; each voice part is free to go its own way, and the whole group dances in and out of harmony with cheerful abandon and occasional grinding dissonances. Like the gargoyles on a cathedral they introduce a secular, joyful, irresistible human element to leaven the solemnity of religion and bring the larger structure to life. Pérotin's wonderful piece is saying, in effect, that it is possible to be a part of a higher spiritual enterprise and at the same time susceptible to the pleasures and conflicts of human weakness. To have created a music balancing two completely different timescales and temperaments simultaneously is a remarkable artistic and intellectual achievement.

Notes

1. Denys Parsons, *Directory of Tunes and Musical Themes* (Cambridge: Spencer Brown, 1975).

2. Diana Deutsch, "Memory and attention in music." In *Music and the Brain: Studies in the Neurology of Music*, ed. Macdonald Critchley and R. A. Henson (London: Heinemann, 1977), 112.

3. Vern O. Knudsen, "Architectural acoustics." In *The Physics of Music*, ed. Carleen Maley Hutchins (San Francisco: W.H. Freeman, 1978), 79.

4. Vitruvius (Marcus Vitruvius Pollio), *The Ten Books of Architecture*, tr. Morris Hicky Morgan (1914. Reprint, New York: Dover Publications, 1960), 141.

5. Vitruvius, *Ten Books of Architecture*, 147-8.

6. *Chants de la Cathédrale de Benevento*, Ensemble Organum cond. Marcel Pérès (Harmonia Mundi HMC 901476, 1993).

7. Michael T. Clanchy, *From Memory to Written Record: England 1066-1307*, 2nd ed. (Oxford: Basil Blackwell, 1993), 269-86.

8. Ian Rodger, *Radio Drama* (London: Macmillan, 1982), 98-9.

9. Martin Codax, "Mandad' ei Comigo." Sinfonye cond. Stevie Wishart, *Bella Domna* (Hyperion CDA 66283, 1988).

10. Pérotin, "Viderunt Omnes." Early Music Consort of London dir. David Munrow, *Music of the Middle Ages* (DG Klassikon 439424 2, 1993), track 8.

CHAPTER SEVEN

Space

MUSIC occupies space as well as time. The space may be large and purpose-built, like a concert hall for a symphony orchestra, or just large, like a cathedral. Music is also designed for smaller spaces, as Haydn and Mozart designed their music to be performed in eighteenth-century private salons or ballrooms. The lute songs of Dowland and J.S. Bach's solo partitas are examples of *chamber music* composed for the private enjoyment of a small number of performers and listeners in a room such as a domestic study or drawing room.

What is the purpose of a musical space? We tend to think of a room as a container for people, an audience. That is true to the extent that a live musical performance today involves other people, and it is they who pay to listen, and thereby provide the musician with a living. Bear in mind, however, that although a work of music may be designed with one particular space in mind, that ideal space may not correspond to the space in which a live performance actually takes place. When the performing space is not ideal, the music changes. An intimate song by Schubert designed to be heard in a private house may instead be performed in a large public concert hall, under conditions that make the expression of intimacy difficult or paradoxical, simply because in the larger space solo voice and piano have to work harder to make themselves heard.

The Symphony No. 3 for organ and orchestra by the French

composer Camille Saint-Saëns is designed to be performed in a nineteenth-century town hall on a town hall organ. A town hall has the right acoustic for this unusual combination of grand organ, an instrument traditionally associated with a cathedral acoustic, and symphony orchestra, an ensemble suited to a less complex and reverberant acoustic. Organ soloists who get the opportunity to make a recording of the Saint-Saëns symphony naturally want to make the most impressive noises on a very grand and important instrument, and the best noises in terms of performer satisfaction will always be associated with a cathedral organ in a cathedral acoustic. Most organists of necessity tend to grow up in a church environment, as a result of which many don't always realise that while a cathedral acoustic is fine for an organ, it is not the best acoustic for a symphony orchestra. But you can't simply unplug a cathedral organ and take it into a recording studio where the acoustic is more controllable. The organ is *built in*; the cathedral acoustic is part of the package. That being so, a producer has to juggle with two very different acoustic requirements when recording an organ symphony or concerto in a cathedral, in order to ensure that the same rolling reverberation that makes the organ sound magnificent does not make the string section of an orchestra sound foggy and sluggish. The reverberant acoustic of a cathedral is not designed with an orchestra in mind. It is great for slow, grand sequences, but no good at all for the fast-moving music that an orchestra is normally expected to play.[1]

Music itself tells us what acoustic it is designed for. This is as true of birds as of the music of human beings. An interesting study of vocalization in birds has shown that in a dense forest habitat birdsongs are found to concentrate within a narrow 1.5-2.5K bandwidth and they also incorporate more pure tones. In less dense habitats there is not so much interference by trees and foliage, allowing for a wider variation in pitch. However, air turbulence in more open environments can also distort signals, hence birds avoid pure tones in their songs and tend to rely more on temporal coding (i.e., rhythm).[2]

The guiding rule for a listener is: Is the musical image clear? Can I hear what is going on in detail? If the sound effect is blurred or muddy, then it is likely that something is not quite right. Example: You have a camera. The point of having a camera is to take pictures. You don't need to know about lenses and focal lengths and film speed in order to understand that if a picture comes out blurry something was not done right. The function of music is to illuminate an acoustic environment, but music is also about an ability to hear and appreciate complex *sound*, and if the images in a performance are not clearly defined, so that an audience cannot clearly hear what is going on, then questions arise. Among the tracks from a recent sampler CD issued by a respected early music label is a robust but puzzling excerpt from the medieval passion play *Carmina Burana*.[3] Sampler CDs are highly recommended and always good value because, in addition to retailing at a budget price, they represent a label's own choice not only of quality performances but also of quality recording, a choice that invites the listener, in turn, to make critical comparisons. In this case, however, the performance is rushed, the acoustic muddy, and the text unclear. A recording in which the balance of voices and acoustic appears less than ideal suggests a number of possibilities. Sometimes musicians get carried away, relying too much on their reading and not enough on listening skills. Speed singing is a team reading skill requiring everybody to have their own copy of the music and a conductor to beat time. It is not however the kind of skill generally associated with authentic medieval church practice where there would normally have been only one large-format music book to read from, with the choir following the voice and finger of a lead singer. The same fast tempo might be perfectly appropriate for an outdoor performance where there is little or no reverberation to be taken into consideration, but too fast for singing inside a church where the acoustic of the space is bound to be taken into account.

In this case one is left with a suspicion that the on-location acoustic may have been subsequently tampered with in the studio. It would be unfair to blame singers for singing too fast or not listening if the reverberation a listener hears in the recording is not the original church acoustic. Listening carefully to this particular track with the bass extension engaged, it is possible under all the reverberation to detect the rumble of passing traffic. Traffic noise is a recognized hazard of recording in old churches. Clearly, something has to be done to minimize the intrusion of outside noise in a commercial recording, and one way of eliminating low-frequency noise is to filter out bass frequencies in the studio. However, suppressing the bass makes a recording sound brittle and lacking in warmth, and to compensate for that loss of warmth the producer may subsequently add artificial reverberation in the bass—a treatment now available on most domestic audio players. Any decision to change the acoustic of a musical performance after the fact, however, even to conceal low-level interference in an original recording, will have

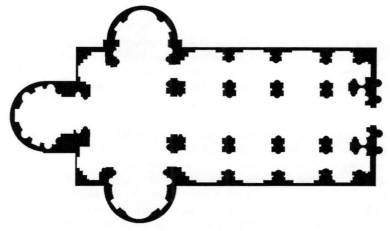

Interior of an imaginary baroque church, showing
the increased surface area contributing to a long
reverberation (after Palladio)

the effect of throwing the originally recorded relationship of voices and acoustic out of balance. For this reason at least a listener should be wary of assigning blame in such matters even when there is clearly a problem in the balance of a recording. It is, after all, a sad fact of life that many authentic and worthy recording locations in Europe and elsewhere are now inevitably to be found on a busy main street in the center of town.

The beauty of medieval chant lies in the balance of voices and the acoustic of the cathedral, a balance of movement and stillness, between clarity of words and depth of sound. Plainchant employs a small number of voices to fill a large space with music. The spatial image created is a *unified space*. A listener can enter a cathedral from any door and feel part of that same musical space. Acoustic unity is a powerful metaphor for social unity. Since plainchant is sung in Latin, which most of the community does not understand, the impression of social unity does not rely on language but on musical and acoustic imagery. The music alternates solo verses and group choruses, expressing leadership and consensus, respectively. The acoustic, in turn, expresses a conception of unity in the uniformity of perceptual experience throughout the reverberant enclosure.

The unusually long reverberation time of a cathedral acoustic is due in part to size and volume, and in large measure to the reflective qualities of stone and glass interior cladding. Absence of sound-absorbent materials in a church or cathedral allows speech and music to continue to resound back and forth, dying away only slowly. Sheer size and volume, however, are not the full story. They only influence how far a sound wave has to travel between walls, ceiling, and floor. Size alone, as in an aircraft hangar, does not create a cathedral-like acoustic. Surface area is more significant, because the irregularities of surfaces determine the complexities of sound reflection that give an acoustic its desirable qualities of full resonance and gradual decay.

You could call it the catalytic converter effect. When laws were first passed regulating exhaust emissions cars were fitted with catalytic converters, that on the outside, on a human scale are compact boxes but, on the inside, on a molecular scale are required to expose a platinum surface area equivalent to two football pitches to trap unwanted pollution. Without that enormous surface area a converter wouldn't last very long. Likewise the interior surface area of a cathedral or any room takes into account every curve, every moulding, every detail—not just walls and windows but fluted columns, statues, sarcophagi, pews, railings, organ pipes. Everything. When sound expands to fill a cathedral space *it touches every exposed surface*. Music is also sensitive to physical size. The higher the frequency, the shorter the wavelength. A 250 Hz tone (250 wavefronts per second, round about middle C on the keyboard) has a wavelength of 54 inches, meaning successive pressure waves are 4 feet 6 inches apart. Music at this order of frequency is therefore scaled to interior features of that order of size, namely people, pulpit, and pews. But the frequency range of music extends far beyond 250 Hz. A 12K tone, of 12,000 wavefronts per second, has a wavelength of just over an inch, and is sensitive at that scale to the detail of stone carving and diffraction by structures the size of a golf ball. It follows that a listener's perception of acoustic space

can and does vary with frequency. The low organ and string bass
tone C0 that introduces Richard Strauss's *Also Sprach Zarathustra*
(title music for Stanley Kubrick's movie *2001: A Space Odyssey*)
generates a hugely powerful low frequency oscillation with a
distance between wavefronts of about 35 feet, way beyond the
human scale. When your unit of measurement is 35 feet, even the
dimensions of a cathedral do not seem vast. Strauss is using the
lowest and most powerful note (powerful in physical terms, but
not loud in human terms) to create an impression of a superhuman presence. The difference in scale makes these low tones
impossible for a human listener to locate. Then, like the great
doors of a cathedral opening a crack to let in the light, an ascending trumpet melody starting at middle C establishes, first, a
sense of location (you can hear where the trumpets are coming
from) and, second, an awareness of movement in space (you can
hear that the melody is rising in pitch and increasing in volume),
both on a human scale. With the entry of the full orchestra the
acoustic space is suddenly expanded in every direction with a
fortissimo blare of trumpets, trombones, strings, and horns, a
musical image of the big bang radiating energies in frequencies
from the lowest to the outermost limits of human hearing. The
impact on the listener of the opening gambit of *Also Sprach Zarathustra* is of an initially vague presence (the organ tone) magic-
ally transformed into a huge acoustic space (the added brass).[4]

Three kinds of space

Around 1600 a group of Venetian composers including Claudio
Monteverdi and Giovanni Gabrieli were pioneering a new kind
of music designed to take advantage of the huge size and multiple chapels and galleries of cathedral architecture, and to exploit
dramatic acoustical effects of distance and direction. The new sensibility coincided with the rise of scientific humanism and important developments in the visual sciences. It led to experiments
in musical perspective analogous to visual perspective in painting.

      We can describe acoustical space as *left-right* or lateral space;
up-down or vertical space; and *front-back* or perspective space.
There are also three kinds of musical space: 1. *real* space, exploiting directional and distance effects that correspond to the actual
locations of performers; 2. *virtual* space, the imitation of distance

and directional effects; and 3. a perception special to music of *pitch* space, the sense of a musical note or line being high or low in pitch.

In his *Vespers* of 1610 Monteverdi juxtaposes the old unitary space of medieval plainchant with brilliant demonstrations of Renaissance spatial ambiguity. The large forces of voices and

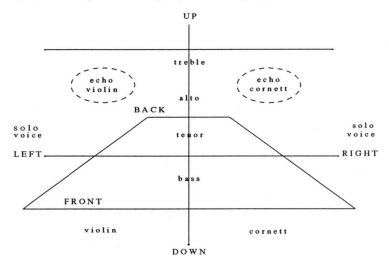

Musical perspective in Monteverdi's *Vespers* of 1610

instruments on display in the opening "Deus in Adjutorium" confront the listener with an impressive show of power. The simplicity, clarity, and fidelity to the text of traditional plainchant is suddenly swept away in a music of overwhelming richness and complexity, a veritable *wall of sound* in which everything seems to be happening at once, and the sacred text, no longer in control, is submerged in a teeming mass of vocal and instrumental life. It is music too complex for the ear and mind to take in, and yet this apparent chaos is held in check by a powerful bassline harmony and a beat of slow, majestic pace implying a regal procession.[5]

But this is a celebration more in the spirit of "Hail, Caesar!" than "Hail, Mary." In a reversal of roles the church is made to acknowledge the glory of Renaissance temporal power. How so? Because the new music the composer is acknowledging is manifestly an image of diversity, not unity, a diversified life held together by images of *temporal power*. A regular beat is a measure of time, time being a human dimension, and the distinctly unchurchlike rhythms of courtly and secular dance that alternate with the striding choruses are clearly more expressive of temporal activity than religious contemplation. Equally symbolic is the imagery of corporate control. The Turkish ambassador at the

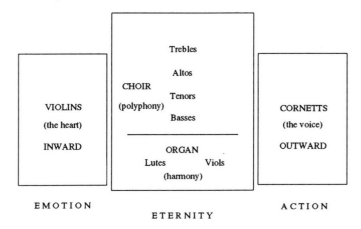

Monteverdi: *Vespers*. The triptych formation expresses
a complex interplay of symmetries in real and imaginary space:
heaven and earth, eternity and time, inner and outer being

premiere performance understands this music as a show of force intended to intimidate as well as to excite. Monteverdi is saying "This is the image of a diversely talented society functioning cooperatively under one authority. If you find this impressive, think what our armies can accomplish."

In alluding to movement patterns of formal procession and dance, Monteverdi is also alerting the listener to the dynamics of space. His orchestra of winds, strings, and organ does not correspond to a symphony orchestra in the classical sense. A symphony orchestra functions in principle as an acoustic unit, *in mono* (as the word *symphony* implies), whereas the mixed ensemble here is staged in such a way as to express concepts of difference and diversity. The players are laid out in triptych formation, winds stage left, strings stage right, and organ, choir and bass lutes in the center. Just as a painted triptych is a representation of symmetrical but complementary realities, as in heaven and hell, so here the wind instruments represent the external "real" world of human breath, action, and military communication signals, in contrast to the bowed strings, which represent the inner world of tension, emotion, and personal expression. In the center, between these two groups, keyboard organ and accompanying instruments, symbolizing the absolute realm of number and measure expressed in harmonic proportion, join together with the choir's voices and sacred words representing the divine realm.

Real space and virtual space

Monteverdi's setting of the "Magnificat" that ends the *Vespers* is a textbook demonstration of musical and acoustical perspective. Lateral space is demonstrated in the left-right opposition of tenors in "Quia fecit." It is followed by a beautiful expression of vertical elevation from bass to treble in "Et misericordia," music that, appropriately for a text about being depressed and having your spirits raised, starts at ground level with tenors and bass voices and floats upward into an angelic realm of altos and trebles. Three times the music's center of gravity is buoyed upward, the first time, treble voices hitting a glass ceiling at the note D5, the second time, breaking through that barrier (an effect of astonishing poignancy), the third time, uniting high and low pitches in a magnificent glowing column of harmony. In the mere two or three minutes of "Deposuit potentes de sede," the composer marries left and right, front and back, real and virtual, temporal and eternal, exterior and interior symmetries in a breathtaking exercise in multi-dimensional perspective, achieved with wonderful economy of means. Two instrumental duos, cornetti on the left (high-pitched winds of trumpet-like tone), then a pair of violins on the right, perform the same music in echo canon against the backdrop of a serene cantus firmus sung by the choir. Echo canon (*canon* as in *copier*) is an artificial echo in music, where a signal initiated at one location is reproduced echo fashion from a different location after a fixed time delay. In this Monteverdi example the illusion of space is reinforced by the physical separation of instruments in real space, one of each pair placed prominently in the foreground, the other in a more distant location where the sound is weaker and colored by natural reverberation.[6]

Cornetts and violins in apposition illustrate not only the

distinction of lateral space (left versus right), but also exterior and interior space (voice and heart); simultaneously the duos in echo canon combine effects of real perspective (actual space) and virtual space (artificial echo). Finally, in the contrast between the music for instruments representing the modern style and that for voices representing the old medieval tradition, a listener detects the even more fundamental antithesis of worlds past and future, mortal (canon, time and space) and spiritual (the timeless realm of heaven).

In the dryer acoustic of aristocratic palaces Italian composers employed echo canon for effect. Orlando di Lasso's motet "Hark! Hark! the Echo" is a delightful invention for four-part unaccompanied double choir, the smaller echo choir being concealed from view.[7] This is a witty, almost postmodern coupling of a vocal part that attempts first to carry on a conversation with its echo, and then, irritated by its persistence, asks the echo to desist, reflected in a music that is equally knowing and self-referential:

> Hark! hark! (*hark*) the echo falling (*falling*)
> Far o'er the vale (*o'er the vale*) replying (*...plying*)
> Ha! ha! ha! ha! ha! (*ha! ha! ha!*)
> Rejoice and laughter (*and laughter*)
> Say, my companion (*companion*)
> Where are you? (*Where are you?*)
> We would desire to hear you (*to hear you*)
> Warble (*Warble*) a ditty (*a ditty*)
> Why, Sir! (*Why, Sir!*) ask you why? (*...you why?*)
> I'm afraid (*I'm afraid*) because we wish it! (*...cause we wish it!*)
> Why, will you not? (*Why, will you not?*)
> Why will you not please us? (*Why will you not please us?*)
>
> Silence, now, pray! (*now, pray!*)
> Hold your peace! (*Hold your peace!*)
> You ill-bred oaf! (*You ill-bred oaf!*)
> What? Now, Sir! (*What? Now, Sir!*)
> No more we'll hear (*No more we'll hear*)
> Of you we met (*Of you we met*)
> Farewell today (*today*), farewell today (*today, to...*)
> We must leave you (*We must leave you*)
> Stop, now! (*Stop, now!*)
> Stop, now! Hold, enough, Sir (*enough...*)
> Hold, enough, Sir (*enough, Sir*)
> Farewell! (*Farewell!*)

Upwardly and outwardly mobile

The Venetians may not have invented surround sound, but they certainly discovered a use for it. Their music's exploration of space and time was conducted within the acoustic confines of church architecture and under the patronage of church authorities. In common with other contemporary inquiries into time and space, represented by the mechanization of time-keeping, and the new astronomical observations and theories of Galileo, the development of "music in the round" was seen as a challenge to the authority of religion, eventually leading to a thousand years of ecclesiastical patronage of musical science being brought to an end.

Giovanni Gabrieli's *Canzon del Duodecimi Toni No. 2* (Canzon in Mode 12) for ten winds, composed in 1597, is a purely instrumental showpiece for groups of players in a surround sound arrangement. Solo cornetts (equivalent to modern trumpets), placed in galleries at points around the listener, engage in a three-way echo dialogue, offset by an anchoring ground-level ensemble of four more cornetti (trumpets) and trombones. It is an exciting and dramatic expression of real and virtual space.[8]

The listener might wonder how such music could be considered subversive. How can *any* music be subversive? We are already aware of the dramatic tension created by Monteverdi's opposition of orthodox plainchant, word-settings expressing the traditional view of music as a servant of religion, in contrast to his own composed inventions in a new instrumental style embodying concepts of time and space representative of the new Renaissance spirit.

First of all, the canzon is purely instrumental music being performed in a church environment. There are *no words*. In itself, purely instrumental music was not a new idea in Gabrieli's time, but traditionally the role of instruments had been to support and accompany a vocal line, and the role of the vocal line was to preserve and enhance the meaning of a sacred text. Music without words is music in the abstract. It is not beholden to the limitations of the human voice, nor is it subject to the jurisdiction of a text. Such music is thus outside the control of religious authority. It can go anywhere and do anything. *Such freedoms are dangerous.*

Secondly, plainchant is religious music designed not only to preserve and enhance the meaning of a sacred text, but also to create and maintain the impression of a unified space. Unity is what religion is about. Gabrieli's spatialized music, on the other hand, is about *multiple spaces.* When this music is performed, listeners entering the performance space by different doors are going to hear it from different perspectives. Where you have a music of multiple spaces, performed by a number of widely separated groups of players, you are defending a version of the truth that allows for more than one "point of view." *The church does not like that either.*

Thirdly, plainchant expresses unity of time both in the absence of a beat and in the exposition of a text that in essence is timeless. Whereas plainchant looks back, expressing the idea of continuity with tradition, of the unity of past and present, Gabrieli's canzona for brass represents a radical break with religious tradition. This music looks forward rather than back, celebrating essentially human perceptions of time and space in musical structures based on echo imitation and the dance. This music is not about instruments imitating singing, but the excitement of discovery, of exploration, of extending the range of human expression beyond the limitations of the voice and by implication beyond the control of mere words. It is music celebrating human achievement, technical skill, exuberance, progress, and competitiveness. In celebrating human achievement and potential, this music of the Renaissance turns its back on the past and looks ahead with confidence to a future in which different worldviews and power structures freely coexist. Gabrieli's music is saying, Knowledge is not a given but a process. The church says, *We can't allow that.*

Instrumental music of this complexity is only possible when you have a universal notation of suitable precision to express pitch and time, and players with the skill not only to play their individual parts but also to coordinate their actions. One of the more surprising outcomes of the Venetian school's investigations into musical space is the very modern discovery that simultaneity is a spatial as well as a temporal concept. The more you separate your players into groups twenty meters or more apart, the more the ensemble coordination is affected by time delays caused by the finite and relatively slow speed of sound. Echocanon—that is, imitation at the same pitch and at a fixed delay—offers a way out of the coordination dilemma: it builds a time delay into the music to match the performers' separation in space, and the effect of a controlled delay also takes care of minor differences in timing that occur for listeners in different locations (not forgetting the players themselves). From a conductor's point of view, however, it means that exact synchronization of players in a widely separated multi-group composition can only be true for a single location.

As musical art spread away from the patronage of the church to invade the palaces of the nobility, composers began to lose touch with the reverberant spaces of the great cathedrals. Interest in spatial music traveled far from Venice, to Germany with Heinrich Schütz, and to the Austrian court of Prince Ferdinand at Graz, for whom Giovanni Priuli composed a *Canzona prima a 12* for multiple groups of woodwinds, brass, and strings.[9] This charming music lacks the raw energy and enthusiasm of the Gabrieli canzona, but makes up for it in richness and variety of instrumental color and texture. In its own way, this music is even more complex than Monteverdi or Gabrieli, a musical equivalent of three-color printing, in which images in red, yellow, and blue are exposed separately and then overlapped to create a holographic surround sound image embodying multiple cues of direction, height, and depth. Each group of instruments occupies its own distinct location in both physical space and musical pitch space, the ensemble delivering a message of harmony in diversity, a diversity not only of space and time, but of timbre or tone color as well.

Corelli's little joke

Generations later one finds a humorous echo of Gabrieli's contesting duo in Archangelo Corelli's 1680 Concerto Grosso in D, an otherwise routine piece enlivened by the composer's prankish sense of humor.[10] You need to understand that a Baroque orchestra normally has two kinds of leader. One is the conductor at the keyboard, who keeps an eye on the complete musical score and is able to bring in the various parts on cue; the other, the concert-master or lead violin, whose role it is to guide the melody line.

Corelli begins his concerto grosso with a plain, slightly pompous overture as a curtain-raiser. Then a faster Allegro begins, led by the concert-master at stage right. But what's this?

Suddenly a second violinist on stage left begins playing the same music after a two-measure delay, as if to say "Pay no attention to him. I'm the real leader." The two rivals continue their duel for control while the rest of the orchestra looks on in silence. After a long hiatus the conductor at the harpsichord can stand it no longer and leads in the remaining strings, *but he cannot find his*

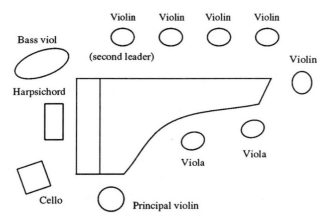

Orchestra as it may have been arranged for the Corelli Concerto Grosso in D, with the opposing violins to right and left of the composer at the keyboard

place and the music lurches from one chord to another and then from key to key without any sense of purpose or direction. Eventually the orchestra does seem to find its place and the music maneuvers toward a cadence, but only to get stuck fast in the key of D, unable to budge. The music winds down to a cadence, the strings slowly sinking as if into quicksand. The payoff, however, is a final chord not based on D as one would expect, but on the note G sharp, an unexpectedly dissonant cadence, after which the alternate solo violin plays a short but elaborately fawning cadenza by way of humorous apology to the audience for having a joke at their expense.

From space to stereo

It is fair to regret the decline of interest in spatial effects, from the exciting and philosophically challenging echo dialogues of a Gabrieli or Monteverdi to the cabaret tricks of a Corelli, but such a decline is inevitable when the magnificent acoustic spaces that gave rise to Renaissance polyphony are abandoned for smaller and lighter secular enclosures. Echo related music survived into the era of Bach and Handel and even to the present day, but more as an intellectual exercise than as an experiment in perception. We can recognize the shift in the *fugue*, a multi-part form of interlaced reflections associated in particular with the composer J.S. Bach, and held up to students of music as one of the highest expressions of musical intellect. In all of the examples of echo canon mentioned to this point, the music imitates nature in that the echo sounds an octave higher, or at the same pitch as the leading voice, though the time of delay may vary. In fugal counterpoint (*fugue* meaning "flight") nature itself is transcended. The answering echoes are transposed not only in time but also in pitch, as though pitch itself were a spatial dimension, so instead of dealing simply in mirror reflections, a fugue is manipulating musical figures in *virtual space,* exactly as a computer transposes, rotates, and inverts visual images on the screen.[11]

For the Italian pioneers of spatial music the performance space was vast and cavernous and their music was structured accordingly, using powerful brass and wind instruments to signal back and forth across relatively great distances. For later generations such as Corelli the environment was smaller, with a livelier acoustic, and the preferred instruments were the violin family, more volatile and less monumental in tone. Unlike the stone and glass and voluminous spaces of a cathedral, the typical eighteenth-century concert venue is a salon or ballroom, a relatively small, resonant structure of wood and plaster with flat walls and ceiling. Such box-like spaces do not require the penetrating tones of trumpets and horns but are ideal for a fast-paced, intricate music that would sound congested and blurry in the long reverberation of a cathedral.

A Marriage of equals

For Antonio Vivaldi, a pioneer figure in the history of the modern string-based orchestra, spatial effects are a rare and special tribute to Italian tradition, reinterpreted in terms of the changed environments and instruments of the eighteenth century. One of the most thrilling interpretations of Vivaldi in space can be heard in an inspired recording of the wedding anthem *Lauda Jerusalem*, a setting of Psalm 147. A wedding is a joining of a couple, and also of two families, and the scrupulous matching of soprano soloists, choirs, and string ensembles on either side of the organ as referee suggests a diplomatic concern on the part of the composer to avoid any hint of bias one way or the other—a delicacy often associated in Italy with marriage between rival families.[12]

The spatial arrangement is left and right, in stereo, rather than front to back, but weddings are like that, with the bride's family on one side of the aisle and the groom's on the other. What distinguishes Vivaldi from Gabrieli and Monteverdi is that this is a music of enormous energy and pace, offering a charge of adrenaline that verges on the indecorous. Matching long and short phrases flash back and forth from left to right in a virtuoso display of mutual agreement that, through a closer and closer overlapping of the two sides, also manages to convey a sense of power rivalries and highly-charged physical attraction. Once or twice the answering echo does not quite reflect the original, but most of the time the agreement is exact, and the balance of power evenly distributed.

Perspective revisited

The late classical era in general has little time for spatial effects, but the *Notturno in D* for four orchestras by Mozart is a rare and fascinating combination of real and virtual acoustic perspective comparable with the "Deposuit de sede" from Monteverdi's *Vespers*. The four orchestras of strings and horns are situated at increasing distances from the audience, producing a rolling echo in imitation of a great *canyon* (the same word as *canon*, by the way). Instead of the single melodies of traditional canon, Mozart, like Vivaldi, echoes entire blocks of harmony, the distance effect most notably expressed in the third-movement Minuetto with horns in call and answer mode.[13]

Notes

1. Camille Saint-Saëns, Symphony No. 3 in C minor, "Organ Symphony." Philippe Lefebvre, Orchestre National de France cond. Seiji Ozawa (Seraphim Classics 7243 5 73430 2 2, 1997), track 2.

2. Clive K. Catchpole, *Vocal Communication in Birds*. Vol. 115 of

Studies in Biology (London: Edward Arnold, 1979), 51.

3. "Pueri Hebreorum," in *Carmina Burana: Passion Play*. Mittelalter-Ensemble der Schola Cantorum Basiliensis cond. Thomas Binkley (DHM 05472 77689 2, n.d. Excerpted in *40 Years Deutsche Harmonia Mundi*, DHM 05471 77820 2, 1999), track 8.

4. Richard Strauss, *Also Sprach Zarathustra*. London Philharmonic Orchestra cond. Klaus Tennstedt (Seraphim 7243 5 73560 2 2, 1999), track 1.

5. Claudio Monteverdi, *Vespers of 1610*. Boston Baroque (on period instruments) cond. Martin Pearlman (Telarc 2CD-80453, 1997), disc 1, track 1.

6. Monteverdi *Vespers*, disc 2, tracks 7, 8, 10.

7. Orlando di Lasso, "Hark, Hark, the echo" (O la o che bon eccho), *Libro de Villanelle, Moresce, et Altre Canzoni* (1581) No. 14. Glasgow Orpheus Choir cond. Hugh Roberton (vinyl, His Master's Voice DLP 1020, n.d.), track 2.

8. Giovanni Gabrieli, *Canzon del Duodecimi Toni a 10* for 6 trumpets and 4 trombones. The Wallace Collection cond. Simon Wright, *Gabrieli & St Mark's: Venetian brass music* (Nimbus NI 5236, 1990), track 1.

9. Giovanni Priuli, *Canzona prima a 12*. Early Music Consort of London (on period instruments) dir. David Munrow, *Monteverdi's companions* (Virgin Veritas 7243 5 61288 2 8, 1996), track 4.

10. Archangelo Corelli, Concerto Grosso in D. Clarion Music Society (on period instruments) cond. Newell Jenkins, *Hidden Masters of the Baroque Vol. 1* (Newport Classics NCD 60075, 1988).

11. Johann Sebastian Bach, *The Musical Offering* BWV 1079. Capella Istropolitana dir. Christian Benda (Naxos 8.553286, 1998).

12. Antonio Vivaldi, *Lauda Jerusalem* RV609. Margaret Marshall, Ann Murray, John Alldis Choir, English Chamber Orchestra cond. Vittorio Negri (Philips 420 648-2PM, 1988).

13. Wolfgang Amadeus Mozart, *Notturno in D Major* K286. London Symphony Orchestra cond. Peter Maag (Decca Legends 289 466 500-2, 2000), tracks 1-4.

CHAPTER EIGHT

Visible sound

IMAGES in art are mute and still, but the world depicted in great art is neither silent nor motionless. A dynamic world is a noisy world, and a noisy world includes music. Artists render the world in all its sensory diversity, not just the visual. The meaning and enjoyment of art works can be greatly enhanced if a viewer is mindful of the ways by which sound and movement are represented pictorially.

The simplest way of representing sound in nature is by depicting people, birds, animals, and machines in the act of speaking, singing or making a noise. If mouths are open, a musical instru-ment is being played, or a wheel is seen as turning, a particular sound is evoked. Small children intuitively make the connection between the image of a farm animal in a picture book and the noise it makes. A child knows that *moo* means cow, *oink* means pig, and *quack* means duck. Animal noises are *names*. Equally, animal images are sounds.

Words are sounds as well. Medieval illuminations, early charts and maps, broadsheets, pamphlets, and political cartoons regularly employ the device of a scroll or banner (whence the term *banner headline*) to spell out the actual words issuing from a speaker's mouth, in anticipation of the speech bubble of today. Words, however, limit understanding to viewers who can read, and the information contributed by a text to a pictorial image has only literal significance. Words are also likely to distract the viewer's attention from other sources of meaning, which is why Western artists in general tend to avoid language in painting.

Sound is implicit in the organization of a picture. The visual congestion of a Brueghel village market scene creates an impression of bustling activity; silence in a still life of fruit or flowers expresses just as eloquently the inexorability of time and inevitability of physical decay. Classical images of the Madonna and Child, or portraits of saints and philosophers, are designed to catch their subjects in reflective mood, and that means the stillness of listening.

Angels as light-bulbs or microphones

The halo as a symbol of holiness is also a visual representation of hearing. We hear things we cannot see, in the real world and in our mind, and the unseen world of the imagination also creates images that cannot be visualized in realistic imagery. Early in the twentieth century engineers had to find a way of visualizing the directional sensitivity of a microphone to allow the user to recognize the difference between microphones that are forward-sensitive (unidirectional), those that are front- and rear- but not sideways-sensitive (figure-of-eight), and those that are sensitive in every direction (omnidirectional). The printed polar diagram that comes with the instructions for your plug-in microphone is designed to tell the user how good it is at picking up sounds through 360 degrees at various frequencies, and therefore where to point it. The halo of classical religious art conveys the same information about human sensitivity to the acoustic environment, indicating an ability to receive and process information from every direction like an omnidirectional microphone. The old comic-strip images of a light-bulb over the head to express a flash of inspiration, or of sun rays to indicate a radio that is switched on, express much the same idea of radiant sensitivity. On the other hand, if we look for an equivalent visual analogy in the literature of classical optics, what we find is an image of dual eye beams pointing straight ahead like the two headlamps of an automobile. Directed light can only see in one direction, straight ahead. The light-bulb, by contrast, casts its light in all directions.

Holiness in a halo expresses a heightened receptiveness rather than an ability to see through walls or work miracles by the power of thought. Just like an omnidirectional microphone, the halo bearer is fully aware of the world around and not narrow-minded or obsessively discriminating. Whereas vision necessarily involves selecting what to look at, hearing is necessarily all-embracing. In order to convey the idea of omnidirectional awareness, the halo is typically represented as a golden sphere surrounding the head but transparent in cross-section, like a bubble. Gold is routinely associated with the power of hearing, both as a sign of the supremacy of hearing over all other senses and as a pun linking the words *aura* and *aureole* to Latin nouns *auris* (ear) and *aurum* (gold). By extension, a golden background in a medieval or Renaissance painting signifies a zone or region dedicated to hearing and harmony.

A halo is not a hat

Not all artists interpret the halo as a sphere or light-bulb. For some it is a badge of office like a wide-brimmed cardinal's hat; others reduce the zone of awareness to a flat disc floating above the head or orbiting around it like the rings of Saturn. As visual representations of reality gain the upper hand during the Renaissance, so the halo is seen to decline in significance, eventually to disappear from the artist's vocabulary of symbols.

Sound is collected and resonates under the dome of a church or cathedral, so, quite apart from the religious symbolism of the sphere as an image of divine perfection, the dome or arch is associated with sound and in particular with divine harmony, which is to say, with music. Images or scenes depicted within a dome or within a frame that is circular or arched are thus acoustical or musical in implication.

A wormhole in pictorial space

Giotto's *The Last Judgement* depicts Christ seated in an oval bubble or mandorla supported by a retinue of flying angels. The arched upper portion of the fresco depicts heaven as an acoustic domain, the rectangular lower half the material world. The curved bubble occupied by Christ glows with light-reflecting gold, signifying the triumph of hearing. The head of Christ is positioned at the center of the arch, corresponding to the acoustic focal point of the heavenly dome or the position of a conductor of an orchestra. The lower part of Christ's body projects into the rectangular half of the picture, signifying his mediating role between the invisible and immaterial world of hearing and the visible and material world of bodily sensation.[1]

Directly above the image of Christ is a triple-arched chapel window. One is tempted to ignore it as an architectural obstacle; certainly if the window were uncovered the light coming through would make the fresco difficult to see clearly. Even this feature can be understood, however, in terms of Giotto's original design, contrasting the two-dimensional painter's image that can only be seen by reflected light with the dazzling manifestation of real sunlight streaming in from another dimension. The window corresponds in today's currency to a "wormhole" in pictorial space, a wonderfully daring conceit emphasized by Giotto's two angels at the top of the fresco peeling away the wall on either side to reveal a heavenly realm.

Christ is floating in space because he is exercising judgment on the visible world from the world of hearing, as a mediator between visible and invisible realms. The world of sound is transparent, weightless, omnipresent, and intangible. Under the dome of heaven choirs of angels intone the music of the spheres; at the half-way border between the rectangular space of the visible world and the curved space of heaven sits a panel of saintly figures, their haloes perched on the back of their heads like cardinal's hats.

The lower portion of the fresco is given over to the consequences of divine judgment. To the right of Christ, the virtuous are marshaled in orderly fashion by winged angels; to his left the damned tumble in confusion to be consumed in a chaos of darkness dominated by a horned and overweight Satan, suggesting the dire physical consequences of excessive appetite (or in today's terms, the perils of unreconstructed consumerism). The two groups are on the same level, so what is being represented here is not an image of virtue as upwardly and vice as downwardly mobile but rather the real world as a level playing field in which virtue and vice jointly coexist. Judgment is being pronounced, therefore, not on the dead but on the living. The difference between virtue and vice can be viewed in today's terms as a matter not so much of religious belief as of lifestyle. The virtuous are distinguished by their degree of self-control, and the damned by their lack of control, exemplified in the involuntary sounds of greed and flatulence that connect non-virtuous humanity with the animal kingdom, as well as the

Giotto: *The Last Judgment*

destructive forces of nature manifested in disease, drunkenness, and death.

In images of this kind the message of divine judgment can often be read as arbitrary retribution by a religious bureaucracy eager for revenge on the unbeliever. Giotto's vision of judgment by contrast seems rather more gentle—lifestyle advice of the kind one is given by a bran cereal ad or a government health warning. The picture declares in effect not that gluttony is sinful but that too much cholesterol is not good for you; not that self-indulgence is a vice but that self-control is self-empowering; not that good thoughts are inherently virtuous but that civilized conversation at table is frankly more agreeable and potentially life-enhancing

Fra Angelico: *The Coronation of Mary*

than spontaneous belching.

Trumpeting

The long trumpets in Fra Angelico's *The Coronation of Mary* point upward at different angles. Nowadays we like our fanfares to be disciplined in military fashion, with trumpets in line, horizontal, and facing forward. It seems a minor detail, but it is, all the same, a persistent feature of musical iconography to depict trumpeters as a rowdy element in our musical life (a reputation they still enjoy in today's symphony orchestras). The impression of disunity is in fact misleading. Here the trumpets are directing sound upward into a dome of gold, filling the pictorial space with harmonious reverberation.[2] The inside of a dome being curved, it makes sense for each trumpet to point in a different direction. Trumpets are long, straight, and loud instruments, and to play them pointed straight at an audience would be both anti-social behavior and awkward for the musicians. Pointing them upward at different angles ensures that their sound is suitably diffused to beam down on the assembled gathering like a heavenly radiance.

Christ and the Virgin float in the center of the composition, as figures in a divine acoustic space surrounded by dancing angels, musicians, and saintly dignitaries. Unlike Giotto, the artist has equipped his saintly characters with genuinely transparent haloes that remain in outline at every angle, full-face, profile, even from behind.

The triumph of science over art

Complementary panels illustrating choir and musicians from the Ghent Altarpiece by Hubert and Jan Van Eyck have become icons of the early music industry. These angels belong to a new musical and religious culture, distinguished on the one hand by an absence of traditional acoustical symbols, and on the other by representations of new musical attitudes.[3] Sumptuously dressed, with no haloes other than their long golden hair, self-evidently appreciative of the sensuous delights of rich fabrics and precious materials, unconcerned at the moral implications of endorsing human craftsmanship expressed in fine embroidery, jewelry, engineering (the cast-iron cantilevered lectern, the pipe organ), woodwork and carving, these idealized singers represent a new attitude that has overcome a pious aversion to human achievement and material goods by dedicating them to a higher spiritual purpose, symbolized by the carved image of St. George slaying the dragon that decorates the base of the lectern. Indeed, the artists celebrate human science and skill to the point of concealing as much carnal human nature as possible beneath multiple layers of what Marshall McLuhan might describe as a protective outer fabric of technology.

The choir of singing angels is eight in number and they are reading music from a book. The image of sight singing is significant, representing a marriage of vision and hearing in place of a traditional reliance on aural memory for music-making. The lectern at which they sing is a marvel of engineering that pivots on a crank-shaped spindle of metal resembling a wall-mounted video monitor on an angle bracket. The eight figures could conceivably

Hubert and Jan Van Eyck: *Angels Singing and Playing Music*

signify the eight notes of the octave. In the matching panel of
instrumental musicians, special prominence is given to another
example of new technology, a bright new portative organ. Associated with St. Cecilia, patron saint of music, the organ is *the only
instrument being played* and clearly sets the pitch standard on
which the singers are having to rely. The novelty of the situation
is indicated by the four musicians standing to one side and looking on. Two are holding stringed instruments—one a plucked
harp, the other a bowed *lyra da braccio*—and the puzzle for the
viewer is why they are not also playing. The harpist has her hand
on the lyra player's shoulder in a gentle admonishment to be
quiet and listen. Seated on her Renaissance Bauhaus-style folding
stool the organist, perhaps a representation of St. Cecilia, appears unaware of their presence. Her eyes are deliberately averted, her attention completely focused on the simple keyboard

where both hands in action suggests a musical accompaniment in two parts. The message is of an earlier custom of tonal variety (harp, lyra, and organ) giving way to a new regime of tonal uniformity (the evenly-matched tones of the organ).

Fixed tuning

What is the first action of a guitar or violin player at a concert? Tuning up. Why does a string player tune up? Because the guitar or violin is not in tune. Why is it not in tune? Because string instruments do not stay in tune. What does this mean for the science of musical acoustics and the study of harmony? That string instruments are unreliable. And what does this imply for a choir of angels? That in order to attain perfect, divine harmony, they need to follow the pitch of a more reliable source than the traditional harp or lyre. So what makes the organ more fitting and more desirable as a guide to pitch?

First, the organ has evenness of tone. Manually bowed and plucked instruments do not, being subject to human uncertainty, emotion, and error. All the pipes of an organ are played from the same reservoir of compressed air, the wind chest located underneath the pipes. Air pressure is maintained by bellows operated out of sight behind the instrument. Second, the organ cannot go out of tune. Strings are made of animal gut, organic material that reacts to changes in humidity and temperature by losing tension and consequently losing pitch. Organ pipes, however, are resonating columns of air enclosed in pipes made of an inert metal compound peculiar to pipe organs, a substance that does not distort under changing atmospheric conditions. Once the pipe has been tuned to a desired pitch it stays in tune, winter or summer, rain or shine.

The temperament paradox

Guaranteed permanent tuning enables the organ to maintain a prescribed standard of accuracy over an entire scale for an indefinite time. Accuracy implies a consistent precision of pitch relations for every two-note combination available in an eight-note octave. This is an image from a time when the morals and mathematics of tuning an entire scale were highly significant issues. Science insisted on a goal of standardized pitch relations, and art represented by music and morals argued in favor of human discretion in such ,matters as intonation and tuning. A harp or violin is tuned by ear, and although suitable for tuning intervals of two notes at a time the ear is not the ideal instrument for tuning an entire scale, a task of reconciling highly compli- cated harmonic ratios. The term for a system of standardized tuning for an entire scale is *temperament*, reflecting the traditional association of a scale or mode with a particular emotional character. The challenge faced by the mathematicians of equal temperament was to devise a system of tuning the scale in which every possible two-note interval and every chord would sound equally harmonious, whatever the key. The problem? *It can't be done.* Temperament is invariably a compromise. As a piece of music *modulates* or moves through major into minor, and from white-note to black-note keys, chords and intervals are heard to vary in harmony. So the solo organist in the Van Eyck image can be interpreted as a statement endorsing the scientific view in favor of a compromise standard of tuning, in contrast to the orthodox view, represented by harp and violin players, that there is no easy way to achieve perfect harmony except by a process of constant adjustment of intonation by ear.

Pitch standards

Celebrating technology and human expertise puts the Ghent Altarpiece firmly on the side of science and standardization. Standard notation is implied in the book of music from which the choir is singing. Eight voices singing from the same book sug- gests eight notes of the scale in harmony. Written notation (the book) implies visual standards of accuracy rather than tuning referred to the sense of hearing. Assigning priority to the key- board pipe organ over the stringed lyra and harp expresses the triumph of objective calculation and measurement over musical instinct. Scientific notation in itself implies agreed standards of pitch, not to mention time. The multi-voice digital keyboard fur- ther suggests a potential to realize music of a complexity only conceivable in visual terms.

In its cool geometrical purity the organ expresses a concep- tion of pitch as a continuum independent of human or material limitations. (The artists are employing poetic license in showing the organ pipes as forming a perfect triangle. In reality the progression of pipe lengths in a musical scale follows a log-arithmic curve and not a straight line.) The organ can be relied upon to produce a tone of uniform quality throughout the scale and throughout the seasons without temperature or human fallibility getting in the way. An organ tone does not fluctuate or die away like a violin or harp, and this consistency of signal also qualifies the organ as a frequency generator for acoustical research, for example, into the resonant behavior of different materials, or into the acoustics of the voice.

In depicting angels as reading from notation and receiving their notes from an organ played with two hands the artist is making a statement of the greater reliability of the pipe organ as an instrument of scientific inquiry over earlier devices such as the harp, not only as a pitch and interval generator, but also in terms of the superiority of the mechanical keyboard and bellows over traditional fingering and manual playing techniques. The organ's importance is directly related to its digital design and permanent storage of pitch information, advantages having nothing to do with the delights of human expression and every-thing to do with scientific precision applied to the study of interval proportions and such pressing issues as standardized tuning.

Apollo's revenge

Jacopo da Palma's double canvas *Apollo and Marsyas* sets pipes once again in contest with strings, but in the mythical duel adjudicated by King Midas, Apollo playing the violin is defeated by the goat-legged satyr's rustic panpipes. Thereupon Apollo takes a terrible revenge on Marsyas, flaying him alive for his temerity, albeit letting Midas off relatively lightly with the gift of a pair of ass's ears for having based his judgment on animal instinct rather than intelligence. On the surface this is a contest between high and low art, the never-ending debate over the relative merits of classical versus popular music.[4] In this sense Apollo's extreme revenge can be viewed in the same way as the appropriately violent retribution that a present-day connoisseur might wish to visit on the perpetrators of elevator music. In such a scenario King Midas, in preferring the rustic pipes to the sophisticated violin, corresponds to the patron of art who has no taste, or the present-day record company executive who prefers the simple music that will appeal to the widest audience and bring in the greatest profit.

Reading the painting as a contest between old and new (or between art music and popular music) still leaves Midas's choice and the violence of Apollo's revenge on the hapless Marsyas insufficiently explained. There is more to the story than the fate of the thick-skinned. Why is Apollo playing a violin, a modern instrument and thus an anachronism? And if Apollo is playing the violin, then what significance can be read into the panpipes? The panpipes appear to be an ironic reference to the Renaissance pipe organ and, in that case, the contest represented here can be understood as addressing the much larger contemporary issue dividing art and science, between the proponents of humanistic expression symbolized by the violin, and the natural scientists' interest in musical acoustics and their preferred keyboard instrument that aims to eliminate human variables in the production of musical tones.

Apollo is holding the violin, not against his chest like a fiddle player, nor even under his chin like a modern violinist, but against his ear. A musical value-system symbolized by the violin is both human, relying on hearing-based skills and intuitions to create a music totally responsive to human emotions, and also god-like, because its subtlety and flexibility of intonation transcend any attempt at mathematical rationalization. In this reading of the picture Marsyas represents the natural scientist who sees music as an expression of physical processes and mathematical relationships rather than human values. A contest between violin and pipes tells the viewer that the underlying debate is between the idea of music utilizing true intonation based on pure harmonic intervals as intuitively practiced by trained musicians, and music embodying a standardized system of intonation as promoted by the scientific community. Already at loggerheads over the representation of spatial relationships in visual perspective, art and science are now witness to the corresponding debate over pitch relationships in music.

A suitable case

At least this interpretation corresponds with the historical facts. Science and the tempered scale, symbolized by challenger Marsyas's natural panpipes, *did in fact triumph* in Midas's public domain, and did lead to a proliferation of keyboard instruments that were easy to play and that embodied out-of-tuneness as a constructional necessity. Marsyas's panpipes are the ancestor of today's midi synthesizer. And today it is as true as ever that because the new keyboards are easy to play, musicianship continues to decline and professional musicians continue to be put out of work. No wonder Apollo takes so dreadful a revenge. The skin is the outer envelope of the senses and emotions. It colors when we feel emotion, and is our layer of contact with the outside world and with other people. Flaying is, therefore, a manifestly suitable punishment for a being whose thick-skinned insensitivity to tuning and expressive nuance are evidence of a fatal lack of feeling for human expression and absolute musical harmony.

Low Country polemic

Hieronymus Bosch belonged to a reformist religious sect that reacted strongly against established religion and the cult of the ear. The panel "Hell" from Bosch's celebrated triptych *The Garden of Earthly Delights* can be read as a violent tract attacking music in all of its manifestations. The scandalous implications of this picture can easily be overlooked or trivialized unless one is aware of their underlying musical theme.[5] The composition includes a genuinely shocking parody of the Crucifixion in which a male figure, arms outstretched, is run through by the strings of a harp; another male is bound to the finger-board of a lute; both victims are tormented by black serpents, and the harp itself is depicted as penetrating the lute in an improbable allusion to sexual congress. The same two victims of crucifixion can also be seen to the right of the panel, the one with arms outstretched being carried off unconscious by a black panther-like demon, the other bearing a wind instrument on his back as a penance.

This is a topsy-turvy world where the hare holds the gaff and blows the hunting horn and humanity is the prey. From inside the snoring belly of a hurdy-gurdy, a nun's face peers out and her arms extend to add the jangling noise of a triangle to a general pandemonium in which crowds on either side jostle and bray in a mixture of pain and religious supplication. Propping up the hurdy-gurdy is a crouching figure dressed in a cardinal's red robes; from his frog-like mouth a string of drool emerges, along

Hieronymus Bosch's "Hell"
panel from *The Garden of Earthly Delights* (detail)

which pearls of saliva descend at regular intervals suggesting a
rosary. His singing (or croaking) is directed by a naked assistant
toward a musical score imprinted on the buttocks of a body lying
crushed beneath the coupled lute and harp. We know it to be

musical notation because the artist has considerately placed an open book of plainsong next to the body and the resemblance is perfectly clear. However, the viewer is evidently meant to understand the marks on the prone figure as an obscene metaphor of notation, a stinking, spotty array of boils and blackheads supposedly appropriate to the lewd music being sung and a suitable pretext for the defiled form of musical worship practiced among the great unwashed.

These images are funnier and also more terrifying than anything else in Bosch. Compared to this work of invective, his *Concert in an Egg* is gentle child's play and *The Ship of Fools* a paragon of restraint. Flatulence and anarchy are everywhere, from the sight of a recorder protruding from the rectum of our earlier victim to the whoopee-cushion sound of the oversized bass pommer he carries on his back, a double-reed woodwind from the same family as the bagpipe chanter and a raucous ancestor of the bassoon. It looks in the picture as though the pommer is being played through a trumpet mouthpiece attached to the center of the instrument, but this is either an error committed by an assistant or a musically uninformed piece of restoration. In reality, the red-faced individual is connected to the alphorn emerging from behind the victim's head.

The world of music and sound is linked to emblems of uncertainty and chance (cards, dice and backgammon); the vanity of trust in a musical liturgy is symbolized by the perils of skating on thin ice. Dominating the entire panel is the hauntingly pale figure of a philosopher reduced to a dried-out hollow upper torso symbolizing an empty vessel, his tree-like arms balanced precariously in boats. On the crown of his hat sits the roseate full bladder of a set of flesh-colored bagpipes, while behind in the background a blood-stained butcher's knife bearing the artist's monogram projects menacingly between a monumental pair of severed ears.

Bosch's extraordinary polemical image signifies a cultural shift from aural to visual belief systems following the introduction of movable type and dissemination of the scriptures in book form. The Gutenberg revolution made mass-produced knowledge available for all to read in the language of the people, taking religion and moral judgment out of the exclusive hands of the church. It was no longer necessary to rely on a priesthood versed in Latin, or on ritualized musical aids to the interpretation of

Matthias Grünewald: *Allegory of the Nativity*

holy writ. The new Protestantism believed in the truth of the written word in direct opposition to older faiths that relied on esoteric codes and rhetorical persuasion. Bosch is saying to his viewers, "Do not believe what you hear: be guided by the truth which is there for you to see in the printed word and before your eyes. "

Complementary realities

Matthias Grünewald's *Allegory of the Nativity* from the Isenheim Altarpiece is constructed like a diptych in left and right halves. But unlike a normal diptych where the compositional symmetries are balanced but divided, here the two wings form a continuous space. The artist intends the two halves of the picture to be understood as interpenetrating and not simply as a balance of opposites.[6]

On the right the Virgin and Child are depicted in an idyllic landscape of light and calm; this is the visual world, a world of light. On the left a choir of angels plays sweet music on viols in a cathedral setting: this is the reciprocal world of music and resonance, the world of hearing. The eye pans smoothly from interior to exterior without encountering any wall, an imaginative stroke anticipating the continuous tracking shot of a movie camera.

Light streams from the visible world into the cathedral interior, illuminating the principal angel on the left, who is playing an idealized and musically improbable viola da gamba with the bow held in reverse (this appears to be artistic license rather than music criticism since the remaining angels in the background are holding their viols more or less correctly). The harmonizing power of light illuminates the musical interior, and we are meant to understand the harmonizing power of music, in turn, as streaming out from the resonant space to influence the outside world. That the music is harmonious is shown by the beatific smile on the angel's face and by the fact that the infant Christ has fallen asleep. As we can clearly see, the baby, after being fed, potted, and bathed, is just about to be put to bed. The music being played is thus a lullaby. The artist has charmingly placed bath and chamber pot in prominent positions at front center stage where they will be readily seen by a small child— the bathtub within the church interior, a layout emphasizing the interpenetration of acoustic and visible worlds, and reconciliation of the real world of bodily functions with the world of the spirit.

The message could not be clearer: the two realms of vision and hearing are complementary and equal, each illuminates the other, and the object of religious devotion is to reconcile the old aural tradition and the new visual understanding.

Why the artist (whose *Allegory* was incidentally the inspiration for Paul Hindemith's opera *Mathis der Maler*) should be so careless in depicting musical instruments and performance practice while at the same time promoting music as a good thing, is a small mystery. Bosch, after all, is painstakingly accurate in rendering instruments and sheet music in exact detail while at the same time suggesting that music is sinful. Perhaps both artists are making the same point indirectly, the inaccuracies of Grünewald emphasizing the truer imagery of musical hearing, and Bosch's visual fidelity making the contrary argument in favor of

Caravaggio: *Rest on the Flight into Egypt* (detail)

the superior truthfulness of the visual sense.

New product endorsement

Both are reconciled, however, in Caravaggio's *Rest on the Flight into Egypt*, an image that combines accuracy of representation with the message of music as a source of inner harmony and repose. Once again music for strings performed by an angel is associated with the portrayal of rest and recovery of a sleeping mother and child. This solo angel is playing a modern violin, a new Italian invention of the period, and reading the melody from standard notation in a contemporary score held up by Joseph, leaning against the dark attentive head of the ass that has carried the holy family on their long journey. Music brings harmony not to humanity alone, but to all of creation. Joseph's gaze is fixed on the angel's face as if attempting to catch a glimmer of meaning from his expression, but the musician is impassive and his attention firmly on the printed page.[7]

This angel is sight reading from the latest music software (a *typeset, printed* score), demonstrating that modern standard notation can be read and interpreted in real time and is therefore an advance on older oral traditions and notations that are more specialized and not designed to be sight read. The painting also declares that the message of this music cannot be reasoned (is beyond the powers of vision), but can only be intuited (as by the mother and child whose eyes are closed). The luminous skin tone and profile suggest the radiant tone of the violin. The angel's averted gaze gives special prominence to the ear as gateway to the soul, which the artist depicts at an angle, allowing the viewer to see right into the ear canal. Caravaggio adds a charming extra touch of his own in suggesting a resemblance between the torso of his gender-nonspecific angel and the curved body of a violin. The dark, narrow wing and long feathers seem to allude to the violin's ebony fingerboard, and the angel's golden curls, to the curled scroll, and loose ends of gut strings. The viewer is reminded of Ingres' famous *Odalisque*, not to mention Man Ray's famous photo image of a female nude as a violoncello body with painted *f* holes.

Invisible torments

Flemish skepticism toward music and the distractions of hearing resurfaces in Marten de Vos's *The Temptation of St. Antony*. Here once again an artist from the same region as Bosch has visualized a landscape of mortal suffering and confusion populated by a cacophony of weird and threatening monsters. Like the Matthias Grünewald panel, this painting is also divided in two halves. But this time the division is lateral rather than vertical, a lower half representing the real world of human affairs and temptations, and an upper half representing the world of the imagination and the spirit.[8] At ground level St. Antony bears the body of St. Paul to a grave dug by friendly lions, while at the graveside a chorus of surreal but relatively unthreatening music-making demons with open mouths hoot, thrum, and parp in noisy disharmony. These demons may be intended to represent the state of music —two of them are reading from musical scores—but in a more

Vermeer: *Young Woman before a Virginal*

general sense they correspond to the noise and distractions of human society from which St. Antony removed himself to live the contemplative life.

The upper portion of the painting shows St. Antony floating in mid-air, surrounded by another menacing crop of demons. Weightlessness and crowded imagery suggest sound and hearing, but this is also an image of other varieties of invisible mental and physical anguish, such as the inner voices of the schizophrenic and the painful tribulations of old age. A boar-faced demon wearing a cardinal's scarlet skull-cap is squeezing the saint's chest, suggesting that he suffered from shortness of breath. An armour-plated lobster pokes at his bare toes, an allusion perhaps to the sharp pain of corns. A third bespectacled demon hints at failing eyesight. The weirdness of these surreal animal creations expresses imagery of painfully acute sensation in both mind and body. St. Antony's temptation or trial is a test of the soul's integrity or power to maintain full composure in a world of excessive sensation.

Music fit for a lady

A sense of complete harmony comes to focus in the serene but knowing gaze of Vermeer's *Young Woman before a Virginal*, one of several works by the artist on the subject of music and young womanhood. It has been argued that the moment captured is just before she begins to play, but it makes more sense to imagine that she has just finished playing and is turning to the onlooker with a smile of quiet satisfaction. The delicate sound of the virginal dying away is a classic image of mortality, completely in keeping with the acoustic of a small, sparsely furnished room with marble floor and plaster walls. A soft light shines diagonally through the window into the room, on to the opposing diagonal of the lid of the instrument from which one can imagine the sounds of music quietly rising in response.[9]

The young lady's back is to the light, her face half in shadow, so it is clear that her attention has been focused on the inner world of music rather than the external world of light. That inner emotional world is symbolized by the picture of a chubby god of Love on the wall behind her, and by the idealized landscape seen on the underside of the lid of the virginal. Music and love are dynamic processes articulated against a built environment of exemplary stillness. By opening the instrument the young woman gains access through music to an idealized but natural landscape of human affections. In a room otherwise dominated by the timeless order of geometry, the human figures and landscapes stand out as images of growth and complexity.

Vermeer is not the first great artist to combine images of music and love, but the strength of his composition elevates the association from a statement of the merely sensual to sublime philosophy. This young woman is in control of her music, and thus of her emotions. The virginal is a simple keyboard instrument that brought instrumental performance within the scope of young womanhood. The implications of this are of a feminine beauty in music that is fresh, artless, and natural in contrast to the contrived and self-seeking rhetoric of the male professional lutenist or serenading guitarist.

Notes

1. Giotto di Bondone, *The Last Judgment*, fresco, 1305-07, Cappella dell' Arena, Padua, Italy.

2. Fra Angelico, *The Coronation of Mary*, panel, 1434-35, Uffizi Gallery, Florence, Italy.

3. Hubert and Jan Van Eyck, *Angels Singing* and *Angels Playing music*, panels from the Ghent Altarpiece, 1432, Eglise de St-Bavon, Ghent, Belgium.

4. Jacopo da Palma, *Apollo and Marsyas, Apollo Flaying Marsyas*, canvas, Herzog Anton Ulrich Museum, Braunschweig, Germany.

5. Hieronymus Bosch, "Hell," panel (detail) from the triptych *The Garden of Earthly Delights*, c1500, Prado Museum, Madrid, Spain.

6. Matthias Grünewald, *Allegory of the Nativity*, panel from the Isenheim Altarpiece, 1512-16, Musée d'Unterlinden, Colmar, France.

7. Michelangelo da Caravaggio, *Rest on the Flight into Egypt*, canvas (detail), 1594-96, Galleria Doria-Pamphili, Rome, Italy.

8. Marten de Vos, *The Temptation of St. Antony*, panel, 1591-94, Koninklijk Museum voor Schone Kunsten, Antwerp, Holland.

9. Jan Vermeer van Delft, *Young Woman before a Virginal*, canvas, c1670, National Gallery, London, England.

CHAPTER NINE

In camera

Dear, if you change
I'll never choose again
Sweet, if you shrink
I'll never think of love
Fair, if you fail
I'll charge all beauty feign'd
Wise, if too weak
More wits I'll never prove
Dear, Sweet, Fair, Wise
Change, shrink, nor be not weak
And on my faith
My faith shall never break.

Earth with her flowers
Shall soon her head adorn
Hell her bright stars
Thro' Earth's dead globe shall gloom
Fire heat shall lose
And frost o'er flames be borne
Air made to shine
As black as Hell shall bloom
Earth, Hell, Fire, Air
The world transform'd shall view
Ere I profess to faith
More strange to you.

John Dowland, "Dear, If You Change"

JOHN DOWLAND is an English composer and contemporary of Shakespeare whose music expresses the down side of the new Renaissance spirit of freedom. His lute song "Dear, If You Change" is an example of chamber music, music to be played in the privacy of one's own home.[1] Dowland specialized in small-scale works of this kind. He was labeled with the catch-phrase "Semper Dowland, semper dolens" (Ever Dowland, ever miserable), a pun on his name (probably pronounced "Dolan") and the Latin word for grieving. His music is intense, intricate, and often troubled in tone, reflections of a wider unease that is found in much Renaissance art and is given the name of *melancholia* or philosophical melancholy. We tend to think of melancholy as overwhelming sadness, but that's not what is meant here. The brooding figure in Albrecht Dürer's *Melencolia I* is overcome not by self-pity as much as by a sense of doubt and uncertainty in the face of an unfathomable universe. The doubt is science in doubt, a science newly divorced from the patronage, emblems, and assurances of religion. Less thunderstruck than the damned soul of Michelangelo's Sistine Chapel frescos painted three years earlier, but certainly looking more doleful than Rodin's *The Thinker*, Dürer's "fallen angel" of 1514 is surrounded by images representing the illusory certainties of a received knowledge having more to do with mind games, paradox, puzzles, and numerical conceits than hard obser-vation.[2]

Not everyone was touched by moodiness, at least not all of the time. Gabrieli's wind canzonas celebrate the emancipation of music from the stifling authority of the church and pointless restrictions of permitted expression to the span of the human voice and meaning of the printed word. He declares for human skills and competitiveness, and uses musical instruments and notation to explore the mysteries of space and time with a sense of energetic cooperation and good humor. Gabrieli is no melancholic. His music celebrates free enterprise and scientific progress rising from the ashes of the Dark Ages.

Cabinet art

A camera is a box where images are stored for later contemplation. The first camera, in fact, was a small room on to whose walls images of the world outside were projected by means of a lens. "A room" is what the word *camera* means. When judges deliberate *in camera* they are coming to a decision in private. Chamber music is music to be played in private for study and contemplation. There is a nice parallelism between the camera as a device for storing pictures and chamber music as a music of personal and inward reflection.

Art historian Herbert Read alludes to the rise of chamber art as the expression of a new humanism and freedom of visual representation from a medieval subservience to religion. With the assertion of personal freedom comes a loss of security. What Read rather dismissively calls "paganism" suggests a relaxation of morals, but this seems a little unfair bearing in mind the anxious intensity of Dowland's or Dürer's imagery. The individual can no longer trust in the comfortable assurances or protection of traditional wisdom. The badge of melancholia catches that sense of awesome isolation and awakening responsibility. He writes:

> Originally the illuminations in a manuscript were conceived strictly as decorations subsidiary to the text, and part of the book as a book. They were incomplete and meaningless when divorced from the book. But gradually during this same period the illuminator began to conceive his decoration *as a page*, complete as a page. It was then a logical step to divorce the illumination from the book, and to paint on separate panels. It only needed the paganism of the Renaissance to make a further divorce between such pictures and any sacred or utilitarian intention. The picture in itself became a desirable object, either as a record of the personality of the owner, or as a decoration to his house, or even as an emblem of his wealth.[3]

Dowland's song for solo voice and lute accompaniment is chamber music in this new spirit. The first line says it all. It is a song in *English*, the language of ordinary people, not church Latin, the language of authority. It is sung by *a woman*, in itself a radical statement of female emancipation, because the thoughts and feelings of women were not taken very seriously in a male world, and women were not encouraged to sing in public. The very first word "Dear" shows this to be a love song, and therefore about individual human emotion, not about collective religious faith. It is music expressing a real, personal, and intimate relationship, not a ritual declaration of public duty. The combination of pure female voice and quiet lute reveal it as music designed for a small domestic chamber, not a vast public space. The voice sings alone, unsupported by any reverberation, with the lute as sole partner and support.

Tablature

The lute player is a specialist and the lute has a special notation called *tablature*. There are many different forms of tablature.

Many still exist. Tablature is an instrument-specific form of notation. It tells the player where the fingers should go, not which
notes to play. That may seem a little odd, like painting by numbers. In fact, it is an intentionally secret notation that only makes
sense if you have the right instrument and are familiar with the
code. In Dowland's time music for voice and lute was becoming

A I R S.

Air for voice and lute in the French style by Gabriel Bataille

available in published form. To protect the interests of lute
specialists and prevent their music being played on any other
instrument, the accompaniment was printed in tablature while
the voice line, a part freely available for anyone to sing or play,
was printed in standard notation.

The above example of a seventeenth-century French lute
song by Bataille makes the distinction clear.[4] At the top of each
system is a five-line stave for the voice with a treble or G clef
indicating the pitch of the melody. The voice line is in *standard*

notation that can be read and performed by *any instrument or voice*. Standard notation is information about pitch and time values—what notes to play, in what order, and how long each note shall last. This notation is not concerned with the instrument or its tuning; those aspects are the player's responsibility.

Beneath the voice line and lyric is a *six*-line stave. This is lute tablature. The six lines correspond not to absolute pitch values, but to the six strings of a lute. But they don't say what pitches the six strings are tuned to—that is a secret. Above each of the six lines is a pattern of letters a, b, c, or d. You might think a, b, c, and d are notes of the scale. You would be wrong. They correspond to index, third, ring, and little finger positions on the string. Once again, the casual reader cannot tell by looking what notes the lute is supposed to be playing.

Between lyric and tablature is a line of standard note values representing the rhythm. Tablature by itself has no way of showing rhythm. It's a notation essentially for a solo performer, and a solo performer can decide on his own timing. However, since a singer is also involved here, and the singer's notation is very particular in indicating rhythm, the lute tablature is obliged to borrow an additional layer of meaning from the voice in order for the two parts to stay in time together. Tablature was the master musician's protection of his investment in skills and his professional income. The imminent arrival of domestic keyboard instruments, such as the virginal in the Vermeer painting, was eventually to put the lute specialist and his fingerwork skills out of business, and make the same music available in standard notation to any young lady who could read a voice line.

Asking the great "What if?"

So what is Dowland's poem about? It is about *the future*. It asks the great "What if?" "Dear, *if* you change". What *if* we should part? What will tomorrow bring? What if you were to fall out of love, or go away to battle, or sicken and die? The lute's slight notes hang in space for only a moment and then die away in a potent image of mortality. And she says that whatever happens, she will forever be true. Dowland's message—her message—is that *love is eternal*—not religion, not society, but the pure love of one person for another. "Love will survive anything." It is the authentic voice of Renaissance humanism: serious, considered, thoughtful, and ultimately positive. If Dowland's melancholy signature is, in part, a consequence of his skills as a lutenist, the constant reminder every time he takes up his instrument to play, that notes once "touched" cannot be retrieved but die away in an instant, then the partnership of voice and lute can be understood as a relationship of hope. *The voice can defy death.* The voice can

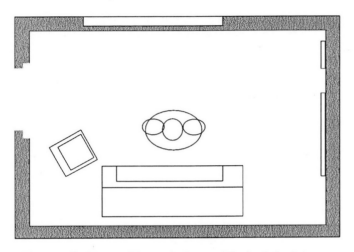

Vermeer's *Young Woman before a Virginal* depicts an ideal acoustic for the intimate clarity of chamber music. Flat surfaces, tile floor, and plaster walls reinforce the delicate sound of an early keyboard, lute, or viol.

sus-tain, hold, project and modulate a melody, even though it may be limited to only one note at a time.

The great "What if?" is about not knowing what the future may bring. Whereas a Gabrieli sees the future from a masculine perspective of unbounded possibilities of technical and human progress, Dowland explores the fragility of human relationships, perhaps from a more feminine perspective. Turning to face the future and away from past tradition means one can no longer rely on old regime values and assurances, in particular the material symbols and structures of belief. In their place the artist proclaims a new set of values in art, music, and human affection, values that are not permanent in the solid material sense that bronze or stone or gold are permanent, but in a transcendental

moral sense of qualities that despite being insubstantial, immaterial, and invisible, are nevertheless eternal.

The voice in close-up

The vast reverberating spaces of plainchant kill off the texture of
language and reduce the voice to a melody of abstracted vowel
sounds. A member of the congregation not already knowing the
words of the service has a hard time discovering exactly what is
going on, being prevented by a combination of screens and security zones from getting close enough to the choir to see their
lips move. The chamber songs of Dowland and his contemporaries bring the voice into close-up and into focus where you can
hear every syllable and every consonant. These are by definition
new ideas and new words. The listener has not heard them
before and can only understand what is being said if they can be
heard clearly. Part of the meaning and music of these verses is in
the sound effects of speech underscoring the ideas the words
convey. Texture is consonant music: the cymbal clash of CH/SH
in CHange, CHoose, SHrink, CHarge; the arrow-swift F of Fair,
Fail, Feign'd, and Faith; the hammer-on-anvil K of shrinK, thinK,
weaK, breaK; the whirl of W in sWeet, Wise, Weak, Wits. No
matter how serenely the song is performed, the mere sounds of
the lyrics express a deeper dimension and inescapable reality of
physical actions and their corresponding implications. To hear
the texture of verse a listener has to be up close and the acoustic
has to be transparent.

Dowland's two verses express a classical symmetry making
the connection between the inner mental world of human emotion and the external physical world of time and the seasons. The
verses' mantra-like repetition of the word *shall* emphasizes the
idea of the future at the same time as the context reassures a
doubting lover that the natural order is as stable and strong as
her love—and by extension, that it is even love itself that sus-
tains the universe.

Logical positivism

Dowland's contemporary Thomas Wyatt is also in love, but his
love has been rejected. His poem "Blame Not My Lute" survives
as a song fitted to "La Gamba," a popular Italian melody, but the

lyric is far wittier than the melody and offers a robustly confident alternative to Dowland's melancholic disposition.[5] Wyatt's song is about the persistence of love and music even when his efforts, as well as his intentions, are roundly rejected. It declares these values to be self-sufficient. They don't even need an object.

Blame not my lute for he must sound
Of this or that as liketh me;
For lack of wit the lute is bound
To give such tunes as pleaseth me:
Though my songs be somewhat strange
And speak such words as touch they change
Blame not my lute.

My lute and strings may not deny
But as I strike they must obey;
Break not them then so wrongfully
But wreak thyself some wiser way
And though the songs which I indite
Do quit thy change with rightful spite
Blame not my lute.

Farewell, Unknown! for though thou break
My strings in spite with great disdain
Yet have I found out for thy sake
Strings for to string my lute again!
And if perchance this foolish rhyme
Do make thee blush at any time
Blame not my lute.

Wyatt is saying, "Fair enough; you don't like my song. You think it's a bit strange, and you are probably right. But *don't blame the lute*. The lute has no mind of its own; it has to play what I tell it to, so if you really don't like my song or me, don't waste your anger in breaking the strings, because new strings can always be found." What does he mean? That the song *will not go away* if the strings, or even the entire lute is destroyed. Even a bad song will last for ever. Wyatt speaks with the newfound authority of the Renaissance wit. In insisting that the lute is not to blame, he is making the deeper point that he accepts that *he is to blame* for his actions, even when they give offense. In fact, giving offense and being rejected are the clearest possible evidence of personal independence. *He is no lute.* "Look at me. I am in control. I am not the mere instrument of some higher power. I am the one who makes the decisions. I am responsible for my actions."

Wyatt's self-assurance is wonderfully encouraging. Dowland is resolute in doubt, only too aware of the unknown perils that lie ahead. Wyatt, on the other hand, is almost recklessly optimistic in his declaration that *Art is eternal. Music cannot be destroyed even though the instruments of music may be destroyed, and Love that springs from the same human values as music is also eternal.* What is true of the music and lute is also true of his affections and his person: "Love endures beyond rejection and even beyond death." The last lines are wonderful: "If at any future time my lines may bring a blush to your cheek, *don't blame the lute.*" Because her blush would be at the memory of the song and the person who composed it, proving the enduring power of music and love over time, rejection, and mortal decay. Dowland and Wyatt stand with Shakespeare's famous sonnet "Shall I Compare Thee to a Summer's Day?" in declaring the supreme values of individual beauty and emotion, and the supreme power of music and verse to preserve that beauty indefinitely.

The power of absence

You can tell a voice line from an instrumental melody. A voice line stays within a limited compass and moves in small, easy-to-manage steps up and down, whereas an instrumental line is able to move more freely over a wider range, and leap awkward distances from note to note. There are many instrumental pieces by Dowland that conceal a voice-like melody in the structure of a pavan or solemn dance. One is actually entitled "Semper Dowland, Semper Dolens;" and it exists in versions for lute alone and for lute with a trio of viols. In this latter version for viols and lute the lead melody is taken by the treble viol. The listener is tempted to treat it as a purely instrumental number, but the absence of a solo voice is perhaps its most dramatic feature: the ultimate expression of mortality. It says, the singer is dead.[6]

The viol is an earlier relative of the violin, and its sound is vibrant and intense. This resonant, highly charged, and emotional music creates its own acoustic environment. As long as the group continues to play the music continues to sound, but when action ceases everything stops. The music stops. There is complete silence. *Time itself seems to stop.* Without will there is no energy, without energy there is no action, without action there is no music, and without music there is no existence. At the end of each verse a silence descends, so black and so profound as to seem like the end of the world. This is music that speaks of *loss*— not just the absence of a voice, it is also the absence of a sustaining room acoustic. The message of this music is that time and value and purpose are wholly dependent on human actions and human responsibility, but also more positively about the persistence of music despite the sad inevitability of human mortality.

Absence can be a difficult concept to grasp. But imagine a band of musicians arriving at the studio one morning to record the last song of a new album. The lead singer is late. While the other players are tuning up and getting ready, word comes from the hospital that there has been an accident and the lead singer is dead. What do you do in this situation? The rest of the band confer, and decide to record the song anyway, but with the lead guitar substituting for the voice. The album will be released as a tribute to the lead singer and with the instrumental as a poignant tribute recorded on the very day of his death. Fans will buy the album and listen to this last song for an overwhelming emotional experience of total and absolute loss.

Conflict resolution

Wherever there is a lead performer and an accompanying group, whether the music is plainchant, folk song, or a song by Dowland, there is manifestly an interaction between the two partners. The relationship is essentially cooperative, but does not have to be friendly. A listener understands interaction as dialogue. A dialogue is not always a meeting of minds. Where a dialogue is at cross purposes, there is potential for disagreement. Conflict of any kind is a test of character and resolve. Not just for the artists involved, but for the audience as well.

In the opposition of solo and chorus in plainchant, for example, a tension is generated between the possibility of freedom (the solo) and the duty of collective solidarity (the chorus). The influence of harmony reaches beyond the musical domain of different pitches in a preordained, stable relationship to notions of teamwork as well as emotional concord. Tension is latent in the very idea of a lead soloist in opposition to the group. So a listener is always primed to the thought that harmony is not a given;

Opening of the *Adagio* from Piano Concerto
No. 23 in A, K488 by Mozart

the soloist may stage a tantrum or go off musically at a tangent.
It may not work out.

In Dowland's "Dear, If You Change" as in any slow ballad
for voice and acoustic guitar, there is a creative tension arising
from the opposition of *melody*, the voice, and *harmony*, the lute or
guitar (or piano, or any chord instrument). A voice line or solo
violin or saxophone creates a sustained musical line that by
definition expresses notions of action, direction, and continuity
of being. A melody expresses a dynamic outlook, but one that is
also inherently uncertain, not to mention unstable. A lute or

guitar, on the other hand, is emblematic of the steady state, of the snapshot of existence in which everything falls momentarily into place.

Planning is thinking, and thinking is a mysterious process. Strategy in a musical sense is control of a process of change and its impact on corporate harmony, the stability of the group, and to express a control strategy you need a harmony instrument like a lute or harpsichord. These instruments map the movement of a melody line with global accuracy in pitch and time, and are able to coordinate multiple melodies that may be going in different directions simultaneously. But they have *no sustain*. Music for lute or harpsichord solo has intellectual precision, therefore, but no independent life. For that harmonic strategy to acquire a sense of the natural dynamics of human existence, it has to be translated into the actions of voices or orchestral instruments that have breath and movement.

Mozart's equivalent to Dowland's lute is the *fortepiano*, an eighteenth-century keyboard, and the slow movement of his Piano Concerto No. 23 magnifies Dowland's melancholic theme to concert proportions. Many hearing this slow movement for the first time imagine that the music is sad, but it is not sad at all. For most young listeners melancholia is an unavoidable part of becoming aware of the extent to which their identity is conferred by relationships—happiness in being with other people, sadness in being alone. A slow movement in classical music is nearly always music for meditation or reflection, more about thinking than action, which makes it interesting, if not exciting. In the spirit of Dowland, the Mozart slow movement is a brooding piece in the dark and remote key of F sharp minor, and it begins with an opening statement by the solo piano.[7]

A solo signifies uncertainty. The soloist may be making it up. It's easy to interpret the mood as one of sadness. There is precision at first but also hesitancy. But the music itself is not depressive at all. How do we know? Because if a person is depressed the world is turned in and the voice shrinks to a near monotone. This music is anything but monotone, though, granted, it is an opening gambit of deliberately few notes. A listener may also describe it as withdrawn, or quiet. But it isn't quiet; it's just thoughtful and when you are thinking you are listening to yourself. The music is just beginning. The orchestra has not said a word, so there is no basis for comparison of loud or soft. *It is loud enough*, and that is all. You can hear everything perfectly clearly.

Hesitancy is not a sign of lack of confidence, because the range of notes covered by the soloist is unusually wide, and the interval leaps from high to low at times breathtaking. The image is of an architect poring over his designs, or a navigator over his charts with set square, dividers and ruler.

The moment of truth for the entire movement comes at the point of entry for the full orchestra after the solo introduction. In a normal concerto a conversational exchange develops between the solo instrument and the orchestra, the solo making a statement or asking a question, the orchestra repeating the statement or formulating a response. Mozart is much more subtle and his idea of harmony is infinitely more profound. The piano, in keeping with its role as a *harmony* instrument, is presenting a model or chart of *relationships*, and the implicit question is how the orchestra as the voice may translate the plan into *action*.

Mozart's orchestra in both published versions consists of strings and woodwinds (I prefer the version with added clarinets myself: Mozart's clarinet writing is wonderful). All of the orchestra are *sustain* instruments: they either breathe or otherwise manifest human measures of continuity, gesture, and sensation. The relationship of the orchestra to the clipped tones of Mozart's pensive fortepiano corresponds in Dowland's song to the life-sustaining voice in relation to the momentary sounds of the accompanying lute, but the voice of Mozart's orchestra is much larger and more powerful, more varied and also more numerous.

The keyboard is *idea*, the orchestra *incarnation*, or "bones and flesh" to borrow a comparison from composer Pierre Boulez. What the listener is conditioned to expect is for the piano to end its opening statement and for the orchestra then to address it in some way. For example, if you were handing out instructions then repetition would show that the message was at least received, even if not fully understood.

What happens here is totally unexpected. The full orchestra enters without ceremony on the piano's final chord, without even pausing to digest what the piano has just been saying. Why not? *Because it already knows.* Into the candle-lit world of the soloist it brings daylight, fullness, color, texture, and confidence all at once. This is no ordinary dialogue. Any impression of the piano as a solitary figure is blown away by an accompanying orchestra whose demeanor reveals that it is completely at home with the piano's mood. The orchestra's serene answering gambit says: *Yes, I know.* However one chooses to interpret the relationship, whether as divine intervention, or the orchestra as the realization of a dream, or Galatea to the piano's Pygmalion, as a vision, or as an epiphany, it remains a poignant and powerful affirmation of absolute harmony of mind and heart.

Equally remarkable is the sequel, because *nothing changes* as a consequence of the orchestra's intervention. The piano, reticent as ever, continues to measure out intervals on the keyboard, but now it is in perfect synchronization with an all-knowing, all-forgiving, compliant, even voluptuous orchestral partner. Apart from this new steadiness of purpose, the solo keyboard shows no change of direction, no inclination to relax, continuing to calculate and plan right down to the last note. This is no sudden passion, but a mature relationship; not sadness or depression, but domestic bliss. You are working late. Your eyes are tired. You reach for the phone to call the one you love. At that moment the phone rings. It is your loved one calling you.

Notes

1. John Dowland, "Dear If You Change," *First Book of Songs*, No. 7, 1597. Emma Kirkby, Anthony Rooley. *The English Orpheus* (Virgin Classics 0777 7595212 4, 1989).

2. Albrecht Dürer, *Melencolia I*, engraving, 1514, Kupferstich-kabinett, Berlin, Germany.

3. Herbert Read, *Art and Industry: The Principles of Industrial Design*, 4th rev. ed. (London: Faber and Faber, 1956), 27-28.

4. Gabriel Bataille, *Airs de Différents Autheurs mis en Tablature de Luth* (1612-17). In Bernard Huys, *De Grégoire le Grand à Stockhausen: Douze Siècles de Notation Musicale* (Brussels: Bibliothèque Albert Ier, 1966), 75.

5. Thomas Wyatt, "Blame Not My Lute". In *The Penguin Book of Early Music* ed. Anthony Rooley (Harmondsworth: Penguin Books, 1980), 135. The Consort of Musicke, dir. Anthony Rooley. *Musicke of Sundrie Kindes* (vinyl, Editions de l'Oiseau-Lyre 12BB 203-6, 1975), side 7, track 2.2.

6. Dowland, "Semper Dowland, Semper Dolens," from *Lachrimae* (1604). Fretwork, *Goe Nightly Cares* (Virgin VC7 91117-2, 1990), track 24.

7. Mozart, Piano Concerto No. 23 in A, K488. St. Luke's Orchestra dir. Julius Rudel (MusicMasters 01612 671649 2, 1998).

CHAPTER TEN

Team players

AN ORCHESTRA is a collection of musicians who perform as a group. It varies in size from a chamber orchestra of around 15 to 20 players to a symphony orchestra of 75 to 150 players. An orchestra performance is not a jam session where the players take turns to make it up as they go along. It is not like choir music where there are large numbers of singers but only a small number of voice parts (soprano, alto, tenor, bass). Instrumental music for large numbers of players demonstrates and requires a high degree of organization. The word *organization*, in fact, means "musical order." An orchestra is first and foremost about people management.

Organizing people is a serious business. In order to run a successful army, manufacturing business, or country, it has to be organized. The history of the world has largely been determined by competing organizational strategies that originated in post-Renaissance Western Europe, many of them originally developed and researched in musical applications.

The simplest form of people management is direct imitation —having the same action performed by large numbers of people, for example, a regimental march-past, a mass gymnastic display, or a fun run. The combination of simple actions and sheer weight of numbers can be effective in generating a sense of corporate identity. Many people enjoy being part of a crowd. It brings a feeling of security and community. But direct imitation of this sort can only work if the action to be imitated is within the reach of the average citizen, which means that the level of skill required by such demonstrations is relatively low. The average national anthem, for precisely these reasons, is not designed to be great music and is rarely performed to a high artistic standard. Unison hymn singing in church or at a football match may have a positive impact on congregational morale, but as a display of organized teamwork it does not prove very much at all. That is as it should be. Imposing a primitive uniformity on group behavior by definition eliminates the distinctions that enable society to function cooperatively in the real world.

Similarity and difference

Higher forms of organization aim at recognizing and reconciling a range of different skills. Taking as their model the various organs of the human body operating in partnership to maintain life, the ancient Greeks and Romans accepted the principle of a social order within which a wide diversity of special functions interact in harmonious cooperation. Organized diversity is a key element of group music, in particular folk music and music of oral cultures generally where each instrument can be heard to play a distinct and functionally appropriate role in the ensemble. The same teamwork audible in the Spanish song "Mandad' ei Comigo" expresses the universal maxim: different instruments perform different functions. The voice lead delivers the text, the harp keeps the voice in tune, the hand drums reinforce the voice rhythm, the bass drum keeps the beat, the drone maintains harmony.

Group music-making results in complex sounds. Organization covers complexity at a number of different levels. Harmony has to do with pitch combinations (what the music is saying at any one moment), orchestration with voice or instrument combinations (how they sound together), and complexity with actions or teamwork (where the music is going). Assume, for the sake of argument, that each member of the group produces one sound at a time (so, no chords) and the resulting complexes can be either homogeneous, when all the members of the group are the same kind of instrument, or heterogeneous, when the members of the group are all different. A choir of voices, a brass choir, or a string orchestra is a more or less homogeneous group, whereas a folk band, a pop group, or a symphony orchestra are heterogeneous in composition. Whether a group is homogeneous or heterogeneous affects the quality of sound it makes, and thus the kind of organization it is capable of expressing. Renaissance music pursues a preference for uniformity of tone, so a consort of recorders was a group of similar instruments of different sizes, from treble to bass, performing as a group to produce a highly coherent and unified combination sound. Likewise, a consort of viols. Uniformity of tone means that the group can attain a high degree of harmonic integration. That same degree of tonal fusion cannot be a realistic goal for a group of instruments that are not all members of the same family. The broken consort, as it is called, aims instead at a different kind of harmony based not on tonal integration but rather on integrated functions. The two roles and their implications are quite different. One is science, while the other is social engineering.

A desire for consistency

Certain instruments, such as the harp and panpipes, are designed to play more than one note at a time, and in this sense resemble the consort of recorders or viols in providing a means of investigating harmony as well as making melody. The earliest forms of keyboard instrument—spinet, organ, and clavichord are one-person mechanisms of tonal inquiry. They are designed with a non-specialist player in mind, the sort of person who is interested in the science of harmony but does not have the knowledge and skill of a professional lutenist or guitarist. Such a person is not interested in the art of music in its human dimensions of hard-won technique and emotional nuance, but rather in the underlying science of tone and tone combinations that makes music possible. For such an inquiry, what is needed is not an instrument designed for art and skill of a high order but one on which a non-expert can make plain demonstrations and observations of combinations of more than one tone at a time easily and with a guaranteed consistency of tone production. That desire for consistency has to mean something.

Pipe organs are another such instrument. The earliest recorded structures for feeding air under pressure to multiple pipes date from Roman times and are based on Ctesibius's invention of an air pump or *hydraulis* using water pressure from a reservoir to deliver compressed air at equal pressure to a number of different pipe outlets. Though usually represented as the first organ, it is unlikely to have been devised with musical applications in mind. Ctesibius himself was almost certainly thinking of using water pressure as a means of controlling a system of air conditioning to ensure the equal delivery of warm air to every room in a Roman mansion. A logical way of demonstrating to your client that the air pressure is the same for every outlet would be by inserting a large whistle pipe into each one. There would have to be a pipe of different pitch in each outlet to enable a listener to assess each one individually. You insert the pipes, apply the pressure, listen, then switch them around. If they all sound at the same loudness, the pressure is the same throughout; if any one pipe sounds louder than the rest, then the pressure is greater at some outlets than others. Acoustic testing of such an invention makes sense in the absence of modern techniques for measuring air pressure or air flow. However, anyone witnessing such a demonstration, for example, a wealthy client, is likely to see the potential of Ctesibius's invention for making music, or better still, for *acoustic modeling*.

Positive interaction

Harmonic science is the study of interacting natural phenomena that oscillate in predictable ways—on the grand scale, the motions of the planets, the tides and seasons, and on the human scale the vibratory behavior of materials such as woods and metals of various kinds (for example, to assay bars of precious metal to determine any adulteration). In ancient Rome that science involved the acoustic matching of strings of different weights and tensions, and analysis of the defining sounds of animal and human speech. To analyze and classify complex oscillating patterns in nature, one needs suitable instruments with which to model and observe harmonic interactions, starting with the simplest two-note combinations in single-figure ratios such as 1 : 2 or 3 : 4, to find out which tone combinations are stable and mutually supportive and which are not. Any instrument or group of instruments exhibiting uniformity of tone across the entire range from extreme bass to extreme treble is a candidate for research of this kind, because it expresses pitch as a continuous dimension and not as an index of human action or emotion. Ctesibius's air pump would be seen to fit a particular scientific need for a reliable tone generator.

The keyboard is an acoustic calculator

The search for order in the universe, for the Romans as for physicists today, is a search for laws governing the relationship of energies and motions in dynamic but stable combinations. Crude but powerful organs based on Ctesibius' model came to the West from the kingdoms of Byzantium where they were used as signal generators to model the orbital motions of the planets, otherwise known as the harmony of the spheres. In keeping with their status as scientific instruments designed to reveal the structure of the universe, organs were regarded with awe, rather like particle accelerators today, and were brought out for display on special occasions to celebrate the divinity of the emperor. The instrument generated a powerful and hypnotic drone, an effect imitated in the harmonization of Ambrosian chant and among lesser instruments such as the bagpipes. However, the general perception prevailing even after many centuries of evolution was that whatever its merits as a religious or scientific instrument, the Byzantine organ did not really fit the description of a musical instrument. "In its liturgical use the organ was limited to high feast days. The extremely crude and clumsy nature of the instrument's mechanism at this early stage, together with its well documented noise and clatter, scarcely made it suitable for extended vocal accompaniment," remarks one scholar, citing complaints of contemporary observers such as Aeldred of Rievaulx at "this awful noise which expresses more the sound of thunder than the sweetness of the human voice."[1] Organs were clumsy, raucous instruments and remained so for hundreds of years, and they were still pretty crude even in the relatively high-tech world of the Ghent Altarpiece in the fifteenth century. For an instrument of such dubious musical merit to have continued in development for so long a time with so little to show for it speaks of a purpose darker and more serious than providing the choir with a tune.

Religious or aesthetic connotations aside, the post-Baroque organ as we know it today is manifestly a keyboard synthesizer incorporating multiple oscillating circuits for the production and analysis of combination waveforms. As such it represents a very high order of technology at a relatively early period in Western civilization. You do not invest that amount of acoustical and technical knowledge and development *for over a thousand years* merely for entertainment or celebratory purposes. It is highly ironic, in retrospect, that the church should have begun to lose interest in acoustical modeling just as development of the organ as a precision instrument finally peaked in the age of Galileo and Gabrieli.

Intellectual inquiry into the nature and laws of harmony persisted outside the church among an informed upper class with access to consorts of recorders, pipes producing very simple waveforms, consorts of viols, whose waveforms are uniformly rich, and keyboard instruments such as the clavichord and virginal which are designed for uniformity of touch. Wherever multi-voice music is performed by a single family of instruments or a keyboard instrument of uniform timbre (or even, in the case of madrigals, by voices in combination) the subtext of such music is intellectual inquiry into the nature of harmony. Chamber madrigals for voices build on the idea of harmony in human relations reflected in musical harmonies and chord sequences. Harmony is as much the issue underlying Henry Purcell's fantasias for multiple viols[2] as in classical favorites such as the slow movement of the Lute Concerto in D by Vivaldi.[3] The emotional neutrality of J.S. Bach's Prelude in C from *The Well-Tempered Clavier* tells the listener that this music is not for singing or dancing but for hearing the same pattern projected on to increasingly remote tonalities.[4] Nowadays the madrigalists' devotion to uniform perfection in harmony is preserved and reflected, albeit in parody, in the exaggerated vocal idiom of the barbershop quartet.

People power

The earliest orchestras were relatively small groups of players employed in the service of powerful families where they provided entertainment in the form of music for concerts, dancing, and theater. In the seventeenth and early eighteenth centuries the orchestra was a more or less *ad hoc* collection of players, usually based around a core collection of violins, a cello, bass viol, and harpsichord. For Vivaldi and his contemporaries, the classical orchestra was a string band with organ or harpsichord, to which other instruments could be added as occasion demanded. A

Keyboard music embodying tonal consistency:
J.S. Bach: Prelude No. 1 in C major

preponderance of strings in early Baroque and classical orchestras was no accident or manufacturing conspiracy. Instruments
of the violin family brought an attractive resonance to the relatively dry acoustic of the eighteenth-century salon. Today's
symphony orchestra is still built around a core ensemble of
multiple first and second violins, violas, cellos, and basses, sup-
plemented by smaller numbers of winds, brass, and other instruments. The string orchestra is a sonority designed to instil a
sense of spaciousness and resonance in environments that are not
particularly reverberant.

Portable acoustic

The orchestra came about because public interest in complex
instrumental music continued to grow in defiance of the church
hierarchy. Institutional and private patronage was on hand to
continue supporting new music but the accommodation available for group instrumental performance—ranging from coffee-
houses to private chapels, ballrooms, and academies—tended to
be smaller in scale, unpredictable acoustically, and very different
in build quality from the vast stone cathedrals of earlier times.
    Composers also had to reconcile themselves with the loss of

the grand organ as the centerpiece of a musical ensemble. The organ in Monteverdi's *Vespers* serves as a focus of attention and source of continuity linking constantly varying combinations of voices and instruments. Along with loss of the organ as a focal point, composers also lost the impressive reverberation of the church acoustic. The conveniently portable harpsichord, chosen to replace the organ as a control keyboard, unfortunately has a much weaker tone, and no sustain. Substituting the harpsichord for the organ is a relatively easy choice, but recreating the ambience of a cathedral has to be a much greater challenge.

Working in a highly reverberant acoustic with an instrument of fixed tuning like the cathedral organ has its attendant disadvantages, nevertheless. The fixed tuning of an organ is not adapted for a music traveling freely from key to key, especially the dark and dissonant keys and harmonies based on the black notes of the keyboard into which Bach ventures in the familiar Prelude in C. A reverberant acoustic, on the other hand, is equally unsuitable for fast-moving or complex music by large numbers of instruments. The great advantage of creating an orchestra around a core string ensemble is that the ample tone and complex reverberation of a group of violins, violas, and cellos all playing together can simulate the warmth of a church acoustic but can also *modify it on demand*. This is exactly what happens when music for string orchestra changes from one key to another. The word for changing from key to key is *modulation*. When small groups of voices change key, the effect created is emotional, internalized, but when a string orchestra changes key it is like changing the reverberation characteristic of a room. We know this is so because when the violin was new, as we see in Caravaggio, it took by surprise architects like Palladio who had been designing room spaces to resonate in particular keys or wavelengths based on a human scale of proportions. For Palladio the musical ideal of harmony was the Renaissance consort of viols, a music that does not modulate freely but stays within the bounds of a particular key that can be tuned to the wavelength of a room. By the time of Vivaldi, however, Baroque architecture had begun to emulate the violin and develop more complex interior spaces suitable for a music of changing key. So the string orchestra can be understood as a complex unit designed to create its own variable acoustic, and we can also understand the characteristic modulation of Baroque music from key to key as an effect intended to be perceived as a manipulation of space.

Time and motion

Mechanical clocks also transformed the art of music. Artificial time, as signified by the ticking of an escapement, brought new order into people's lives, and that order also began to infuse into musical organization. Musical notation became more precise and also more objective. Timing, referred to the clock, no longer needs to refer to human measures of feeling or action. The nature of time itself could be investigated in musical applications. Music could pose such questions as: what is fast? or what is slow? Do we live at different speeds from one another, and is this a reason why some creatures, for instance birds, appear to live faster lives while others, like tortoises, seem to live at a slower pace than human beings? Art and science began to develop their own rules and models of human and universal order. To examine the nature of time, music had to seek alternative environments that allowed composers to experiment with concepts of fast and slow. The orchestra was a part of this new trend.

The Baroque orchestra of Bach and Corelli resembles an organ where the keyboard operator is controlling people instead of pipes. The players stood round the harpsichord in the middle of the room, acting the role of different organ registers or stops. The violins were kept in time by the concert master or lead violin, playing a highly decorated and easy to follow melody line, while the conductor at the harpsichord took special care of the bass line and supporting chord structure, and made sure the various parts came in on cue and sounded well together. In a recording studio today, the engineer at the mixing desk performs a very similar role to the harpsichordist of old.

Because a Baroque orchestra is designed to perform in a smaller room with a less reverberant acoustic than a church, the musical action can be faster and more complex, freer to explore human sensations of speed and motion in dance rhythms and melodies. During the Baroque era, composers began to describe music in terms of fast and slow for the first time, and the word *movement* was introduced to describe a musical number. The term can be understood as referring to motion in the abstract as well as to human scales of movement and timing embodied in the dance repertoire.

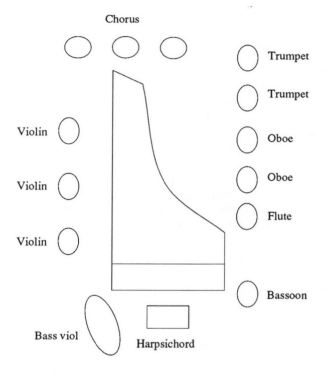

Chorus

Trumpet

Trumpet

Violin

Oboe

Violin

Oboe

Flute

Violin

Bassoon

Bass viol

Harpsichord

Voices and orchestra from an anonymous
German painting c1775

Take, for example, the sprightly first movement Allegro of
J. S. Bach's Brandenburg Concerto No. 2. The first question to ask
yourself when listening to any music for the first time is "Can I
sing it?" and the second, "Could I dance to it?" If you are tempted to sing along with a piece of music, that tells you it has a
melody or guiding line, and that the melody moves in a voice-
like, stepwise progression up and down. By this standard Bach's
Allegro is not vocal in style at all. Instead it starts from a very
high pitch, beyond human voice range, then zig-zags back and
forth like a bugle call—which, in fact, it is—making it hard to
follow. Could you dance to this music? This takes a little longer
to answer but the more you listen, the more you realize that this
music does not, in fact, resemble any dance. It has pace and
movement but no rhythmic structure suggesting dance steps,

changes of direction, nor do we hear the reciprocal patterns or male-female symmetries of traditional formal dance. Nor does the music pause, slow down, or speed up to indicate beginnings or endings, it just keeps going at the same mechanical pace. So it's not a dance, and it's also not a song. What is it?

This is music like a clockwork toy that you wind up and set down on the floor and it rolls across the room. What makes a clockwork toy so fascinating? *It goes by itself.* What people admire in a self-propelled machine—whether it is a clockwork ladybug on wheels or a Mars buggy on Mars—is that it moves unaided. We recognize the Allegro movement as mechanical in nature because the piece does not vary in speed from start to finish, because the speed of the music is unrelated to normal human action (is neither slow nor fast), and because there is neither vocal melody nor expression present (both loudness level and complexity remain relatively constant throughout).[5]

This is Bach the clockmaker, playing Geppetto to the music's Pinocchio. Everything is happening at once and everybody is doing something different, just like the very best animated video games where even the figures in the background have something to do. The group range of pitches extends from an extremely high trumpet down to an extremely low bass. The lead melody is handed from trumpet to oboe to violin to recorder. Nobody takes charge. This is a very even-handed, democratic distribution of leadership. Everybody has an interesting role to play, and everybody whether high or low in pitch moves to the same putt-putting beat—in fact, the movement only comes to a stop by the machine, in effect, hitting a brick wall.

Perhaps you remember the pink bunny playing the drum in the battery ad? Just so. Bach's image of clockwork motion is designed to impress on his audience the idea of musical *propulsion.* He has two sources of energy: harmonic *tension* acting like a powerful spring, and *rhythm* expressing perpetual motion. The combination of a rhythm that is always going round in circles and a chord sequence that seems not to care where it goes as long as it keeps going generates a strong sense of dynamic uncertainty and momentum. This, in fact, is one of the things that Baroque music is all about.

Building a team

The word *symphony* means "playing together." Teamwork is the art of playing together. A symphony orchestra is a product of Western European culture and social organization. Prior to the nineteenth century, organized music was the highest and most complex expression of team management achievable in human terms. For centuries the nearest comparison to a symphony in terms of action by numbers was experienced on the field of battle. No military campaign, however, could compare in precision timing or complexity with an orchestra, not to mention the software control that enables a symphony to be exactly repeated time after time.

Today the organizational implications of music are easily overlooked. It is normal today for the same orchestra to present works by composers as diverse as Handel, Vivaldi, Haydn, Beethoven, Berlioz, or Bartók in the same concert and in the same concert hall. This is convenient and admirable, but it takes music out of context and suggests that the only differences that matter are issues of personal preference or "historical evolution"—a form of cultural Darwinism still reflected in the ongoing debate among musicians over the use of period instruments.

Warfare

Team games as we appreciate them today are a relatively recent invention. On the other hand, music and warfare go back a long way. Apollo was not only the god of music, but also the god of war—his instruments the lyre and the bow, both powered by stretched strings. To win in war with the bow, you have to understand the relationship of force and tension; to win the heart of an admirer, you have to be skilled in tuning and targeting musical notes (the image uniting the two arts of warfare and love being Cupid with his toy bow and amorous darts).

Better to appreciate the implications of teamwork in musical affairs why not go to a football match where in one and the same arena you can witness three levels of team management promoting three levels of organization, namely the cheerleaders, the marching band, and the game itself. Cheerleaders are teams of attractive young ladies performing simple uniform actions for the purpose of building enthusiasm and a sense of community among the audience. This is crowd control. The marching band combines music and movement in a skilled display of abstract pattern formation occupying a large area of the playing field, action embodying harmony at a number of levels of complexity and also the symbolic occupation of territory (where incidentally the widely distributed participants have to deal with the same problems of coordination as Gabrieli's trumpets and trombones). This level of organization stands for good government, its message "By working together we can maintain harmony over our territory and at the same time move forward."

The highest level of organization, and the center of attention, is the game itself. Here we see team management put to the test, facing attack and having to respond to an unpredictable and violent challenge. In a team game the individual player is no longer a mere cog in a wheel but a specialist with a particular role to play. The excitement of a game has to do with imposing order on an unpredictable and potentially chaotic situation, and doing so by a combination of power, tactics, speed, and teamwork involving *delegated functions*. The history of the symphony orchestra in the eighteenth century can also be characterized as an era of increasingly complex organization from cheerleaders through marching band to today's highly specialized football team.

Industry and revolution

Vivaldi worked for most of his life on the problems of building an orchestra as a body of diverse skills united in a harmonious team. Composers joined with players and instrument makers to develop new variants of traditional string and wind instruments more suitable for a large, mixed ensemble. Many of the older instrumental families such as recorders, shawms, and viols were designed as families to perform together and produce a uniform group sound. In embracing a new philosophy of diversity, composers were looking for new qualities of individual expression for which these older instruments were not really suited. Over time new forms of oboe, bassoon, clarinet, and later the trumpet and fortepiano came on stream and were gradually incorporated into orchestras of increasing size and expertise.

Spring action

Team management is at a relatively early stage in Vivaldi's *The Four Seasons*, music every bit as popular as the Brandenburg

Concertos of J.S. Bach, but organized in very different ways. This
is music for a core string orchestra, supporting solo violins, and
directed from the harpsichord. It reflects a situation of building a
team of little leaguers around a pair of star players who know
their way around. When you are working with a young team, the
emphasis has to be on getting everybody to pull together, so

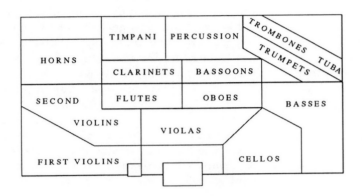

Layout of a modern symphony orchestra

much of the orchestral action consists of marching in step while
playing different notes. The music expresses the *temperament* of
each season in music of appropriate key and timing. In "Spring,"
for example, the pace is upbeat and the more elaborate writing
for solo violins expresses the idea of new growth in up-thrusting
melodies of vigor and vivacity, and imitation birdsong. The relationship of soloists and orchestra follows traditional church practice, and uses the formal repetitions of echo canon to convey an
image of pastoral space, much in the tradition of Monteverdi.[6]

As teamwork grows and develops, the team itself becomes
not only more capable but also more aware. Along with discipline comes prestige. The political and cultural status of leading
cities such as Mannheim, Vienna, and Paris is reflected in an
orchestral music of abundant confidence and skill. An eighteenth-century orchestral concert had not yet become the public
entertainment it is today, but was more like a staged event at a
shareholders' meeting of a large company, an event designed,
that is, to impress an audience of the organizational skills of the
current CEO. As corporate sensibilities begin to prevail, the focus

of attention shifts from team skills to images of speed, efficiency, and power. With the trend moving away from J.S. Bach's mechanical model of integrated design to the more overt displays of managerial authority of the later eighteenth century, one observes a corresponding change in tone from the intellectual complexity of the Baroque era to the simpler and perhaps shallower determinism and outward charm of a new Rococo era.[7] Embracing a lighter *galant* style in keeping with the age of Voltaire's *Candide*, Leopold Mozart's Concerto for Trumpet in D major caters to the complacent sensuality of well-to-do eighteenth-century society, a music written for elegant entertainment, not job satisfaction, treating the musicians as factory workers doing routine work to produce a disposable consumer product. How can a listener tell? Listen to the way the music is structured. All the interesting parts are taken by a few star players. Everybody else plays *really boring stuff*. It's like elevator music: jaunty, shallow, uninvolving, the kind of music that makes people want to talk rather than listen. You can't sing to it, the melody is too skittish; you can't dance to it—it's too fast, you would fall over. Especially after dinner. There is an underlying ruthless production-line insensitivity to a music that seems to be saying of the rank and file players: "They're only workers." The Rococo era is marked by a willful simplification of the world's complexities.[8]

Storm and stress

Anybody who thinks composers are simple creatures should listen to Haydn and think again. Haydn's more than 100 symphonies chronicle the evolution of the symphony orchestra from the harpsichord era of Vivaldi to the industrial age of Beethoven. For many years the genial composer was also conductor and manager of the Esterházy palace orchestra, a position he filled with great tact and good humor exemplified in the well-known story of the "Farewell" Symphony, written to persuade the prince to allow the hard-worked orchestra members time off, in the final slow movement of which the players one by one blow out their candles and tip-toe out of the room.[9]

The vivid contrasts and unpredictable changes of late eighteenth century music express not only higher levels of musicianship and teamwork but also a burgeoning sense of social and political unease. The natives were getting increasingly restless. The American Revolution had challenged Europe and prevailed under a banner of individual freedom. It was a time of *Sturm und Drang* (storm and stress) when the heirs of the Ages of Reason and Enlightenment were becoming uncomfortably aware of their own vulnerability in the face of unfathomable higher forces. Haydn, the most reasonable of composers, traveled to England by sea and was greatly intimidated by the power of the waves. But it was not only nature in the sense of rain, wind, and tidal forces that created unease among a privileged aristocracy. To the artists and composers of Haydn and Mozart's generation, the idea of nature embraced *human* nature as well, and human nature included the anger of an increasingly outraged and oppressed working class.

His own good nature and sense of humor do not disguise the underlying signals of social unrest in Haydn's "Surprise" Symphony No. 94, composed in 1791, in the shadow of the French Revolution. He delights in lulling his well-fed audiences to sleep with quiet music, only to jolt them awake with a loud bang, as famously happens in the second movement Andante; but something more disturbing is happening in the fourth-movement finale, a music light and cheerful on the surface but prone to dark and unpredictable mood-swings, culminating in a literal bolt from the blue as a sudden and awesome clap of thunder from the timpani interrupts the music, a divine warning that the storm of revolution is about to break.[10]

Rage against the Enlightenment

During the brief but intense lifetime of Leopold's son Wolfgang Amadeus Mozart, the oppressive simplicities of Voltaire's Age of Enlightenment declined into social and intellectual doubt and disorder which came to a climax with the French Revolution of 1792, a year after the younger Mozart's death. In the final Allegro of one of his last major orchestral works, the Symphony No. 40, classical elegance and charm are shot through with imagery of overwhelming anger, violence, and desperation.[11] This Mozart can see the revolution coming, and his music expresses a terrible warning. The image of him as an obnoxious child in the movie *Amadeus* is a travesty not only of the composer as a person but also of classical music as an intelligent activity.

To understand just how intimidating music can be, imagine being in the audience at the premiere of Mozart's Symphony No. 40. Mozart works for the aristocracy but he is regarded as a servant, an artisan; in consequence, he is undervalued and underpaid. He has already composed a number of operas on themes of social unrest and the revolutionary consequences of

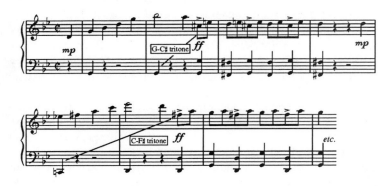

Mozart: Symphony No. 40 in G minor K550,
finale. Note the emblematic tritones G-C sharp, C-F sharp

unbridled sexuality, for example, *Don Giovanni*. He knows exactly what is happening philosophically and politically.

This late symphony is a staged confrontation between the old regime and the new, disguised as palace entertainment. A modern concert takes place in a specially-designed concert hall where the audience is comfortably seated on a different level, in a different zone, and *at a safe distance* from the orchestra on the platform. Not here. Mozart has composed his symphony for a large orchestra and it is seated at one end of a private ballroom on the same level as an invited audience representing civil and economic power. As for a concert in the White House, orchestra and audience are drawn up facing one another like opposing armies, only ten or so feet apart. Theoretically, it is the members of the audience who are in control. They are the ones who own the house. They are the ones who pay. But for the brief moment of the concert performance, it is the orchestra who speaks and the audience that is obliged to listen.

Mozart's orchestra has come a long way from the *ad hoc* team of Vivaldi's day. This is a powerful and highly disciplined army of virtuoso players, and the message—delivered in a spirit of high drama and with a sense of powerful forces of Nature ready to be unleashed, is "Watch out!" You know the scene in *NYPD Blue* where the two detectives take a suspect in for questioning: there is a nice one and a nasty one and the constant change of tone from friendly to menacing is intentionally destabilizing. Just listen here. The strings begin the interview, speaking in a light and elegant tone, a bit nervously, perhaps. Scarcely have they begun before the rest of the orchestra interrupts without any warning or ceremony with a series of violent slaps across the listener's face: *bang-bang-bang-bang-bang-bang!* The violins carry on as if nothing had happened; the sequence is repeated; a pause, then repeated again.

The most terrifying moments are when cellos and basses cascade and slide like an avalanche in an image of overwhelming and unstoppable power. This orchestra is no longer just a team but a highly-trained commando squad in complete control, behaving in a manner to suggest exactly what the aristocracy can expect from a professional class whose compliance can no longer be taken for granted and who are liable to explode at any moment. It is a very noisy, headlong but at the same time *entirely disciplined* movement combining imagery of anger pushed to the edge and the determination of a fearless and motivated underclass with the skill and resolution to win.

Orchestrated progress

It has been said that the Industrial Revolution began in England when Josiah Wedgwood, witnessing a performance of one of the "London" symphonies of Joseph Haydn, realized that he could apply the organizational principles of music for orchestra to the manufacture of quality chinaware. Prior to that time, the industry was organized as a collective of craft workers, each of whom took charge of the entire manufacturing process for a particular item, such as a cup or dish. It was a slow and inefficient process leading to inconsistent quality and wastage. Wedgwood reformulated the manufacturing process as a sequence of distinct operations following a fixed timetable, adopting the orchestra model of sharing production responsibility among a team of specialists, each assigned a particular role. Haydn's symphonies are lengthy and extremely complex acoustic processes working to a precisely timed master plan that gives every member of the production team a role appropriate to his or her particular skill. By adopting a similar approach, Wedgwood was able to introduce the same high standards of consistency and reliability to the mass production of fine china. The new production line system greatly improved the speed and uniformity of chinaware production, so that more complex patterns and processes were able to be introduced to the catalog.

Standardization

Standard notation enables an orchestra to function, and an orchestra in action is a model of production line organization. In 1900 the nineteenth-century orchestra was described by sociologist Max Weber as the ideal model or paradigm of Western European industrial society. He may have been thinking of the music of Beethoven, Bruckner, and possibly Mahler, but Weber was not talking about aesthetics but about the orchestra as a model of social organization, and social organization as a manifestation of economic power. The British humorist Miles Kington once commented in the London *Times* that Yamaha was the only company manufacturing both motorcycles and pianos. He went on to describe road testing a concert grand. The remarkable world success of Japanese and Malaysian industry in the second half of the twentieth century is directly linked to the encouragement and cultivation of Western classical music in these countries. We may wonder why these Pacific Rim societies with their own long and distinctive musical traditions have chosen, in what the rest of the world may see as a form of reverse colonialism, to embrace the heritage of Bach, Mozart, and Beethoven, in particular, concert orchestral music of the eighteenth and nineteenth centuries. It is because in their own cultural traditions there are no equivalent models of social and corporate organization. In order to compete successfully with Western industry, these countries had to embrace the classical orchestra model of people organization. Even the introduction of karaoke —a form of musical entertainment that many in the West regard as a peculiarly Japanese exercise in humiliation—can be understood as part of the same assimilation process. If you watch a karaoke party in progress, you see that it involves regular team players, individuals not used to taking control, having to assume a leadership role in a musical performance. It is no coincidence that companies like Yamaha and Sony have become such powerful players in a classical music industry dominated by Western codes and traditions which, being foreign to traditional Japanese culture, are for that reason much easier to understand from an organizational perspective.

Notes

1. Edmund A. Bowles, "The Symbolism of the Organ in the Middle Ages: A Study in the history of ideas." In *Aspects of Medieval and Renaissance Music: A Birthday Offering to Gustave Reese*, ed. Jan La-Rue (New York: W.W. Norton, 1966), 33-6.

2. Henry Purcell, *Fantazias for Viols*. Rose Consort of Viols (Naxos 8.553957, 1995).

3. Vivaldi, *Lute Concerto in D* RV93, arr. Bream. Julian Bream, Monteverdi Orchestra, cond. John Eliot Gardiner. *Concertos and Sonatas for Lute* (RCA 09026 61588-2), 1993.

4. J.S. Bach, Prelude No. 1 in C from *The Well-tempered Clavier*. Vladimir Feltsman (MusicMasters 01612 67105-2, 1998).

5. J.S. Bach, Brandenburg Concerto No. 2, BWV 1047. Academy of St. Martin in the Fields, cond. Sir Neville Marriner (EMI 7243 5 69877 2 2, 1987).

6. Vivaldi, Concerto No. 1 in E, RV 269 "Spring" from *The Four Seasons*. Felix Ayo dir. I Musici (Philips 438 344-2, 1999).

7. For example, *The World of the 18th-Century Symphony* (Naxos 8.554761, 2000).

8. Leopold Mozart, Concerto for Trumpet in D major. Wynton Marsalis, English Chamber Orchestra, cond. Raymond Leppard (Sony SK 557497, 1995).

9. Franz Joseph Haydn, Symphony No. 45 "Farewell" in F sharp minor. The Hanover Band, cond. Roy Goodman (Hyperion CDA 66522, 1991).

10. Haydn, Symphony No. 94 "Surprise" in G. The Hanover Band, cond. Roy Goodman (Hyperion CDA 66532, 1991).

11. W.A. Mozart, Symphony No. 40 in G minor K550. The Cleveland Orchestra, cond. George Szell (Sony SBK 46333, 1990).

CHAPTER ELEVEN

Resonance

THERE is a harmony of likeness, and a harmony of difference. If you like Southern fried chicken and fries, that's a harmony of likeness: everything is covered with a crispy coasting on the outside and deep fried to cover up any differences in taste and texture. A string orchestra or a piano is the musical equivalent of a harmony of likeness, and so is a choir, because in a harmony of likeness the ingredients are *blended* into a uniform texture and sonority. Sushi, on the other hand, is a harmony of difference. A folk group or a rock and roll band, even a modern symphony orchestra, is a harmony of difference, because here, instead, the ingredients are *combined and contrasted* for effect. Uniformity is relatively easy to achieve. In boot camp you have no option but to do what everybody else does, and nobody cares what you think. It takes a different approach to reconcile an ensemble of contrasting elements and styles. In traditional Western music, consistency is the key and classical tonality the regulating principle, the fact of everybody being tuned to the same scale of pitches and playing to the same beat.

Sushi, now, that's something else. It's a little of everything: sweet, sour, salt, sharp, spicy, bland, crunchy, chewy, soft, glutinous. This is the harmony of difference, a harmony of *taste sensations*: everything is distinctly different, and in balance with everything else. When you eat sushi, the whole of your mouth resonates in a chorus of tastebuds. The harmony of difference is

175

expressed in the "Edo Lullaby," an arrangement of a traditional Japanese melody for a small ensemble of Japanese instruments.[1]

In this music there is a mode but no harmony in a Western sense, no major or minor key, and no chord sequences to give the music a sense of directedness. The lead instrument is an end-blown flute, the *shakuhachi*. The other instruments are a lute, the *biwa*, that makes resonant sounds that also bend in pitch; the abrupt sounding *shamisen*, a pick instrument with a small skin-covered box like a banjo; two *koto* or long zithers that play a harp-like role but with a much sharper and more intense tone that can also bend upward in pitch; and the glistening sound of a string of bells. This is chamber music in terms of numbers and intimacy, but the Japanese vocabulary of sounds is vividly physical and somewhat disconcerting to Western ears, and it is not easy at first to associate these very sharp and sudden sounds with an air of tranquility. I think of it more as a philosophical or meditative music, even though my Japanese students assure me that this is a song they used to hear as children at bedtime.

A useful image for comparison purposes is a formal Japanese garden with a wooden bench to sit on and a pond with fat koi carp, water lilies, frogs, and attendant dragonflies. Tall reeds stand at the water's edge and a weeping willow leaning over just grazes the surface. A cherry blossom tree is host to birds and bees. It is a warm mid-afternoon, partly cloudy with a hint of approaching rain. A Japanese garden is a world in microcosm. It combines all the elements of fire (warmth), earth, air, and water; it is a place where stability and change, life and not-life, coexist. A garden is a haven for creatures that live in the water, in the ground, on the ground, and in the air, a place of light and shade, of sounds, images, textures, scents, and sensations of heat and cold, dry and wet, hard and soft. Such a garden is also a *sensing device* by which an observer can monitor the cycles of nature, a harmonious composition in which every element is connected and in balance with every other.

Like the traditional art of Japan, Japanese music is based on direct observation of nature. The lullaby begins and ends with the shakuhachi whose leading role may be likened to the solo flute in Debussy's orchestral *Prélude à l'Après-midi d'un Faune*, music composed in an antique pastoral style, influenced perhaps by Japanese tradition in just the same way as French art was heavily influenced by Japanese imagery in the latter half of the nineteenth century.[2] The flute resembles what, in fact, it *is*: a breeze, a freshening but also a *disturbing* element. It is also by nature a *breathing* instrument, so represents the breath of life, a dynamic living force; it is also a *restless* impulse like the inquiring mind of the contemplative observer, and since it is the only *sustaining* instrument of the whole ensemble, the shakuhachi could even be said to represent time itself.

The other instruments respond to the flute as elements in a garden to a breeze. Shamisen, kotos, and biwa are plucked instruments of varying hardness of tone; like Dowland's lute or Bach's harpsichord, sounds of this kind once struck die away, so the combination of instruments in "Edo Lullaby" resembles a lute song in that the flute has a sustaining function and the string instruments play incidental roles. Western music, however, has little use for tones that *bend*. In the Western tradition, a gliding or bending tone is uncertain, imprecise. For the Japanese listener, on the other hand, a tone that bends expresses a bending in nature. Look out of the window at a tree and see how it bends in accordance with the strength of the breeze, compared to the swaying of standing reeds or grasses. Japanese musical imagery is very direct. If a sound is hard, it expresses hardness; if abrupt, suddenness; if it bends, it expresses a bending action.

Japanese art likes to incorporate a discreet element of sound —a cricket on a branch, a frog on a lily pad, a brightly colored bird amongst the cherry blossom—characters that chirp, croak, or trill to dispel the image of silence and bring a picture to life. Perfect stillness in a garden is just as disturbing and unnatural as silence in a picture. In a garden there is usually very little movement, but what movement and sound there is, is in harmony with nature however random it may at first appear. A frog croaks. A cricket chirps. A dragonfly hovers over the surface of the pond. Slow-moving fish glide and turn and occasionally break surface to feed. The small but deliberate sound actions of living creatures are accompanied by spontaneous incidental sounds that express no intention but the effect of time: the falling of a leaf or petal, the trickle of water, the rattle of wind chimes, the rustle of foliage. These incidents seem random and unrelated at first, but they are essential to the experience of tranquility, and the philosophical point of a music designed to imitate the outward randomness of a natural setting is to draw attention to an underlying harmony. In music of the Western tradition, harmony is an imposed conformity: a beat, a rhythm, a group of tones in mathematical proportion. In this music, however, suddenness and unpredictability are more significant because the purpose of the exercise is not to organize and control but to contemplate nature, and unpredictability in nature is both perceptually necessary and philosophically significant. When the unexpected happens, it reminds us of the existence of the real world. Events that are not anticipated strike our awareness with greatest clarity, with "the shock of the new."

When a breeze blows through a Japanese garden, things happen. Water drips, leaves fall, blossom scatters. The surface of the pond loses its sheen and transparency. Reeds and branches bend. Birds and insects stop chirping. And everything that is disturbed *is disturbed in a particular way*, and every disturbance is revealing of the *nature* of the thing disturbed: of water, that it flows; of reeds and branches, that they bend; of rock, that it is solid and unyielding. Everything reacts differently *but to the same cause*— the breeze itself. In the middle of "Edo Lullaby" there is a brief flurry of kotos and shamisen playing rapid tremolos, as though a rain shower is passing through. In nature a gentle breeze causes each element to move more or less independently; the leaves and end branches of a tree, for instance, do not move in synch, nor do standing reeds sway in rhythm. When the pace quickens, however, and the breeze becomes a wind, its greater force tends to impose a uniform swaying and bending on trees and plants, and disturb the entire surface of the pond at once. Similarly, as the energy level in the music reaches a peak, a common rhythm momentarily unifies the ensemble. Eventually calm is restored and the music fades out with the solo flute.

What is the message of harmony in such a music? It says something rather profound, first of all about nature being a system, a resonating structure like a musical instrument. The same view was held by the Pythagoreans of ancient Greece, who understood the universe as a structure in motion held together by forces of tension and oscillating in regular cycles. So the image of nature represented in "Edo Lullaby" is of an organic system that is not a loose collection of parts but a structure of interrelated forces and counter-forces in dynamic equilibrium. In the cause and effect interaction of flute and other instruments, a listener understands that the harmony of nature is normally hidden and only revealed by contemplation or action that, paradoxically, disturbs the balance—just as we only become aware of the real world through some disturbance affecting the sense organs. However, it is not just the breeze but *the act of observation itself* that disturbs and reveals the essential qualities and distinctions in nature and mysteriously brings a sense of harmony to their apparent disorder.

Chance and determinism

It seems like poetic justice that at the same time as Japan and other Pacific nations were embracing Western-style industrialism and a Western music of team management and production-line organization, artists and composers in the West were starting to cultivate approaches to music and art of a more contemplative kind. Faced with a music of such stillness as "Edo Lullaby" it becomes all too obvious how Western music is focused on regulation and action. Counter-examples are rare, John Cage being a noteworthy example. Inspired in part by abstract expressionist art, the New York school of Cage and associates Earle Brown, Christian Wolff, and Morton Feldman attempted in a variety of ways to creat a music of natural incident uncorrupted by aesthetic or artistic intention. It was an enormously difficult task, both philosophically and practically, since the whole apparatus of Western music—instruments, keyboards, concert-halls, social rituals, training methods—militates against spontaneity and innocence.

Since the final decade of the twentieth century, stimulated in part by the example of early music's cultivation of what might be called a more holistic awareness, there have been signs of a renewed appreciation of the contemplative virtues of music-making and a willingness to accept the challenge of a music of seeming randomness. For an increasingly stressed civil population, the experience of a sympathetic and well-recorded music such as Cage's *The Seasons* is a natural restorative.[3]

Structural resonance

Why does a grand piano have legs?
Various answers spring to mind:
- As a means of support.
- To raise the keyboard to a convenient height.
- Tradition—it's always been that way.

Yes, but apart from all that, is there a *musical* reason?

Consider the neighbors

If you own a piano, you have to consider the neighbors. Music travels not only through the air, but through the fabric of a building as well. Walls and floors act as membranes, radiating low to middle frequency sound. Floorboards and joists conduct higher frequencies. Most instruments of music don't have to touch the ground. A violin, flute, or saxophone is supported in mid-air, so the sound it makes is purely airborne. But there are a few larger and heavier musical instruments that either stand on the floor on legs like a harpsichord or grand piano, or are held in an upright position by the player with the base of the instrument supported on the floor by a spike or stud.

These instruments are heavy because they are big, and they have to sit on the floor because they are too big and heavy for the musician to hold unaided. Big, in musical terms, means low notes. A bass tuba is bigger than a trumpet, and it plays the low notes; a bassoon is bigger than an oboe, and it plays the low notes; the cello and double bass (bass viol) are bigger than the violin, and they both play the bass line. What is interesting is that if you study pictures of an orchestra in Bach's time, you see a group of players not sitting but *standing* around a harpsichord: violins, oboes, flutes, trumpets, horns, etc. To the harpsichord player's immediate right is a lone cello; to his left, a lone double bass. Harpsichordist and cello player are the only players in the group who are sitting down. All three string instruments share the bass line. All three are in contact with the floor. They are the *only* instruments in direct contact with the floor.

Bass response

Audio equipment manufacturers know all about spikes. Studs and spikes are big business. Audiophiles—people who are willing to pay large sums of money to acquire top-quality equipment in their quest for perfect music reproduction—agonize over the shape, weight, composition, position, and pointiness of objects that should come between their prize speakers and the polished floor. Spikes influence sound quality by channeling cabinet vibration into pressure points. Instead of large patches of speaker vibrating against the floor (or sucked up by wall-to-wall carpet), the same amount of vibration is concentrated in usually three points per speaker, and the weight of the speaker drives these points right into the floorboards, feeding cabinet vibration directly into the room structure. Sub-woofers for home theater hi-fi come from the store with spikes already in place. These people know what they are doing. Spikes or studs improve bass response. They make your classical or jazz music collection sound more realistic. An extended bass makes any group on record seem to perform tighter, with better integration. It also opens up the sound, making it appear more spacious and lively.

Could it be that the seventeenth- and eighteenth-century pioneers of classical orchestra music knew about spikes? These things don't usually happen by accident. If the pictorial evidence shows that the entire bass section of a Baroque orchestra is coupled to the floor by legs or spikes, while all the other players are holding their instruments in mid-air, that is a pretty good indicator of deliberate practice. Composers and musicians were aware even then of the effect of coupling the bass section of an orchestra to the floor. This was no accident. It was done to improve bass response and tighten the ensemble sound.

An orchestra whose bass section is coupled to the floor behaves like a very expensive hi-fi system. It uses the same acoustic principles to create a similar effect. Most of the sound of an orchestra comes through the air. Airborne sound travels at a certain speed and is reflected off the interior surfaces of a room or concert hall. The remainder of the sound energy, for bass instruments coupled to the floor, goes into the room structure. Structure-borne sound travels a great deal faster. Just what effect this has on an audience depends on the structure of the concert hall and the position of the orchestra within that structure. We need to take another look at the documentary evidence. If the orchestra is located in the middle of a large room on a hardwood floor with an open space, such as a cellar or pantry, directly beneath it, the floor acts as a membrane; if, on the other hand, the orchestra is mounted on a raised platform at one end of the room, the stage area and the airspace beneath will behave in a different manner, like an enormous speaker cabinet.

Cellos and keyboards did not always have spikes or legs. A delightful pen drawing by Bernard Picart dating from 1701 shows a standing solo cello player in action with his instrument resting on a cushioned footstool. The tenor viol, the cello's immediate precursor, is played sitting down, but with the instrument held in place between the knees, not resting on the floor. The tenor viol has no spike. Clavichord and dulcimer, predecessors of the harpsichord, were originally suitcase instruments designed to sit on a table-top. If there is no direct contact between an instrument and the floor, the sound produced is only airborne, light and airy in tone. Look at a cello up close and see how the metal spike and its mount are connected to the instrument. It is a crude and brutal treatment of a beautifully sculpted and finished all-wood structure to have a thick iron prong piercing it at the base. It looks like body-piercing. The two materials don't go together. It doesn't look right. What it tells you is that this is not how the instrument was designed to be. The spike is an afterthought—a musically useful afterthought, but an afterthought nonetheless.

A coupled system

So how does it work? It works like a tuning fork. A tuning fork is a pocket implement of tempered metal with two long, thickish prongs that musicians use to tune their instruments. You tap the prong end against the table edge to set it ringing, which it does at a high pitch, and then you set the pointed handle end on the table top to produce a clear tone. The sideways motion of the prongs moving in opposite directions is converted into an up-and-down motion at the point, and it is the slower and more powerful up-and-down motion on the wood that becomes audible as the tuning pitch. It is not just the tuning fork by itself but the fork and table-top vibrating *as a coupled system* that produces the lower tone. Take the fork off the table and all you hear is the airborne vibration at the ringing pitch. You can't tune a violin to the ringing pitch.

Longitudinal vibration

Back in 1867, Professor John Tyndall gave a famous series of lectures on sound to the Royal Institution in London that included a demonstration of how sound can travel through a solid rod, such as a piano leg or cello spike. It was the audio equivalent of demonstrating how light travels down optical fibers, and it caused a sensation. Down in the basement of the Royal Institution, a hired musician played a grand piano continuously for the duration of the lecture. On the soundboard of the grand piano rested one end of a twenty-foot long wooden rod extending up into the lecture theatre two floors above, passing through holes specially drilled in the floors and ceilings and soundproofed to prevent leakage. (The holes are still there.) To his audience, the vibrating tip of the wooden rod made no audible sound, but as soon as Tyndall placed various instruments—a violin, a cello, a harp—in contact with the rod, the sound of the piano playing in the basement could clearly be heard. With the violin acting as a resonator the music sounded rather thin, in today's terms, like a cheap portable radio. But as soon as Tyndall brought the larger-bodied cello or harp in contact with the rod, instruments of a range and size closer to the piano, the quality of reproduction dramatically improved.[4] This was a highly effective demonstration of music conduction, proving not only that wooden struts and floorboards were capable of conducting sound but that music conducted through solid wood *retained its original structure and coherence* across the frequency range from bass to treble, meaning that a medium much denser than air did not slow down or otherwise interfere with the transmission of weaker high frequencies.

Experimenting with legs

Adding legs transformed the early keyboard family of instruments from small table-top devices into grander machines suitable for use with an orchestra. Their role changed as well. As the harpsichord family grew in size, it also grew in range of notes, but adding legs also improved the bass response, and this was critically important because the original keyboards were not designed to make powerful sounds. Cristofori's hammer-action fortepiano with controllable dynamics made its first appearance in the late seventeenth century, but it was not until the late eighteenth century that it succeeded in overtaking the harpsichord as the composer's instrument of choice. In the interim, clavichord and harpsichord makers experimented with different shapes and styles of leg to find out which was best for conducting keyboard and sound-box vibration to the floor. Some surviving early instruments are mounted on flat feet resembling a violin bridge; some rest on legs tapered like the business end of a tuning fork.

In other instruments, the legs are connected by struts at mid-point, perhaps in an attempt to equalize the stresses of discharging cabinet vibration to the floor, since the energy distribution in a keyboard instrument is heavily loaded in favor of the bass. In other keyboards of Mozart's time, long and slender legs connect independently to the perimeter of the case and soundboard.

All of this effort seemed to bear very little fruit. The instruments did not appear to gain significantly in loudness or in strength of tone. Even the grandest harpsichord today, authentic or otherwise, sounds relatively feeble in a concert hall, compared to a piano. Conductor Sir Thomas Beecham, who could be very rude about a lot of things, expressed contempt for the tonally challenged harpsichord. Writing in 1851, the concert pianist Sigismond Thalberg, a contemporary and rival of Franz Liszt, was equally scathing in his criticism of the fortepiano:

> In 1760 the piano was successfully competing with its more established rivals . . . as is sufficiently shown by Haydn, who left sixty sonatas composed expressly for it. Gluck also adopted the piano; and we have seen the instrument on which he composed his Armida and other works, made for him by Johannes Pohlman in 1772. It is but 4½ feet in length, and 2 feet in width, with a small square sounding-board at the end, the wire of the strings being little more than threads, and the hammers consisting of a few plies of leather over the end of a horizontal jack working on a hinge. The instrument, compared with a fine piano of the present day, is utterly insignificant and worthless; and it is difficult to conceive how it could have been used for the purposes it certainly served.[5]

Energy transfer

Playing a piano, or even a harpsichord, is hard work, like operating an old-fashioned manual typewriter, but with the difference that a musical keyboard is a great deal more massive than a standard Imperial typewriter, and is designed to bang out lists of notes rather than letters of the alphabet.

All that energy has to go somewhere, and since only a very small percentage is radiated into the air as sound, the excess is either going to accumulate, in which case the instrument heats up and eventually bursts into flames, or it has to be dissipated by conduction through the legs to the floor and thence to the walls, the ceiling, and the apartment next door. However, surprisingly little attention (or none at all) has been paid in the literature of instrumental acoustics to the question of what happens to all that excess energy that is not radiated as sound.

> The eighteenth-century harpsichord . . . has a rich and clear sound, a reasonable volume level, and a precise and flexible action. Its shortcomings, to the ears of musical romantics, lay in its inability to produce an expressive crescendo or accent, and in its lack of volume level adequate to balance the instrumental forces of the romantic symphony orchestra . . . .
> The soundboard is a roughly triangular plate, about a meter in longest dimension, stiffened by light ribs. Its function is to receive energy from the vibrating strings and, because its coupling to the air is very much better than that of a string, to radiate the sound. As with most vibrators, however, we expect that most of the energy will be dissipated in internal losses and only a few percent will be radiated.[6]

Of a modern grand piano soundboard the same highly-respected authors note "Losses do occur, and these result in a decrease of sound radiation in the upper treble range. At the bass end, below the critical frequency [i.e., the frequency at which the speed of bending waves in the soundboard equals the speed of sound in air], the radiation efficiency drops dramatically. This results in a favored region for acoustic radiation between about 200 and 2000 Hz."[7]

Where has all the power gone?

We know perfectly well what happens to the excess power. It goes into the floor. "Vibrating machines in contact with the structure can impart a great deal of energy to it, and in a continuous structure noises may be carried great distances. . . . The source of the loudest noises, such as those from the radio or the piano, is usually the living room or the parlour." So says an official post-war guide to English architects and builders.[8]

If a harpsichord sounds weak and feeble compared to the modern grand piano that is because it is being judged in hindsight, and hindsight assumes that the two instruments are performing an equivalent function. They are not. A modern grand piano is designed for the *audience* to hear. The harpsichord in a Baroque orchestra is designed primarily for the members of the orchestra to hear, not the audience. And the orchestra can hear the harpsichord perfectly well. Why do they want to hear the harpsichord? Because the harpsichord, aided by the cello and double bass, has the conductor's role of keeping everybody in time and on cue. Why is it acceptable for the harpsichord not to be heard by an audience? Because that's not the point. The point is to coordinate the louder instruments and to do so discreetly.

Sound travels in air at around 1,100 feet per second. Through floor-boards along the grain, the same sound travels at around 14,000 feet per second. A listener sitting thirty feet away from the harpsichord will sense the impact vibration from the floor after only *two* thousandths of a second, but the associated wavefront takes approximately *twenty-two* thousandths of a second to travel the same distance through the air. It follows that the player or listener is going to be aware of the impact vibration first and the airborne sound as the after-effect. You might think that a difference of a few thousandths of a second is not significant, but remember, it is differences of this order of magnitude that the brain is processing *all the time* to enable ordinary human beings to tell what direction sounds are coming from and where they are located in an acoustic environment. So, far from being insignificant, such fine differences are actually routine—both when we listen and when we speak. Indeed in speaking we are already monitoring the difference between low- and mid-frequency impact vibration (bone-conducted sound) and high-frequency airborne reflected sound.

For a small-scale Baroque orchestra, the impact sound perceived via the floor is real, easier to follow, and more reliable than a conductor's hand waving a stick. That makes sense. But for players in a modern orchestra, a conductor like Sir Thomas Beecham is a remote figure somewhere in the distance waving a baton. A modern conductor doesn't make any sound—or rather, shouldn't (some conductors do). A conductor in the center of an ensemble, playing a keyboard instrument and controlling the bassline from close by, means that the musicians playing flute or horn or violin *can keep their eyes on the music* and not have to keep glancing up to check the silent beat. The harpsichord's relative transparency of tone can be understood, therefore, not as a weakness of design but as a positive feature, allowing the players to keep in time without looking and a Mozart or Haydn at the keyboard to monitor the full orchestra without the sound of the instrument getting in the way. Consider Duke Ellington conducting at the piano. He doesn't play loudly. He doesn't hog the limelight. He's part of the band, but he's also listening and in control.

Symphony orchestras are a lot bigger in size today than they were in the late eighteenth century. They perform in special concert halls with seating for two thousand or more, and are no longer conducted from the harpsichord. For the members of a modern symphony orchestra, the guiding principle is *Watch, don't listen, because if you listen you get lost*. A modern concert hall is not designed with a flat floor and plaster walls to conduct structural vibration. On the contrary, it is designed to *eliminate* structural vibration. Orchestras vary in their layout from country to country and from conductor to conductor but for violins and double basses on opposite sides of a 20 meter-wide modern concert hall platform, it means that whereas the violins can synchronize directly with the double basses by floor conduction, the double basses have to rely on airborne sound just to hear the violins, which for a distance of twenty meters amounts to a twelfth of a second delay, or fraction of a beat. So coordination by listening no longer works for a modern orchestra as it used to in Mozart's day. Today's widely separated rank and file players now have to rely on visual cues from the conductor in the middle. And that means a loss of precision.

Architectural acoustics, the science of concert hall design, dates from just over a century ago when the American Wallace Sabine was asked to make improvements to a small lecture theater at Harvard University. The science he was instrumental in founding was based on audible speech, not music, and on optimizing airborne reflections, not utilizing structural resonances.[9] With a few exceptions (such as the Berlin Philharmonie), today's acoustically doctored concert halls where great symphony orchestras entertain large audiences are not the places where these same orchestras make recordings. For that they queue up to hire places like Amsterdam's Concertgebouw or London's Watford

Town Hall, nineteenth-century structures where a combination of classic shoebox design, flat floor, and *structural vibration* help musicians to perform at their very best.

Notes

1. "Edo Lullaby" traditional melody, arr. Minoru Miki. Ensemble Nipponia, dir. Minoru Miki. *Japan Traditional Vocal and Instrumental Music* (Elektra Nonesuch 9 72072-2, 1976), track 4.

2. Claude Debussy, *Prélude à l'Après-midi d'un Faune*. Belgian Radio Television Philharmonic Orchestra, cond. Alexander Rahbari (Naxos 8.550262, 1989), track 1.

3. John Cage, *The Seasons*. American Composers Orchestra, cond. Dennis Russell Davies (ECM 1696 465 140-2, 2000), tracks 2-5.

4. Charles Taylor, *Sounds of Music* (London: British Broadcasting Corporation, 1976), 77.

5. Peter and Ann Mactaggart eds., *Musical Instruments in the 1851 Exhibition* (Welwyn: Mac & Me, 1986), 95.

6. Neville H. Fletcher and Thomas D. Rossing, *The Physics of Musical Instruments* (New York: Springer-Verlag, 1991), 296-7.

7. Fletcher and Rossing, *The Physics of Musical Instruments*, 326.

8. Great Britain. Department of Scientific and Industrial Research. Building Research Board. *Sound Insulation and Acoustics*. Post-War Building Studies no. 14 (London: His Majesty's Stationery Office, 1944), 22-7.

9. John R. Pierce, *The Science of Musical Sound* (New York: Scientific American Books, 1983), 140-1.

CHAPTER TWELVE

Leadership

A CONCERTO is a piece of music for orchestra with a featured soloist or soloists. The great majority of concertos are exhibition pieces for one star performer, but it was not always so. The period from 1600, the time of Monteverdi and Gabrieli, through the lifetime of J.S. Bach ending in 1750, was an age of experiment and consolidation for group instrumental music. Composers divided their loyalties between religion (the choir) and the state (instrumental music). Instrumental music dealt with scientific and intellectual matters, such as time and motion studies and the nature of key relationships. The orchestra was the composer's laboratory, and a great deal of creative energy was spent in deciding what instruments should become permanent members of the ensemble. Vivaldi and his contemporaries composed numerous test pieces for auditioning purposes, often rearranging the same piece for different solo instruments. New secular concert halls built of wood and plaster provided a vibrant and flexible acoustic for team music-making, compared to the solemn reverberant stone churches of earlier times. In time, the uniform sound quality of recorders and viols gave way to a new aesthetic of color contrast, and the pure meditative style of old also yielded to a new age of fast-paced action music.

Massed violins set the standard to which other instruments of the orchestra had to conform. Unlike Renaissance viols, the violin family of instruments is not designed for tonal purity and consistency but rather for their extended range, subtlety, and powers of expression. Violins are persuasive signal generators, with a novel unpredictability and unstable quality of tone that holds the attention in the manner of a speaking or singing voice. Traditional forms of wind instruments were modified, sometimes drastically, to achieve greater expression. Instead of the cool tonal purity of the upright recorder, the expressive and warmer-toned transverse flute; instead of the Renaissance shawm, a double-reed with a wind-cap covering the reed for consistency of tone, composers welcomed the Baroque oboe, a more eloquent sound produced by the lips in direct contact with the double reed, shown to great effect in Alessandro Marcello's Oboe Concerto of 1717.[1] Trumpets, trombones, and horns also acquired new expressive skills. The plucked-string harpsichord assumed a leadership role in place of the pipe organ. Each change accorded with a new Baroque aesthetic of creative instability, a dynamic in direct contrast to the static tonal uniformity of earlier times.

Making the team

Putting an orchestra together is no different from assembling a band today. What you are looking for is the group sound. You also want the personalities of the different instruments to work together. There is also the question of style, the kind of music the band is going to play. An orchestra can be designed for action, for speed and precision, or for cruising in comfort (easy listening). All have their place. Whatever option you choose, there has to be a unanimity of purpose on which the composer can depend. The composer is the person who writes the music on command and the one therefore who makes the decision on the image the orchestra is going to project. That unanimity of purpose is expressed most powerfully in the musical score, a visual command language that everybody can read, and that is the same for everybody. Software integration is the key.

The Baroque era concerto grosso is typically a prototype and not a production model. We are interested in the composition of the orchestra as much as in the music that the orchestra is actually playing. It is about what instruments go together and what they can do. Often you find different rank and file players in the group taking turns to stand up and play solo, just as in a jazz band. In the opening movement of Bach's Brandenburg Concerto

No. 2, for example, the lead is passing constantly among high trumpet, violin, oboe, and recorder. In Corelli's tongue-in-cheek Concerto in D first movement the lead is fought over by rival violins like players in a singles tennis match.

What Nita did

Remember Nita and her encounter with Benedetto Marcello's Trumpet Concerto? What did she hear, and how did she come to a decision about interior design from her clients' musical taste? First she listened to the combination of instruments. There is a trumpet soloist and an orchestra. The orchestra is a small body of strings with a harpsichord. There are no other instruments, no flutes or clarinets. So the music makes a clear distinction between solo and chorus. The trumpet is a brass instrument and has the solo role, the strings together with the harpsichord form the accompaniment and have the supporting role. There are no ambiguities here.[2]

The trumpet is a military instrument of commanding power and range formerly used in the heat of battle to communicate instructions to the troops. By design it is an outdoor instrument with a bright, clear tone that can be aimed in a particular direction. The military trumpet today is largely ceremonial, used for parades and rituals like "taps," but these outdoor functions still employ the signaling capabilities of the instrument.

A traditional military trumpet is like a bugle. It has no valves and only a restricted range of notes—which is actually an advantage from a signaling point of view. During Marcello's lifetime, however, the trumpet acquired a system of valves that made it possible to play normal melodies as well as sound the attack or call soldiers to attention. In the Marcello trumpet concerto, Nita could hear that this normally bright and forceful instrument, traditionally the embodiment of leadership, was making statements of a restrained and polite character, thereby showing its civilized and domesticated side. Clearly this concerto is not music designed for an outdoor setting like a march-past; rather, it takes the trumpet out of its usual context to feature as solo in an *indoor* performance for a *civilian* audience in a *private* house for enjoyment as a party piece. Orchestra and harpsichord defer to the lead of the trumpet, and the trumpet responds with an evenly modulated tone that shows it is quite capable of making intelligent conversation. So a listener is able to under-stand this music as expressing *authority, discipline,* and also *self-restraint.* It is *sociable* but not *wild* or *over-stimulated.* This is not music you would leap up and dance to; indeed, it is not de-signed to spur the listener to much action of any kind.

The classical era is a period of consolidation. Composers are still working out how to combine different instruments to make a successful orchestra. Like many Italian concertos of the period by Vivaldi, Corelli, and others Marcello's trumpet concerto is an ex-ercise in *harmonizing* two different kinds of tone color, in this case the bright timbre of the trumpet and the warm, reverberant sound of a string orchestra, the latter highlighted just a touch by the sparkle of the harpsichord. Nita understood this in interior design terms as a limited range of muted but rich and har-monious colors, essentially only two colors (the harpsichord more resembles gold thread in the weave than a separate color). Such restraint tells you this is music acknowledging the virtues of *consistency* and *control.* There are no dramatics, and none of the violent and unexpected *Sturm und Drang* contrasts that add spice to a Haydn or a Mozart concerto. This is not a palette of colors or materials that leads to any surprises. The music is not composed with any intention of playing tricks on the audience, and there are no unexpected dynamic contrasts, no sudden changes of pace, and certainly no vivid color contrasts. What you have, instead is a subtle play of *texture* in the string orchestra, at times thicker, at other times smoother, but always done with great finish, elegance and attention to quality, and with the comfort of the listener in mind.

This is civilized music, generating a sense of security. The trumpet delivers a firm, clear outline. The movement is organized in a regular sequence of balanced and symmetrical phrases that are easy to follow. It is music to set a listener at ease.

Follow my lead

The idea of creating concert items for a featured soloist and or-chestra may have been provoked by audience interest in the role of the concert-master or lead violin in the Baroque orchestra. The concert-master shared responsibility with the conductor at the keyboard for keeping the ensemble together. His function was to make himself visible by standing apart from the other players, moving his body in time to the melody and embellishing or decorating the melody with extra notes called *graces*: quivers, hesitations, and runs to dramatize the line and draw attention to the important beats. A plain melody is always an opportunity for a skilled performer to show his paces. Leading the melody gave the concert-master a chance to display his improvising and technical skills. Display techniques are good advertising as well as good management. Audiences liked the idea of artists showing a combination of leadership and technical skills. We shouldn't be surprised to find the role of the concert-master helping to feed a popular demand for display pieces for a number of different stand-up solo instruments with orchestral accompaniment.

Magic

Luigi Boccherini's late eighteenth-century Cello Concerto No. 9 in B flat opens with a cheerful but routine overture, as though for a theatrical act.[3] The curtains part and the cello launches straight into a dazzling display of fast fingerwork, trills, double-stopping (playing two notes at once), and wide leaps from low to high and high to low. The cello had been a bit of a dark horse in the Baroque orchestra, sitting quietly by the harpsichordist and reinforcing the bass line. Here Boccherini, a cello player himself, seizes the opportunity to present the instrument as a brilliant solo with a radiant tone and enormous range—a virtual Pavarotti among instruments. A little way into his opening solo the cello begins to soar like a Montgolfier balloon above the orchestra into the musical stratosphere, leaving everybody else behind. Floating even higher than the violins, seemingly turning a corner, the solo seems to vanish into thin air, momentarily dropping out of sight before re-emerging from a cloud to make a rapid but controlled descent back to ground level. Wonderful stuff.

This is music designed to celebrate human achievement and technique in a classical spirit of wonder and entertainment. It says, "Human science and skill can do anything." The smallish orchestra of strings and horns plays a discreetly supportive role. The cello holds the stage and the limelight. There is no competition for attention and little dialogue. The soloist is firmly in command.

Solo and orchestra in a concerto reinvent the complementary

relationship of lead singer and choir in medieval plainchant. As before, the soloist plays on the impression of making the music up on the spur of the moment, modifying and embroidering the music to express his individual personality and technical mastery. In Boccherini's opening solo, we witness a brilliant display of technical fireworks, but for most of the movement the orchestra continues to play—only a couple of violins perhaps, but enough to tell the audience that there is in fact a script, that they know what is going on, that it is not all being made up. Later in the movement, however, the soloist gets a real opportunity to improvise. The *cadenza* is a special feature of a concerto, a moment where the solo has completely free rein. The orchestra stops. The conductor stops beating time. *Time stops.* In this moment without time, all the attention of orchestra and audience alike is focused on what the soloist is going to do. The idea is to take elements of melody and harmony from the existing script and transform them by the magic of superhuman technique and creative vision into a mysterious transcendental experience. At this point the soloist is completely alone. Nobody can second guess what is going to happen. It is a moment of inspiration, a moment of creation, a moment of success or failure, a moment of direct communication with the gods. You hold your breath.

At the end of the cadenza the soloist plays a musical formula that signals to the conductor that enough is enough. With a firm down-beat from the conductor, time and the orchestra begin once again, and the movement draws to a controlled conclusion.

A concerto is about leadership. As the culture changes, so the idea of leadership also changes. The Age of Reason valued technical skill. A soloist's expertise aroused admiration and demonstrated leadership, representing intellectual as well as physical prowess. A concerto in which the soloist completely outclasses the rest of the field is also making a statement about how society is organized. Boccherini's soloist is like a magician. His artistry combines an understanding of human psychology with a knowledge of hard science. This is a leadership that has to be earned, but the image of leadership is aristocratic. Some are leaders, and the rest are followers, is what the music says. The role of the orchestra is subservient and enabling.

Prima donna

Sixty-six years after Boccherini, the Cello Concerto in A by

Robert Schumann inhabits a completely different world.[4] The slow movement is an essay in Romantic introspection. The soloist has nothing to prove. The movement is not a showcase of technical expertise. Indeed, its melody is markedly vocal in style, simple enough for a nineteenth-century middle-class audience to hum along to, and so undemanding technically, in fact, that it could almost be played on a child-sized instrument by a gifted eight-year-old.

What is going on? This music is not concerned with strength and prowess but about mental attitude—not about brilliant fingerwork but about *the soloist as prima donna.* The image of leadership Schumann expresses here defines the artist as a person *entitled* by virtue of superior taste, style, and sensibility to offer emotional and spiritual guidance to the *hoi polloi.* And what is the soloist doing? Enacting a soliliquy, but not the one from *Hamlet.* Pacing elegantly back and forth across the stage in a silk robe, the leading lady pauses every so often to reiterate, Garbo style, "I want to be alone." Dear me. The audience is supposed to be drawn, fascinated, into this endless loop of melodic neurosis. Schumann was a gifted songwriter, and we can be sure the movement is based on personal experience. His attention here is focused almost entirely on the serene flow and beautiful tone of voice of a solo cello melody with nowhere to go. An insistent, erratic bass-line heartbeat hints at some dark mystery or threat lurking in the background.

The orchestra is behaving very circumspectly during this display of artistic temperament, but one has the impression that it is playing the role of a private secretary with the uncomfortable duty of reminding Madam that the taxi is waiting. A discreet cough from the clarinets. "Go away!" says the cello. "I want to be alone." The orchestra waits, then tries again, more urgently, with oboes and horns now sounding impatient. Madam improvises a little tantrum and stamps her foot. "How dare you," she cries; "I want . . ."—But then, in the blink of an eye, the mood changes: "Oh, all right then . . ." and the action quickly segues into the final movement.

Action model and strategist

Entertaining a crowd, whether by athletic display, as in the Boccherini concerto, or by charismatic moodiness, as in the

Schumann, is not really true leadership. Perhaps this is why these two works are not in the front rank of classical concertos. The eighteenth-century orchestra provided two models of leadership: the concert-master or performing artist, and the harpsichord playing conductor or strategist. These two cello concertos belong to the first, concert-master category. The soloists' role here is to be visible and to inspire the troops. The troops include the audience. The audience is inspired by demonstrations of action and charisma and is not expected to inquire about motive or cause.

The harpsichordist conductor coordinates and controls the musical action. The composer at the keyboard is the brains behind the entire campaign. He understands the objective. He has written the script. The essential difference between the field commander or concert-master and the keyboardist or general is that the field commander does not have to be a strategist. A concert-master executes strategy as a series of short-term maneuvers, but is not required to demonstrate long-term understanding of the campaign. This is leadership on the surface, giving directions and building morale. The harpsichordist, however, is working through a long-term process with a clear understanding of how each piece of the action fits into the overall pattern. His role is to coordinate, delegate, and observe. The sign of good generalship is not frontline charisma but backroom resolve and ultimate success, not pleasing the crowd but facing up to conflict and working things out. So music demonstrating leadership has to do more than dazzle and entertain. It has both to acknowledge and then resolve conflict, and that means expressing conflict in musical terms.

The hero as outcast

The French Revolution has passed. It is 1806. The people are in control. The Age of Reason is over, and the old social order has gone. In this uncertain time, power, authority, and leadership no longer coincide directly. A generation after Boccherini composed his cello concerto Beethoven confronts the new situation in the slow movement of his Piano Concerto No. 4 in G. Once again the piano is the voice of the artist, this time at loggerheads with the personality of the orchestra, which consists in this movement entirely of strings.[5] The contrast is stark. The strings open the dialogue with a brusque, energetic rhythm. It is a rasping, brutal *unison*, hence *unanimous* (signifying the absence of harmony), expressing the old revolutionary slogan "Unity is strength." The piano replies with studied eloquence and self-conscious beauty of tone, making a counter-statement consisting almost entirely of chords, in a sequence conforming to the logic of classical tona-

Opening of the slow movement of Piano concerto No. 4
in G by Beethoven

lity. The mood of the orchestra is aggressive and dynamic. It says, "Come on! Let's get going." The piano's mood is *harmonious* and sweetly reasonable; it says, "Okay, but first let's decide where we are going, and then how to get there." The orchestra allows the piano to speak, showing respect, but then refuses to listen. How do we know? Because at first it *waits for the piano to finish*, but then carries on where it left off as if nothing had happened. There is no meeting of minds. One is conscious that the solitary piano is no match physically for the might of the orchestra. This is the image of an artist from the Age of Reason having to defend what he represents against a new people's government that threatens to do away with every last vestige of class privilege.

Beethoven grew up in the last decades before the French Revolution and survived to become the first great composer of the Romantic age. He studied with Haydn, another survivor, and in the persona of the piano solo is defending the values of eighteenth-century classicism as worth preserving. This is leadership with a difference. One is always aware that the orchestra represents *power*, even though the piano represents *authority*. Their dialogue is confrontational and reminiscent of the McCarthy era interrogations of suspected un-Americans. After a further exchange which takes a different tack, the pace quickens and both sides begin to interrupt one another. The orchestra, uniformly loud, maintains its aggressive, irregular (dotted) rhythm, while the piano becomes increasingly assertive but maintains its harmonic character. At a certain point the piano, seemingly provoked beyond endurance, completely loses its classical "cool" for a moment and launches into a fierce and passionate tirade that leaves the orchestra without any response. The outburst climaxes in a tearful cascade of trills, a display more of hysteria than emotion, but certainly a show of utter defiance. *And it works*. The orchestra becomes very subdued. It loses its earlier aggressive rhythm. It begins to show a sense of harmony. The piano quiets as well, and the movement ends with a harmonious cadence by the now compliant orchestra, over which the piano traces an imperious, over-arching right hand arpeggio. The message? *Reason has triumphed*—or at least, old-style reason in combination with new age (Romantic) emotionalism.

Alienation and acceptance

Beethoven worries a lot about leadership, which is understandable. After the French Revolution it was left to artists, poets, and composers—figures like Schiller, Goethe, and Beethoven himself—to assume moral authority. Beethoven expresses authority in a new grandeur of orchestral rhetoric. Though not a concerto, his "Coriolan" overture is a study of the artist as hero. An overture does not give away the play, but it does lay out the tragedy, and the tragedy for Coriolanus is that the hero is a person who does not fit into society. The very same qualities that make a successful leader—persistence, single-mindedness, uncompromising dedication to a goal at the expense of everything else, an inability to enjoy life or to relax—all conspire to make the hero impossible to live with. There is a touch of Coriolanus in Beethoven himself.[6]

What you notice about Beethoven's use of the symphony orchestra that makes him different from the Mozart of Symphony No. 40 is that with Beethoven the orchestra is a powerful force, but *a force of nature*. During the classical age of Gainsborough humanity is depicted as bigger than nature: the landowner, his wife, and the dogs occupy the foreground of our attention while the mansion and parklands are depicted on a much smaller scale and in the distance. That all changes under Romanticism, when suddenly a huge disparity of scale appears in genre landscapes where ruins or mountains become the dominant feature and humanity is reduced to tiny Lilliputian figures dwarfed by their surroundings. *Beethoven makes you wait.* His music is on a larger scale. He can shift from quiet to loud in a moment, not to intimidate an audience, but because he is conscious of dealing with a natural power that has its own rhythm, a force, like the sea in the art of J.M.W. Turner, that is simply and awesomely big. The human tragedy of Coriolanus is summed up in the opening and closing gesture of a long, horizontal unison ending in a violent vertical hammer-blow of a chord. It is so simple yet so rich: an uncompromising musical right-angle, an irresistible force meeting an immovable object—and also an image of the guillotine. It says: *I am what I am.*

More diplomacy

Violin concertos don't carry quite the same heavy emotional baggage as piano concertos. The violin solo plays a concert-master role, and, if anything, the rivalry is not between the solo and orchestra as much as between the solo and conductor at the podium. Of course the violin has the advantage. He (or she) can at least sing. The third movement Rondo of Beethoven's Violin Concerto in D begins with a jaunty tune in the style of a folk dance.[7] This is a deliberate gesture. Beethoven was the first major composer to be an independent. He is no longer writing for an élite group of aristocrats. He composes the music, books the orchestra, hires the hall, handles the advertising. His music is for a mass public, and the public wants tunes it can relate to. Folk tunes are about national identity—a new sensation in Romantic culture and public awareness. The opening is non-threatening, in fact, ingratiating. This is a tune anybody can sing along with. Then all of a sudden the same melody reappears, but *two octaves higher*, far out of reach, in the musical stratosphere. One can imagine members of the middle-class audience marveling: "Is that really the same melody? It's so high!" The gesture says: Popular music is just as inspirational as art music. This simple tune can also lift the spirits.

Technique and control are just as much a part of Beethoven's style of leadership as they were for Boccherini. The difference is that, whereas Boccherini's cello had no problem whatsoever in dominating the orchestra, in Beethoven's case the violin is a relatively small sound competing with an orchestra that is massive *and conscious of its power.* What this conveys, as we become aware in due course when the orchestra enters en masse dancing to exactly the same jaunty tune despite its great size, is the message that human skill and ingenuity are no match for the sheer weight and power of the massed orchestra. The symbolism is quite explicit. Human science is wonderful, but nature is bigger. The artist may lead, but the people have the power.

There may be a similar hidden message in Aram Khachaturian's 1940 violin concerto. What makes the comparison with Beethoven especially interesting is the use once again of folk music for the purpose of ingratiating an audience. Again it was a time of conflict. There was a people's regime in power, and Soviet composers had to balance the moral authority of modern art against the power of the state. A composer whose music was deemed too intellectual or an artist whose style was considered too abstract risked losing the right to make a living, or even the right to life itself. In Beethoven's time folk music was cultivated as music of the people and of the nation. In incorporating folk melodies into his concertos and symphonies, Beethoven was declaring his solidarity with the underclasses. In Khachaturian's Soviet Union, the use of folk music themes was a government imperative, so ironically could be construed as an expression of capitulation to a repressive regime.[8]

This is not great music. It goes through the motions and outwardly, at least, aligns itself with the music for pleasure aesthetic of a Leopold Mozart. But wait, now. The composer of *Sabre Dance* is Armenian, and the folk idiom he employs in this concerto is gypsy violin music. But the gypsies under Stalin were a persecuted minority, a culture that did not recognize national boundaries or loyalties. In alluding to gypsy music, the composer might therefore have been sending a coded message—not of national unity under Stalin, but of freedom and defiance of oppression. Now *that* would really be moral leadership.

Notes
   1. Alessandro Marcello, Concerto for Oboe and Strings in D minor. József Kiss, Ferenc Erkel Chamber Orchestra (Naxos 8.550475, 1993).
   2. Benedetto Marcello, Concerto for Trumpet and Strings Op. 2, No. 11. Miroslav Kejmar, Capella Istropolitana, cond. Petr Skvor (Naxos 8.550243, n.d).
   3. Luigi Boccherini, Cello Concerto No. 9 in B flat. Steven Isserlis, Ostrobothnian Chamber Orchestra, cond. Juda Kangas (Virgin VC7 59015 2), 1992.
   4. Robert Schumann, Cello Concerto in A Op. 129. Jürnjakob Timm, Gewandhausorchester Leipzig, cond. Kurt Masur (Curb D2-78028, 1995).
   5. Ludwig van Beethoven, Piano Concerto No. 4 in G Op. 58. Anthony Newman, Philomusica Antiqua of New York (on period instruments), cond. Stephen Simon (Newport NCD 60081, 1991).
   6. Beethoven, Overture "Coriolan" Op. 62. Slovak Philharmonic Orchestra, cond. Stephen Gunzenhauser (Naxos 8.550072, 1987).
   7. Beethoven, Violin Concerto in D Op. 61. Schlomo Mintz, Philharmonia Orchestra, cond. Giuseppe Sinopoli (DG 463 064 2GH, 1988).
   8. Aram Khachaturian, Violin Concerto in D minor. Hu Kun, Royal Philharmonic Orchestra, cond. Yehudi Menuhin (Nimbus NI 5277, 1988).

CHAPTER THIRTEEN

Time and motion

A CLOCK is about telling time, and also about *marking time*. Look at a watch, look at a clock, and you know what hour and minute it is. For most people knowing the time in itself does not mean very much. If you are flying overnight between continents, it may not mean anything at all. If you are flying long distance you are between time zones, no longer at the time your watch was set on departure, and not yet arrived at the time of your destination. That's an extreme example, but it tells you something. The significance of telling time is intimately connected with *where you happen to be* at that particular moment.

Before the clock era, time for the traveler was measured in distance or work. "That village is an hour's drive from here." "I'll see you when that field is ploughed." The time of day was estimated by the sun, and the time to cook an egg by an egg timer. Work-time could be coordinated in song, based on the rhythms of digging and hauling: the blues, the sea-shanty. Rest time was given in the sound of the church Angelus bell ringing morning, noon, and night. Time was not the unified dimension as we understand it, but a composite of completely different layers of experience, some of which were linked to internal processes such as hunger and fatigue, others to social activity, and still others to the rhythms of nature.

Time is supposed to measure change, but in reality change measures time. The ancients contemplated the various motions of the planets and stars in the heavens and sought to reconcile them with patterns of change on the ground: the days, the crops, the tides. Cycles in nature, being stable and predictable features of the universe, were distinguished from non-recurring, cataclysmic events that were both unpredictable and also potentially destabilizing. In order to conceptualize and possibly explain the forces that maintained the universe in a dynamic but essentially stable condition the ancients resorted to musical models. In the twang of a bowstring and the howl of the wind they perceived a relationship between force and tone and recognized that a musical tone of stable pitch was also cyclical in nature. By constructing measuring equipment containing multiple signal generators of different pitch, and observing the interactions of different combinations of signal these early theoreticians were able to discover laws of harmony that could be used to explain the observed stability of the universe. *These signal generators were musical instruments.* Some, like the lyre or kithara, equated pitch or frequency with force (tension ratios), while others, like the harp or panpipes, equated frequency with length. Harmony is the name given to a combination of different frequencies that manifests a strong and stable inter-relationship; disharmony results from combining frequencies that do not interact in an agreeable or supportive manner. This is not an aesthetic issue of the ear deciding which combinations are harmonious and which are not, but a matter of physics since a harmonious combination of tones actually produces a stronger and longer lasting vibration. To the ancients, a musical performance on an instrument such as the harp or kithara was also a reminder of the natural laws governing the universe.

A kind of order

Telling time and feeling time are two different things. Universal time is cyclical, but human time is linear. Spring and taxes are cyclical events; they come regularly every year, but the date on the newspaper and each new birthday only happen once and never again. An individual experience of time has to do with human sensations and emotions. We move on two legs, so music in double time (two or four beats per measure) is natural for linear patterns of motion and by association for linear musical forms such as sonata form. The universe rotates, and triple time for two-legged human beings translates into circular motion, so music in triple time tends to be universal in implication, cyclical in pattern, and repetitive in form, like the minuet and trio of a symphony—or indeed, the triple rhythms of Monteverdi's "Deus in Adjutorium." Every dance is an affirmation both of a kind of order, and of a relationship to time—even the dissolution of time, as for rave music of today.

Dancing and dance music are perfectly natural developments of human behavior. They have symbolic meaning in that uniform actions communicate a sense of social unity. Unity of action is unity of time and unity of purpose; harmony of movement expresses stability and efficiency, allowing for complexity of social interactions. But dance is also enjoyable because the patterns of movement conform to the body's natural rhythms, and working out is a pleasant way of improving body tone and passing the time.

Body tone

Why should classical dance be of interest to anybody in the twenty-first century? Well, if you want to be a fashion model, you have to learn how to move. If you want to be an actor, you have to learn how to stand, as well as how to move gracefully. An actor or a fashion model has to do a lot of standing. If you don't know how to hold a pose without getting cramp, or tired, or your foot getting numb, you won't be successful. Moving gracefully is not only good to look at, it is controlling your body in the most energy-efficient way possible from the body's center of gravity through the limbs to the tips of the fingers and toes. Athletes know this, which is why they gear up in reflective striped body- suits so that their running, leaping, or throwing actions can be filmed and computer analyzed and any inefficiency noted and corrected. (Computer games employ the same technology in order to model human actions more realistically—which means reproducing body movements that are more harmoniously integrated.) The military understands the value of properly coordinated actions and movement. In addition to promoting corporate integrity and discipline, group marching alleviates fatigue, enhances endurance, encourages the under-performer to keep up, and makes time appear to pass more quickly. For the younger person, classical dance promotes good posture and body awareness, keys to a healthy diet and lifestyle. Finally, classical dance encourages young people from an early age to learn habits of courtesy and confidence in their dealings with the opposite sex.

Three kinds of dance

Classical dance forms in music correspond to three types:
- Social dance, in which anybody can join;
- Ballet, which tells a story and is danced by professionals; and
- Virtual dance, which is concert music based on a dance idiom.

Standard musical notation is the written expression of clock time, which is time in the abstract. Folk dance and courtly dance were long established by the time standard notation came into existence. Notation made it possible for differently styled dances, in different rhythms and speeds, to be written down and examined. The point of doing so was to discover why some actions are perceived as fast, and others as slow, and why some styles of movement are felt to express stress, and others relaxation. Clock time, which is neutral, and standard notation, which is universal, bring together the means of describing and coordinating the dynamics of human behavior.

The eighteenth-century dancing-master went from house to house in the better-off neighborhoods to teach the children to dance. In one pocket of his cloak he carried rolls of music, and in the other a thin-bodied violin called a kit or pochette. The children assembled in the salon, standing in rows and holding hands. The dancing-master played his kit, tapped his foot and called out the numbers like a square-dance caller.[1]

Learning to dance is learning to listen with your entire body. A dance is movement in space, and that movement may be broken up into alternating segments in which one partner leads, then the other. The dance form, its melody and phrasing, represents a young dancer's navigational chart describing the various moves and changes of direction. The tapping foot provides an underlying beat that regulates the speed and keeps the dancers all together. A dancer needs a strong form in order to follow and eventually anticipate the moves, and a strong beat in order to remain in time with his or her partner and everybody else.

The well-known Gavotte from Bach's Partita No. 3 for solo violin is a good example of virtual dance. Outwardly it sounds

like a dancing-master playing his pocket violin, but this is not
music for learning to dance, but a model to show off the violinist's technique.[2] A young learner trying to dance in time with
this music quickly loses track of the beat. Bach has converted a
rhythmic pattern into a *linear* pattern. The musical line has a good

"Gavotte en Rondeau"
from Partita No. 3 for solo violin by Bach

sense of direction and momentum, but not a strong and stable
beat that you can follow with your body. This gavotte is for
listening to, and for admiring the violinist's ability to combine
several parts into one.

Town and country mouse

Mozart's German Dance No. 3 is social dance, but social dance
with a vengeance. Mozart the Austrian is making fun at the expense of his German cousins. It is like the tale of the town mouse
and the country mouse. Although they share a common language, the Austrians have traditionally considered themselves as
the merchant aristocrats of Europe and regarded their Germans
neighbors to the north and west a tad condescendingly as simple
farmers, over-bearing, and rather clumsy. The music says it all. It
begins with the full orchestra in unison playing a loud bugle
summons to the dancers. Elegant, it ain't. Trumpet, fifes and
drums declare that this is music for the kind of people who are
only galvanized into action by military orders.[3] The unison
(absence of harmony) means what it says, and the orchestra's
untoward loudness and ebullience tell you that the people for
whom this music is written are neither light on their feet nor
very quick to respond. This lumpen parody of a country dance
continues with a contrasting verse featuring rhythmic jingles. It
is the ladies' turn. The bells clearly indicate that this is, in fact,

music for a rococo pastorale or costume ball in which the men are dressed up as stout huntsmen and their partners as plump shepherdesses wearing jingle bells on their sleeves and ankles. A pastoral entertainment means that the male dancers are probably wearing boots—another reason for the orchestra to be playing so loudly, so as to be heard above the clatter of heavy feet on the dance floor.

For all its comic exaggeration this is good music to dance to. Mozart has given his dancers a clear structure, vivid contrasts and a strong beat. The dance ends, all the same, in utter chaos. The music reaches its appointed final beat, but we hear the dancers carrying on regardless. The orchestra goes around one more time, making the cue to stop even more emphatic, but on the very last chord a flurry of jingles tells the listener that the ladies are not where they should be and are having to scamper back to their partners to the high sustained note of a trumpet that says "Enough! already."

Surprise, surprise

Haydn's Symphony No. 101 in D ("The Clock") is virtual dance music for couch potatoes. The third movement minuet and trio sound very danceable, but it is music for dancing at your peril. Why? Because Haydn delights to play tricks on the listener. He will lull you to sleep only to waken you with a bang. He will begin perfectly properly with a four by four opening gambit, only to follow it with an answering phrase that carries on for so long that by the end of it your partner has drifted right across the ballroom floor and is lost in the crowd. Symmetrical formations and repeats are a given in classical dance forms that involve male and female partners. If you move to the left, expect an answering move to the right. A composer like Haydn can take advantage of a listener's expectation of symmetry to create imbalances that raise a smile in the audience. Classical symmetry is about stability in the social order. Imbalances carry a different message about the dynamics of social progress.[4]

Ballet

Franz Schubert's *Rosamunde No. 2* ballet suite is music to accompany a staged dance that tells a story. Ballet differs from classical

Minuet from Haydn's "Clock" Symphony No. 101

social dance in a number of important ways. A minuet is a step pattern, not a particular piece of music. There are many musical compositions that are minuets, but there is only one step pattern that you have to remember. Once you learn the steps you can dance to anybody's minuet. Classical dance is like that. It's a pattern of moves. At the end of the pattern you are back where you started, so it's also a cyclical pattern. The role of the music is to

keep time and cue the moves. You already know the moves, so
the music is not designed to teach you anything new, only to
keep you on track.

A ballet, on the other hand, has a plot line. It is a new configuration of moves. It has its own special music; every ballet is
linked to a particular composer. Every dance narrative, like any

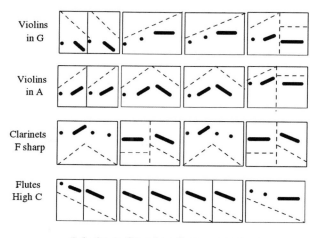

Schubert: Opening theme structure
of *Rosamunde* ballet music

story, is a linear process. It does not go round in a circle. A plot
line involves growth and change, and dealing with the unexpected. Little Red Riding Hood goes off to see Grandma and
encounters the wolf, and at the end of the journey the situation is
changed, there is no longer any Grandma and no wolf. At the
end of a story things are no longer the same. By definition, every
story is a one-off. You don't know in advance what is going to
happen. So an audience at a ballet performance, like the first-
night audience at a movie, is being invited to participate in a
growth process with an unknown outcome. Since the process is
revelatory, it has to be learned; since it involves the unexpected
it is also a test of stamina and individual skill. For that you need
specialists. And once you involve specialist dancers, their other
skills inevitably come into play: how high you can leap, dancing
on points, spinning like a top—the usual concerto solo stuff.

What we are interested in is the composer's role. Schubert

understands that the dancers, solo and chorus, rely on the music
to tell them when to start, how far to go, and when to stop. This
is not an event where everybody is on the dance floor from the
word go and stays there. It is a spectacle and an audience is
watching. A ballet has entrances and exits: people come on and
go off, and many spend a lot of the time waiting in the wings for
the signal to make a new move. Since there are no words being
spoken and it is difficult to see everything that is going on from
the wings, listening to the music is absolutely crucial. Schubert
has composed music of amazing lightness and charm for the
Rosamunde ballet interludes. From the audience's point of view
the music sounds effortless, simple, folk-like. To the dancers it is
a chronometer.[5]

Even ballet has its symmetries and repetitions. One group
enters stage left, then another stage right; the women circle in
graceful patterns, then the men leap about in competitive athletic
displays; the hero's solo daring to go to the limits of the physically possible, the heroine's response expressing the most intense
physical emotion. There is endless cross-referencing of solo and
group, left and right, male and female, young and old characters,
tragedy and comedy, action and passion. The music has to coordinate all of these balancing features and make the composition work as a unity.

Schubert builds complex linear patterns from the tiniest of
motifs, endlessly recycled and recombined. The rhythm units
(long, short-short; short-short-long; etc.) derive from classical
poetry and serve the same function of imprinting pattern on the
memory. This patterning is a form of *counting*, and the counting
process involves a range of switching processes. A three-note
group flicks down the first time, then up the second time around.
A long and two short beats becomes two shorts and a long. A line
given by the violins is answered by the clarinets, then by the
flutes. There is always a new combination for the dancer to latch
on to, to know when to start, how many steps to take, when to
stop, and how to stop. This is very sophisticated cue-giving, easy
to listen to, but amazingly complex when you look at it in detail.
In fact, *nearly every second* of the movement can be tied down to a
specific musical cue.

The musical continuity, on the other hand, is also masterfully
controlled, so much so that a casual listener may not notice how
expertly *Rosamunde* is edited together. At the half-way point and

again at the climax of the movement there is a change of tempo into a fast triple-time where all the dancers seem to join hands and dance in a ring, faster and faster. What is astonishing is that Schubert achieves this change of tempo without a change of beat. He goes from a rhythm in twos to a rhythm in threes, but the underlying beat remains the same. This is important for the dancers because maintaining the beat is essential to the counting process and to the dancers staying together.

In many ways this music anticipates the movies of a century later, in the sense that the dance narrative is constructed out of a sequence of very short actions placed end to end, relying on key and rhythmic momentum to connect them. Schubert is using essentially the same tactics as a movie editor to create the impression of a seamless flow, though of course in the auditory rather than the visual domain. (In the ballet it is the visual element, the dancing, that gives continuity to the music, whereas in the movies, it is the music that provides continuity to the visual action). Music of this quickness and kaleidoscopic immediacy is rare, even among movie composers of the twentieth century. Had he been born a century later, Schubert might have been one of the great Hollywood animation composers.

A German joke

Brahms is a formidable German composer of symphonies of the mid-nineteenth century. His Hungarian Dance No. 5 is a popular concert encore item, a short exhibition piece to round off an evening of serious entertainment. Originally one of a collection of twenty-one such pieces composed over a number of years for piano four-hands (one keyboard, two players), No. 5 has become famous in an ingenious arrangement for full orchestra by Ernst Schmeling.[6] This is dance music with a certain wild erotic charge, introduced into the normally polite forum of a public concert, so part of its humor for a nineteenth-century audience is the humor of embarrassment. The other part is the humor of scale—the spectacle of a volatile and sexy music normally performed by a five-piece gypsy band, being interpreted with improbable vigor by a committee. This music has all the entertainment potential, thrills, and danger of grand prix truck racing. Even if you think the joke is in poor taste, you have to admire the control that can make so large a sound mass accelerate, brake, and turn with such precision, changing gear in an instant from high drama to tears, and from high ecstasy to heavy breathing.

Brahms is already exaggerating the mood swings of the style in the original piano-duo version,[7] perhaps alluding to the Romantic cliché (as depicted by Burne-Jones, for example) of music lessons as an easy route to the seduction of a young lady piano student by her teacher (crossing of hands, etc.). What we have here is a piano piece that can only be appropriately performed by male and female duo-pianists, since it is, after all, music about male-female relationships, but with the added irony that the mere undertaking of a performance could be interpreted by polite society of the day as acquiescence to the idea of seduction itself. Not, of course, by a full symphony orchestra. But then, the version for orchestra is also not a dance one would seriously wish to attempt, for if you did, you would fall over. Perhaps for that very reason audiences year after year are able to enjoy the humor, appreciate the skill, and overlook the sex.

Navigating in space

Human progress is the subtext of Stanley Kubrick's celebrated movie *2001: A Space Odyssey,* and the director's choice of music to accompany the action and underscore the film's moral and political message is as fascinating and provocative as anything in the screenplay. One thinks in particular of the early scenes of the earth shuttle docking with the space station, and later of the moon shuttle on its journey to the moon base at Clavius, with the music of Johann Strauss II's waltz "The Blue Danube" in the background. This famous dance music was published in 1867, a century before the movie which, on its release in 1968, in turn anticipated Neil Armstrong's first step on the moon by more than a year. The audience's first reaction is one of disbelief. What does social dance music from Europe at the height of the nineteenth-century industrial revolution have to do with space travel in the twenty-first century?

In-flight music—the kind you listen to on those flimsy headsets—is designed to make the passenger feel at ease. The first impression that Kubrick's choice of "The Blue Danube" may have been intended to awaken in the audience might well have been precisely a sense of anti-climax, of the very ordinariness of the scene of travel in space unfolding before the viewer's eyes. It says, this is so routine, and indeed, when you get to see the only passenger in the shuttle, he has fallen asleep, his fountain pen quietly floating above his shoulder. A stewardess awkwardly maneuvers forward on velcro shoes, retrieves the pen, and tucks it into his breast pocket.

Navigating in time

Viennese waltz is celebratory social dance with very particular associations. Viennese society is prosperous, civilized, and well-bred. The waltz is a social ritual. Everybody observes the protocols: formal dress, elegance, manners. The dance itself, as the poet Byron makes clear in his epic *Don Juan*, is a revolutionary innovation in social manners for the new nineteenth century. In classical times, guests at formal dances were organized as strictly as diners at an official banquet. You were assigned a place according to rank, and not only your place at table but even your place on the dance floor was determined by your position in the social pecking order, a precise location in space from which you were not sanctioned to move. So you sat and you danced with your partner in one part of the room, and there only. The waltz, however, introduced the radical notion of *free-range dancing*, corresponding perhaps to a new dynamic of social eating where food and drink circulated among the guests instead of the guests sitting down at table. Dancers were now permitted to move freely throughout the ballroom. Instead of rigidly enforcing a social hierarchy, the waltz promoted a new industrial image of social dynamics.

The waltz is a very uncomplicated dance in triple time. Three beats per measure means circular motion, and a ballroom floor packed with couples waltzing is an image of multiple rotations and effortless precision. "The Blue Danube" waltz begins cautiously, first luring couples out on to the floor with a horn serenade, then winding up tension like a clock. Once the dancers are all in place and ready to go the actual dance portion begins with a gradual release of the brakes. As the waltz comes up to speed, a momentum is generated that seems as though it could go on forever. Unlike traditional formal dances such as the minuet, the waltz flows in a continuous movement *around* the floor, without any of the back and forth moves, symmetries, or sudden changes of direction of old-fashioned dance forms. *The*

waltz has no form. It expresses pure momentum. Its cycle of movement implicitly has neither beginning nor ending. And for a
movement of such mechanical simplicity and total absence of
pattern, the Viennese take the waltz very seriously, holding the
upper body rigid, the partner at arms' length, and maintaining a
smooth continuity of flow at all costs.

Rotation, precision, discipline, order: these images come to
mind when a waltz is seen or heard. In dedicating his waltz "to
the beautiful Blue Danube" (its full title in German is *"An der
schönen, blauen Donau"*) Strauss clearly intends his listeners to
make the connection between the flow of the dance and the flow
of the river. The Danube is naturally emblematic of Vienna, but
like any other great river, it is primarily an image of energy and
time. Always there, always changing, like time itself. From our
first-floor gallery overlooking the dance floor, we can see and
interpret the dancers' pattern of motion as an image of wheels
within wheels, or even, equating orbital motion with stability
and power, as a representation of the music of the spheres. Certainly Kubrick's intention in using the waltz as a backdrop to the
scenes of space-craft in space is to suggest a parallel between the
nineteenth-century social ritual and its image of a dynamic but
stable universe, and the twenty-first century technological ritual
of maneuvering in space and its corresponding image of high
civilization. In Vienna the dancers are silent, moving with coordinated precision; in the movie the spacecraft also navigate in
the silence of space, orbiting bodies in a universe of perpetual
motion.[8]

Orbital motion

In space, every move is a curve. Left to their own devices, the
free-floating pen and dinner tray turn slowly in circles, obedient
to laws of motion that govern the entire universe. Ironically, the
human participants in Kubrick's spatial choreography are unable
to behave with the same grace and precision as a fountain pen,
either in space or even in the artificial gravity of the space station
itself. There, while the music stops, the blue earth outside the
telephone booth window continues silently to revolve. Consider
the dialogue as music. Like the waltz, it is stylized, formal, correct. The carefully staged interactions convey the same message
of civilized behavior—and yet, as Kubrick delights in pointing

out, human relationships are marked by an underlying disso-
nance of motive expressed in persistent failure to cooperate and
inability to agree. No, he cannot be there for his daughter's birthday. Sorry, I am not permitted to give you that information. The
scientists circle metaphorically round one another, trying to
connect but not succeeding.

These scenes of docking and onward travel to the moon are
silent movies with orchestra accompaniment. There is no audible
dialogue, and there are no audible sound effects. The observer is
not just a passenger inside the craft listening to Strauss on a com-
plimentary headset, but a régisseur choreographing objects moving in space to the sound of music. The music in turn signifies
high technology, orchestrated movement, and harmony, echoing
the ancients who first explained the harmony of the universe in
interacting frequencies. Because it is uniformly continuous, the
circular motion of the waltz can match movement of any speed,
whether of a space bug, space station, or even the earth itself.[9]
Circular is cyclical, and cycles are routines, and routines are self-
perpetuating laws. Strauss's waltz is more than just the dancers
on the floor. They are only a visible outward expression of the
dynamics of Western industrial civilization, as manifested in the
complex interactions of the expanded nineteenth-century symphony orchestra and encoded in the digital software of standard
notation.

Paradoxical motion

Tchaikovsky's ballet suite *The Nutcracker* is a popular favorite,
composed for the great St. Petersburg choreographer Petipa and
combining narrative elements of traditional ballet with some of
the formalities of conventional social dance. In adapting Tchaikovsky and other music for the animated movie *Fantasia*, Walt
Disney was interested not just in expressing the movement of the
dancers in pictures but also with developing the relationship of
music and film in new directions. Premiered in 1940 at the height
of the second world war, *Fantasia* introduced innovatory animation techniques and revolutionary "Fantasound" surround-sound
recording to public attention fifty years before essentially the
same audio technology reached the home video market.

It was a triumph for RCA who had been working with conductor Leopold Stokowski and the Philadelphia Orchestra for a

decade to develop new technologies of sound recording and reproduction, with movies rather than the domestic audio market
in mind. Disney himself wanted animated film to be respected as
a creative art, and he saw collaboration with Stokowski and RCA
both as a marriage of convenience and as a logical partnership of
advanced research in the visual and auditory domains.

In conventional ballet the music remains in the background,
and the dancers in the spotlight. Normally the dancers move and
the music stays put. *Fantasia* is not like that. In this animated
movie the relationship of image and music in the movie is called
into question by the new audio technology. The moment the
decision was taken to introduce surround sound the question of
what to do with it should have become awesomely apparent.
With rare exceptions, such as multiple orchestra projects by the
French nineteenth-century composer Hector Berlioz (the *Grand
Messe des Morts* of 1837) and by the then little-known American
innovator Charles Ives, orchestral music up to the late 1930s had
always been regarded as a static and essentially monaural experience. How were Disney and conductor Stokowski to justify the
expense of RCA's advanced technology if the musical items were
recorded in a conventional acoustical setting? It was decided that
the music itself would have to move, the orchestra dividing,
separating and dancing about the room in imitation of the animated characters on the screen.[10]

This is exciting, certainly. It is also bizarre. On reflection,
you have to wonder if Disney fully understood the implications
of three-dimensional sound for animations on a two-dimensional
movie screen, that success in the auditory domain could well be
at a cost of making the visual animations, however brilliantly
achieved, appear flat and confined. It is interesting to see the
ways in which Disney's animators contrive to blur the boundary
between image and musical space to make audiences less conscious of the limitations of the screen, for instance, by the use of
sculptural highlighting, by taking the action in and out of the
light within a scene, and in particular by the use of black space to
conceal the edges of the screen to give the effect of a spotlight, in
the style of a circus or nightclub act.[11]

Space notation

Like Kubrick's shuttle choreography in *2001,* Disney's original

Fantasia is also movie-making in the grand tradition of the silent movie era. In between concert items, spokesman Deems Taylor and the Philadelphia Orchestra do their best to create an impression of a live concert, but as soon as the music begins, no other sound is heard and the music itself is transformed into complex patterns of movement. Normally in an animated movie, the animation comes first and the music is created to fit. In this case the music comes ready-made and the animation artists have the very difficult task of exactly coordinating the visuals to a pre-existing soundtrack. *Fantasia* deserves credit for some bravura examples of animated motion reflecting a considerable depth of understanding of natural movement and fluid flow: leaves falling and being blown by the wind, water spilling and falling, ripples, bubbles erupting and spiraling in water, the weight of seeds floating in air—an entire catalogue of closely observed and analyzed movements. The viewer may wince to see lipstick and eyelashes on goldfish, and not notice how beautifully the movements of the goldfishes' tails are executed—all of it done, by the way, without computer assistance. In the sphere of human choreography, the movie draws recognizably on pre-filmed imagery of ice-skating, folk-dancing, ballet, synchronized swimming, even at times the geometric patterns and overhead camerawork of Busby Berkeley. Building on the discoveries in motion photography of Eadweard Muybridge and Louis Marey in the last decades of the nineteenth century, the contribution of classic animated movies to our understanding of natural motion deserves to be acknowledged as entertainment with a solid scientific foundation, just as the dance suites for harpsichord of the seventeenth or eighteenth century deserve proper recognition as studies in encrypted motion in digital code.[12]

Notes

1. Anon."The Wonder," "Slingsby's Allamande." Susan Baker (*Susan Baker's Fiddles and follies*, vinyl, Argo ZK86, 1979), track 2 .

2. Bach, Partita No. 3 for solo violin. Yehudi Menuhin (EMI CHS7 6035-2, 1989).

3. Mozart, German Dance No. 3. Philharmonia Orchestra, cond. Sir Colin Davis (Seraphim 7243 5 68533 2 4, 1990), disc 1, track 7.

4. Haydn, Symphony No. 101 in D ("The Clock"). La Petite Bande, cond. Sigiswald Kuijken. *London Symphonies* (DHM 05472 772451 2, 1994).

5. Franz Schubert, *Rosamunde* Suite No. 2. Orchestra of the Age of Enlightenment, cond. Sir Charles Mackerras (Virgin VC7 91515-2, 1992).

6. Johannes Brahms, Hungarian Dance No. 5, orch. Ernst Schmeling. London Symphony Orchestra, cond. Neeme Järvi (Chandos CHAN 8885, 1991).

7. Brahms, Hungarian Dance No. 5. Silke-Thors Matthies, Christian Köhn (*Brahms Four Hand Piano Music, Vol. 2*, Naxos 8. 553140, 1997), track 5.

8. Johann Strauss II, "The Blue Danube" waltz ("An der schönen, blauen Donau"). Vienna Philharmonic cond. Herbert von Karajan (DG 439 104 2GDO, 1968).

9. *2001: A Space Odyssey*, dir. Stanley Kubrick (VHS MGM/UA M-203103, 1994). Kubrick's use of pre-recorded music is defended by Hollywood composer John Williams in Irwin Bazelon, *Knowing the Score: Notes on Film Music* (New York: Van Nostrand Reinhold, 1975), 111, 200.

10. Peter Ilyich Tchaikovsky, Ballet suite *The Nutcracker*, arr. Stokowski. Philadelphia Orchestra cond. Leopold Stokowski (*Fantasia: Remastered Original Soundtrack Edition*, Walt Disney D STCS 452 D, 1990), disc 1, track 7. From the 1940 original "Fantasound" optical sound-track.

11. John Culhane, *Walt Disney's Fantasia* (New York: Abrams, 1983).

12. *Fantasia* (1940) with Leopold Stokowski and the Philadelphia Orchestra (Disney VHS 1132, n.d.).

CHAPTER FOURTEEN

Noise

HARMONY is the dream, but noise is the reality. After the final shootout, the lone cowboy rides off into the sunset to the wordless keening of an invisible choir. *Doesn't it make you sick?* Understanding music is knowing that it isn't noise. It's the only definition—completely subjective. But in fact, you *don't* know. If you can't be sure what noise is, you can't be sure what music is either. You won't find a satisfactory definition of noise in any dictionary. Go on, look it up. "Unwanted sound." "Excessively loud sound." "Sound without harmony or structure." Noise is not a term. *Noise is an attitude.*

The doctor says, "Say aah." Why *aah*? Why not *ooh* or *eee*? Because if you say *ooh* your lips are pursed, and if you say *eee* your teeth are clenched. Neither of which is good if you want to look into somebody's open mouth. You say *aah* because it opens your mouth wide, drops your jaw, and flattens your tongue. No other vowel does that. The doctor is not asking to *hear* you say *aah*. It's not the sound of your voice that counts—what matters is the action the vowel signifies. If you say it, you've done it. *Aah* is an interesting vowel, nonetheless. Singers practice their scales on it. Choirs chanting "Alleluia" make a meal of it. Philosophers and problem-solvers express delight with it: the "Aah!" effect. Diners express satisfaction with it. Now take *ooh* for contrast. If you are frightened or curious you say *ooh*. If you don't like your food you say *ooh* (ugh). If you are uncertain, you say "ooh . . . I don't know." If you are *in pain*, you say: "ooooh . . . that hurts."

What we do understand and need to acknowledge is the connection between what a sound intends and what it actually does. A child says *aah* willingly because the gesture is associated with agreeable sensations. It is non-threatening. You close your eyes and say *aah*, you might get a gumdrop on your tongue. Opening your mouth is laying yourself open. The meaning of *aah* is taking something in, as in food and drink, or letting something out, as in a sigh of relief or an "Allelu-*yah*!" That's what the vowel does, *and that is what it means*. The difference between *aah* and *ooh* is that *aah* is unconditional whereas *ooh* is conditional. *Ooh* says "I'm not sure." As long as your mouth is not fully open, you are protected against unwanted invasion, not least by the doctor's wooden tongue depressor.

There are five vowels in the alphabet. The ogre missed one: "Fee, fie, foe, fum!" should be "*Fah,* fee, fie, foe, fum." What about all the other letters in the alphabet? What do they signify? Well, they're noises. Are they important? You bet. Vowels get most of the attention, but consonants do most of the work of speech communication. A secret message containing all consonants—

T: b: , :r n: t t: b: , th: t :s th: q:: st:: n

— is a lot easier to read than one with only vowels:

:o :e , o: :o : :o :e , :a : i: :e :ue :: io:

Okay, so you are reading it, not listening to it. If you could *hear* the vowels, you would be able to distinguish the three different sounds represented by the letter [o]. But it's the same even when you read the two versions aloud. In the first case you are whispering, and the words are not loud, but they are distinct. In the second case, you sound like a person with a speech impediment, or somebody in pain. The way a listener construes all-consonant and all-vowel speech tells you something important about the relative functions of noise and harmony, in speech and also in music. Vowels are voiced harmonious resonances. They carry the sound of your voice and the emotion of your argument. Consonants are noises. They are not voiced in the larynx, they do not convey your personal identity (the sound of your voice), but they are essential for articulating what it is you have to say. Consonants are tactile: you can *feel* them in a person's speech, face to face. At a distance, consonants become less and less tangible until, at a rally or opera performance, all the audience has left to latch on to is vowel sounds and rhythms (the primal scream thing). That, of course, is music to your ears, even if you don't understand a word of it.

Consonance

Musicians understand consonance as harmony, but the word actually means "sounding with" or "sounding together." Consonants are *sounded with* vowels. They introduce, they sign off, they actually *modify* the vowel, something you learn if you are editing a speech or a song on tape or disk. Conso*nance*, in the sense of chords in music, is a harmony of resonance. Conso*nants* express a different attitude to harmony, a harmony (consistency, continuity) of action.

Music's bias in favor of melody and harmony and its rejection of noise, and a similar bias among speech scientists in favor of vowel sounds and vowel combinations and rejection of consonants, is totally fascinating to read about because *more* meaning can be found in noise, disharmony, and consonantal elements of speech than in harmony and vowels. It is a very ancient prejudice. Vowels are more highly regarded because they are historically easier to localize and to quantify. Vowels are harmonious, meaning they have a definable frequency structure that can be analyzed: consonants have no internal structure, only boundaries. Vowels carry the song, so they have substance, duration, and weight. Consonants define boundaries, are unpitched, and have position but no duration. Vowels are mid-frequency voice elements, therefore are historically easier to record; consonants include the highest-frequency elements of speech that the pioneers of speech analysis could not register on their primitive cylinder acoustic recorders.

Why the prejudice? Consonants are the rhythm section of the human voice. Every letter of the alphabet has a musical instrument equivalent. Hi-hats and buzz cymbals: *tiff, tiff; tit, tssh, tssh; tssss, fssh;* tom-toms: *bdim, bdam, bdom;* bass drum: *thng, duh;* cowbells, wood drums: *toc, kik, gnk;* brushes: *tzzadah, tzzadah, tzessh!;* sticks: *diga diga bim diga bam.* Ignore consonants and it's like listening to your favorite group without the drums. It makes no sense.

Hush!

Of course, it's not true to say consonants have no duration. You can't use them to hold a melody, right enough, but some vocal noises do function as communication signals. The quickest and clearest way of telling somebody to keep quiet is to look at them and say *Sshhh!* If you want to say hello to a baby, you click your tongue against the back of your front teeth: *Tst! tst! tst! tst!* —or make kissing noises: *Pwst! pwst!* (To make this noise, you have to suck, not blow.) Disapprove of what somebody says at a meeting, you audibly catch your breath: *Hffff! Hssss!* or you hiss: *Ssssss!* Greet the neighbor's cat: *Hey, pss pss psss!* or whisper *Kidikidikidi!* Show disappointment: *Tsk tsk tsk!* (intake of breath). These noises mean something. They are what they mean, and they mean what they say. A hiss or a shush is a blanket of noise that you throw over something unpleasant. It blots out the sound and protects you from having to listen to it. It also helps to erase the memory of the sound. Greeting noises to small creatures are high-frequency clicks that are easy for the target listener to locate, and are not threatening because they don't carry the full weight of the voice.

Zones and vectors

To the extent that speech and music demand a listener's complete attention, they are *all noise*. Noise in the blah-blah sense that it masks or distracts attention away from other sounds in the environment that might be more interesting. Masking sound can be soothing as well as offensive. Elevator music is deliberately unrewarding sound that because of its resemblance to real music is an acceptable feature of the environment; however, its function is not to entertain but to create an auditory presence that is reassuring, either because it alleviates silence, and silence in a supermarket or shopping mall is intimidating, or because it raises the threshold of hearing above the level of air conditioning, refrigerator cabinets, cash registers, and other paraphernalia of mainstream commerce. Conversation functions in much the same way, meaning that whether or not it is intended to communicate anything useful, the speech act is designed to *defend* the speaker from interruption by external sound, and to create a privacy zone for the speaker's own thoughts. Speech is often used aggressively by talk-show hosts to prevent their guests from communicating directly with the viewing audience. When a host talks very fast or interrupts, the subtext is "Hey, this is *my* space."

In more civilized times, before the arrival of shopping malls and talk shows, the city fathers installed fountains about town at places where passers-by could rest for a while and talk. The water spray cooled the summer air, and the sound of a fountain also created a privacy zone in which a person could relax. The sound of a fountain is a natural form of white noise. It's nature's way of saying "hush."

We have now isolated at least four signal categories in every-day speech: the low frequency *hum* or fully enclosed sound representing an internal state (lips closed), the *aah* category or open middle-frequency sound that *ex*-presses or connects what is inside your mind with the world outside (lips open), the *tsk tsk* or abrupt high-frequency click that *stimulates* a response from the environment, and the diffuse *sshh* type of high-frequency noise that *neutralizes* the acoustic environment. Logically what this implies is that spoken language, *whatever its intended meaning*, is a patterned interplay of externally and internally directed acoustic signals, and that even individual words are directional indicators as well as socially meaningful gestures. Speech vectors are most easily recognized in one-word expressions such as *Hey! Wow! Whoops!* and the more abstract and colorful terms in the swearing lexicon. Each of these words describes a geatural trajectory from inside to outside, or from outside to inside, and that trajectory is part of its meaning. The gesture "Hey" projects tone into space; "Wow" opens up an inner buzz to the outside world, then closes back up again; "Whoops" describes a vector from inner to outer space, as in *throwing up*, etc. These directional vectors in word gesture create a natural turbulence in spoken language that one could describe as a *texture*. We already know that high-frequency speech sounds are spatial in function, while middle- and low-frequency sounds are perceived as interiorized and self-assertive. Take away the middle- and low-frequency components of speech —that is, whisper—and the sense of personality is removed while the words remain comprehensible. This tells us that functional language communication operates in a *spatial* domain that is essentially *high-frequency, high-definition* orientated. As soon as you reverse the process, pinch your nose, and take away the high-frequency component of speech, you are left with a low-definition tonal residue that conveys presence, personality, and emotion but very little objective meaning. The texture of normal speech, therefore, reflects a characteristic balance of environmental and internally directed signals. At the very least, it tells you *how the speaker is listening*.

Noise in music

Doting parents hang mobile chimes over the baby's cot—bright, clangorous, high-frequency sounds that the infant can hear and locate easily—spatial information. At eighteen months the child is playing with toys that whirr, click, and buzz, is banging spoons on plates and lids on saucepans—all good practice in physical coordination. Music has always had an element of percussion: drums bang, cymbals clash, bells chime. Percussion plays a similar role in music as it does in speech, for positional and spatial definition, and for punctuation and emphasizing features of a musical line. The audience understands the action manifested in drum rhythm, and the directional and spatial functions of bells, cymbals, and triangles.

To understand why melody and harmony are accorded a higher value than percussion in music (the same as in speech) it may be helpful to reflect on the essential difference between percussion play and speech play in infant development. Vocalizing by an infant, whether it is regarded as intentional or unintentional, is a form of exercise that has the production of *internal* sounds as a byproduct, and that generates an *external* residue as environmental information. Percussion play (for example, banging a spoon) on the other hand is essentially *silent* gesture that generates no internal sound as a byproduct but only environmental information. Percussion play is ideal for monitoring the environment because there is no sound in your head to interfere with the listening act. In vocalizing, the listener is comparing an internal act with its external reflection. In percussion play, both the act and the sound are perceived externally; the act is not the cause of the sound, the sound is simply a byproduct of the act.

In-tension and ex-tension

Strictly in a dictionary sense, therefore, vocalizing and its derivatives melody and harmony arise from inner tension or *intension* (personal character and motivation) and are therefore able to communicate *intention* or will. (The two words derive from the same Latin root.) Since percussion is also a byproduct of action *in extenso* (in space), it can only communicate action. Since there is no direct link between the sound of an action and an intention, cymbals and drums are logically incapable of expressing personal character and/or motivation. That could be one reason why percussion instruments and their sounds are generally accorded lower status in the musical pantheon.

Let's go to some real examples. The drum sounds in the song "Mandad' ei Comigo" previously cited are typical of the rhythm and beat highlighting functions of percussion in classical and folk music traditions. Noise also refers to loudness and complexity in music without accompanying percussion, as heard in the introductory "Deus in Adjutorium" of the Monteverdi *Vespers*. The offensive clamor to which Hieronymus Bosch refers in his depiction of Hell in *The Garden of Earthly Delights* includes drums and jingles, but his attack also alludes indirectly to the sheer intellectual and audible complexity of *ars nova* multiple-part vocal music of the Flemish and Netherlandish schools of Ockeghem, Busnois, and Willaert.

Fireworks

Handel's *Music for the Royal Fireworks* combines space, grandeur, and percussion in the purest sense of exploding fireworks. It's a great shame that the piece isn't available on disc with the sound of fireworks included, not only because it is no longer a difficult operation to incorporate pre-recorded fireworks into a digital recording, but also because the music and the occasion *demand* this vital ingredient of incidental noise, intended from the outset to convey the excitement of the battlefield without any of the danger that usually accompanies the sound of real musket fire and exploding gunpowder. Handel assembled the music from earlier pieces, and for that reason it might be argued that it was not really composed with fireworks in mind, unlike for example Tchaikovsky's "1812" Overture with its attendant clamor of steeple bells and cannon fire. But as an outdoor exhibition piece with pyrotechnics, Handel's original *Fireworks* would be hard to beat. It is both more effective than Tchaikovsky, whose orchestra is not designed to perform satisfactorily out of doors, and more exciting than Stravinsky's *Fireworks*, a bravura exercise seemingly designed to convey the dazzling effect of a movie image rather than the noise and chaos of the real thing. Handel deserves due credit for tackling the logistics of a music for outdoors to offset and accompany fireworks in a realistic and pragmatic manner.[1]

Nowadays the cannons and fireworks in outdoor exhibition pieces are timed to the second, a tribute to the precision of today's expertise in solid fuel rocketry, mining, demolition, and munitions. But having fireworks exploding on cue is not what Handel planned for. His music for massed brass is loud, strong, rhythmic, and disciplined. It conveys majesty and control in the field. Fireworks in Handel's day could not be timed to the second, so instead of trying to coordinate music and fireworks, the composer boldly superimposes the two to create a philosophically more interesting opposition. On the one hand, the music presents harmonious effects that manifest *order* and *intention*, and on the other, the fireworks emit percussive effects that are *incidental* in nature and *spatial* or environmental in connotation. Handel's fireworks should certainly be enjoyed as a fireworks display, but—which is even more important—at the same time, they should also be understood as an image of randomness, of gunfire on the battlefield. Movie audiences today are excited by unexpectedly loud noises, especially random explosions, and in Handel's England of 1749 audiences felt exactly the same way. Unpredictability is the key, not only to the music itself and the excitement of the occasion, but also to the hidden role of percussion in human speech and perception.

Modern times

A movie sound track is composed of three elements: music, dialogue, and "foley" or sound effects. Each plays a specialized role and is understood in a special way by the audience. The role of music is psychic; background music tells you what the movie is "thinking" (meaning not what the character is thinking, but what *the movie as a whole* intends the *audience* to think). The role of dialogue is to progress the action and explain *motivation*. The role of movie sound effects is to add a very selective *realism*: footsteps, gunshots, the squeal of brakes. You should notice that each of the audio elements in a movie requires a different degree and kind of expertise. Music is composed and recorded and is intended to create, consolidate and anticipate audience reaction. Dialogue is scripted, then improvised—to make it appear *made up on the spur of the moment*, then possibly re-recorded in post-production—to make it sound scripted again (and thus allow the screenwriter to claim credit for it). Sound effects perform the role of punctuation that clarifies and articulates intention. Ironically, sound effects tend to be symbolic rather than real (the punch on the jaw), and in practice are highly specific rather than environmentally realistic, since a truly natural sound environment would, in fact, be much too noisy and distracting. There is a charming scene in the movie *Singin' in the Rain* that makes the same point with a great deal of wit as part of a sequence illustrating the trials of silent movie actors coming to terms with the new world of sound and the all-hearing microphone.[2]

Music of noise

Before the twentieth century, classical music in the Western tradition had very little to do with percussion alone. A drumroll at public events tells the assembled public to pay attention, but there is no musical content in the action, so nothing on which a composer could claim a royalty. A drumroll is simply using military drums to create a bandwidth of noise in the region of speech, masking normal conversation and in that way imposing silence on the gathering. Once the trumpets begin, of course, the drums fall silent.

At a parade, a solo drummer at the back taps a beat to keep the marchers in step. That too has no content other than the beat. Drums and jingles in Mozart's German Dance No. 3 perform a pictorial function in the music's comedy narrative. Timpani reinforce the heavy beat of the dance and convey an impression of weight and bulk among the dancers. This amounts to artistic license because you don't *see* the dancers Mozart has in mind —you have to *imagine* them. Likewise the jingle bells, which in a concert performance are played by one or more percussionists in the orchestra. Often referred to as sleigh bells (as if one would ride in a sleigh in triple time) the rhythmic jingling just as easily conjures up an image of generously-proportioned female dancers dressed as shepherdesses with bells stitched into their costumes or wreathed around their shepherdesses' crooks. Since you don't see any dancers in a concert performance, the real purpose of Mozart's added percussion has to be inferred from the orchestral score and its other mannerisms.

In the late eighteenth century, Turkish and oriental percussion instruments were introduced to the European musical world of Haydn and Mozart. This new apparatus of heavy cymbals and triangles (named *janissary music* after the Turkish word for militia) added color, excitement, and a sense of the exotic, but not much else. Increasingly during the nineteenth century, composers dared to incorporate unusual percussion and noise elements into their repertoire of sounds. Opera composer Richard Wagner was particularly drawn to special effects such as anvils, thunder, mysterious bell sounds, and the rather unconvincing *ding* of the celestial hammer on the entry of the gods to Valhalla.[3]

With the coming of silent movies in the early twentieth century, the trend toward more realistic sound effects accelerated. In *Eine Alpensinfonie*, a Byronic evocation of a day's excursion into the mountains, German composer Richard Strauss does not hesitate to borrow the sound of a pipe organ to signify sunshine and grandeur, and to employ special effects theatrical engines (thundersheet and wind machine) to illustrate, well, bad weather on the way.[4] Crude it may be, but orchestral music of the late Romantic era was simply following operatic taste for direct imitation of naturalistic effects by mechanical means rather than indirect allusion by conventional means (as, for instance the storm scene for orchestra in Beethoven's "Pastoral" symphony,[5] or the thunder for massed timpani in Berlioz's *Symphonie Fantastique*).[6] Sound effects, in the opera and subsequently the movie age, needed to make the point more quickly, and the late nineteenth-century composer, unlike Beethoven, had a different attitude to wind and thunder, treating them as environmental effects and not as events of psychological or emotional consequence. At the opposite extreme from Strauss and his inflated Romantic vision, we find French composer Erik Satie employing sound effects in a witty and ironic context. Satie's *Parade*, music for Jean Cocteau's *ballet réaliste* of 1917 (décor by Picasso), an entertainment destined to do for ballet what Marcel Duchamp's "readymades" came to signify for art, incorporates sounds of revolver fire, ship's siren, and typewriter into a charming and purposely aimless sequence of musical scenes.[7]

Hold that siren

Credit for developing the concept of a "music of noises" is usually accorded to the Italian Futurist movement, whose members included artists Giacomo Balla and Gino Severini, poet Filippo Marinetti, and composer Balilla Pratella. The Futurists were great propagandists for a music of sound effects and made some useful suggestions for notating noises, which is no mean achievement. But their musical efforts, partly for lack of know-how, partly for lack of facilities, bore little fruit. It was left to a genuine talent, the Burgundian-born composer Edgar Varèse, to pioneer a new art of spatial composition employing wind and percussion orchestras as signaling rather than storytelling devices. In *Composers, Conductors, and Critics,* her memoir of the American League of Composers, co-founder Claire Reis recalls how Varèse discovered the musical potential of the siren. In 1923 while rehearsing a new work, *Hyperprism,* in his New York studio apartment, Varèse heard a fire engine pass by with its siren wailing. Fire engines prior to that time carried bells on their hoods. The siren was a new and powerful sound effect, and the composer was fascinated by what he heard.

> Rushing to the window, he leaned out and listened intently. Then, turning back to the musicians, he announced, "The rehearsal is over; I'll call you soon." Without waiting for them to pack up their instruments and leave, he clapped on his hat and coat, and dashed out the door and off through the streets to the nearest fire station. There he announced flatly to an astonished fireman, "You must lend me one of your sirens for a concert!"[8]

Fortunately, a sympathetic Department of Acoustics at Columbia University agreed to lend Varèse a siren and the concert was saved. Sirens had been employed since the mid-nineteenth century as a laboratory item for acoustic research into the nature of musical pitch and timbre.[9] The musical value of a siren lies in its ability to create a sense of melodic line and direction, but free of any classical or tonal implications. Varèse subsequently employed the theremin and *ondes martenot,* early electrical instruments, in similar fashion. In turning away from the conventional subjective elements of music (harmony, melody) in order to focus on the signaling qualities of wind and percussion, the composer echoes artist Wassily Kandinsky:

> In music the line supplies the greatest means of expression. It manifests itself here in time and space just as it does in painting. In measuring tonal pitch in physics, special apparatus is used which projects the vibrating tone mechanically on a surface and which thereby gives the musical tone a precise graphic form.[10]

The siren's sinuous line functions in Varèse's music as a dynamic feature connecting patches of brass and percussion texture, in much the same way as the black whip-like line in Kandinsky's abstract compositions is intended to lead the eye and provide some sense of continuity and connection to otherwise static patches of color.[11] Sirens figure prominently in the genuinely noisy *Ballet Mécanique* by George Antheil, a score composed a year later than Varèse's *Hyperprism* for an experimental movie of the same name by artist Fernand Léger, and featuring multiple keyboards, airplane propellors, and tuned electric doorbells.[12]

Do it yourself

Foley artists at the end of the twentieth century sample and play back noises from a digital keyboard. The purpose of noises is still to emphasize movie action and create spatial realism, but the sounds themselves often bear very little resemblance to real life or real space. What seems to matter nowadays is for *action* sound effects to correspond plausibly with the dynamics of a plot with which the audience is vicariously engaged, while at the same time giving *spatial* sound effects an alien quality to reassure the audience that the physical screen environment in which these actions are supposed to happen is only an illusion. There is an involvement in action, but a distancing from its effects.

In John Carpenter's wry 1974 science fiction movie *Dark Star*, a disgruntled crew member played by Bruce Dern seeks solace in his cabin by making percussion music on a made-up instrument constructed of empty bottles and spare parts. The spatial function of such noises serves therapeutically to release emotional tension. A listener is reminded of the music of composer Harry

Partch, exploiting new scale divisions through a variety of strange, hand-made percussion instruments employing unusual timbres of glass and wood to create images and soundscapes of disorientating freshness.[13] John Cage, another musical inventor, deconstructed the piano, the very embodiment of Western musical thinking, to refashion its image of tonal and digital uniformity as an exotic percussion band. Cage was perhaps influenced by the piano transcriptions of Balinese music by Colin McPhee,[14] but by his own account he wanted to create a percussion band that he could carry round in his pocket. The *prepared piano* is a piano with various materials such as wood, metal screws, and rubber wedged between the strings to transform a scale of uniform tones into a sequence of contrasting percussion effects, some bell-like, some drumlike.[15] The effect is initially disconcerting, not only to the listener but also to the pianist, who no longer gets the expected feedback from playing a given key or melodic shape. The prepared sounds no longer express uniformity or continuity of tone and touch (loudness). Each prepared note is connected to a specific timbre effect. The resulting music avoids a sense of narrative flow or emotional development, reflecting not only a Western perception of Balinese and other non-Western traditions, but also the aesthetic of contemplation Cage wanted to convey.

Dreams and trances

Random activity in the neural pathways generates an internal "ringing in the ears" that has powerful associations with sickness, delirium, and fantasy. The age of special effects music for radio and the movies was responsible not only for infusing a new tactile imagery into classical music, but also for extending the emotional range of harmony and melody into previously uncharted areas. This tendency was also influenced in part by reports of the disorientating effects of experiments in sensory deprivation conducted during the early 1950s and involving prolonged exposure to white noise. Dense masses of tone with broad presence but little or no gestural content can be very effective in generating a sense of mystery or apprehension in an audience. Musically the technique is aligned to abstract colorfield painting, in which an entire canvas is filled with more or less uniformly saturated color, as in the paintings of Barnett

Newman, Mark Rothko, and others during the fifties and sixties. An *absence* of form combined with the enveloping size and involving *presence* of color can be quite hallucinatory in effect. The Hungarian composer György Ligeti, a refugee of the 1956 Soviet invasion, created his *Atmosphères* for orchestra in 1960. It was among the first of a new genre of abstract composition to manipulate the orchestra as a source of "colored noise" and the effects of the masking phenomenon for musical expression. Other compositions in the field include *Anaklasis* for strings and percussion by Polish composer Krzysztof Penderecki,[16] *Konx-Om-Pax* by the Swiss Giacinto Scelsi,[17] *Trans* by the German Karlheinz Stockhausen,[18] and *Coptic Light* by Morton Feldman—a music of dreams and trances, though the Feldman also resembles the sounds an unborn child might hear in the womb.[19]

Waitlessness and unexpectancy

Two curiously apt neologisms apply to cluster music and its effects. *Atmosphères* functions in purposely disorientating mode as the mysterious "Overture" to Stanley Kubrick's movie *2001*. The stock response to music of this kind is "horror movies." If it were genuine white noise delivered under experimental conditions (now discontinued, incidentally), fear would perhaps be a normal response, but in reality none of this music is anything like a sensory deprivation experience. So where does the horror come from? It's an alarm response, for sure, but based not on what the music is, but what it isn't.

Cluster music is alarming because it does not do what the listener expects and wants. It tells us more about ingrained habits of listening than specific cause and effect. *Atmosphères* is music without apparent boundaries in pitch or time (although a listener to the soundtrack recording can hear perfectly well that the orchestra is made up of real instruments and recorded in stereo). Contributing factors in a movie context are 1. darkness: the house lights are down, and in darkness a listener has to rely on hearing rather than sight; 2. the listener cannot escape: you have paid money for your seat, and what is more, you want to see the movie; 3. there is no beat and no melody, thus no dialogue on a human scale and no movement on a human timescale.

By having this music performed as an introduction in front of a darkened screen with only the word "Overture" visible,

Kubrick is holding his audience in a state of heightened expect-
ancy, which is frustration for many. Shrewdly, the director is
also ensuring that the movie theater house management is not
going to set an inappropriate mood by playing bubblegum music
instead.
The truncated "Overture" version of *Atmosphères* suggests a
static presence that over time grows more mobile, evoking a
more definite sense of location and movement in both pitch (up
and down) and dynamics (loud and soft). After three minutes,
this movement converges to a focal point of laser-like brightness,
abruptly changing to a stomach-churning rumble on bass viols
(as in Beethoven's storm scene from the "Pastoral" Symphony,
there are no drums in Ligeti's orchestra), then reform as a see-
thing mass resembling a swarm of bees. So, in fact, the music
describes a progression from stasis to movement, and from an
undefinable to a definable mass that nevertheless remains without shape or form. This is music that slows you down, blows
your mind, and promotes a heightened sense of expectation.[20] It
is interesting to note Kubrick's earlier adoption of Beethoven's
Symphony No. 9 (in *A Clockwork Orange*) and "Ride of the Val-
kyries" from Wagner's *Die Walküre* (in *Dr. Strangelove*) as music
emblematic of human arrogance and *disorder*, whereas in *2001*
Ligeti's cluster music is intended to represent superhuman intelligence and transcendental order.

Notes

1. Handel, *Music for the Royal Fireworks* (original version 1749).
The English Concert, dir. Trevor Pinnock (DG Archiv 453 451-2,
1997).
2. *Singin' in the Rain* (1952), dir. Gene Kelly and Stanley Donen
(MGM/UA M202539, 1993), VHS.
3. Richard Wagner, "Entry of the Gods into Valhalla" from *Das
Rheingold*. Cleveland Orchestra, cond. George Szell (Sony SBK 48175,
1992), track 1.
4. Richard Strauss, *Eine Alpensinfonie* Op. 64. Staatskapelle Dresden, cond. Karl Böhm (DG 447 454-2 GOR, 1996).
5. Beethoven, Symphony No. 6 "Pastoral." Berlin Philharmonic
cond. André Cluytens (Seraphim CDL 7243 5 69017 2 8, 1995), track 4.
6. Berlioz, "Scène aux Champs" from *Symphonie Fantastique*. London Classical Players, cond. Roger Norrington (EMI CDC7 49541-2,
1989), track 3.

7. Erik Satie, *Parade: Ballet Réaliste*. Orchestre Symphonique et Lyrique de Nancy, cond. Jérôme Kaltenbach (Naxos 8.554279, 1997), tracks 1-3.

8. Claire R. Reis, *Composers, Conductors, and Critics* (New York: Oxford University Press, 1955), 9.

9. Hermann Helmholtz, *On the Sensations of Tone as a Psychological Basis for the Theory of Music.* 2nd rev. ed., tr. Alexander J. Ellis (1885. Reprint, New York: Dover Publications, 1954), 11-14.

10. Wassily Kandinsky, *Point and Line to Plane,* tr. Howard Dearstyne and Hilla Rebay (1947. Reprint, New York: Dover Publications, 1979), 99.

11. Edgar Varèse, *Hyperprism* (version with siren) rev. Richard Saks. Asko Ensemble, cond. Riccardo Chailly (*Varese: The Complete Works,* London 289 460 208-2, 1998), disc 2, track 4.

12. George Antheil, *Ballet Mécanique,* 1925 version. Rex Lawson, The New Palais Royale Orchestra and Percussion Ensemble, cond. Maurice Peress (MusicMasters 01612 67094-2, 1994), tracks 4-6.

13. Harry Partch, *And on the Seventh Day Petals Fell in Petaluma.* Gate 5 Ensemble, dir. Harry Partch (CRI CD 752, 1997), track 5.

14. Colin McPhee, *Balinese Ceremonial Music.* Peter Hill, Douglas Young (*East-West Encounters,* vinyl, Cameo Classics GO CLP9018(D), 1982), track 1.

15. Cage *Daughters of the Lonesome Isle* for prepared piano. Boris Berman (*John Cage: Music for Prepared Piano, Vol. 2,* Naxos 8.559070, 2000), track 8.

16. Krzysztof Penderecki, *Anaklasis* for strings and percussion. London Symphony Orchestra, cond. Krzysztof Penderecki (EMI CD MS 65077-2, 1994).

17. Giacinto Scelsi, *Konx-Om-Pax,* Cracow Radio Television Orchestra, cond. Jürg Wyttenbach (Accord 200402, 1988), tracks 9-11.

18. Karlheinz Stockhausen, *Trans.* (1) Southwest German Radio Symphony Orchestra, cond. Ernest Bour; (2) Saarbrücken Radio Symphony Orchestra, cond. Hans Zender (Stockhausen-Verlag SV19, 1992).

19. Morton Feldman, *Coptic Light.* Deutsches Symphonie-Orchester Berlin, cond. Michael Morgan (CPO 999 189-2, 1997), track 9.

20. György Ligeti, *Atmosphères.* New York Philharmonic, cond. Leonard Bernstein (*Music of Our Time,* Sony SMK 61845, 1999).

CHAPTER FIFTEEN

Mechanical music

REPRODUCED music—that is, music on compact disc or tape—is the medium of preference for a majority of music listeners in many parts of the world. A sound recording documents a performance that, to all intents and purposes, is both live and real, more so perhaps in the case of classical and world music, which is largely acoustic, than for popular music which is frequently electrically generated and modified. For listeners to recorded music, the medium is one of *recall*: vocal expression, sentiment, action, body rhythms, and so on. A recorded performance is ostensibly a recollection of an acoustic event connecting the real life of the listener with the real experience of the recorded artist.

Since the music industry is unthinkable without hi-fi equipment, it is tempting to classify all recorded music as mechanical, since it is all reproduced by machine. But for a music recording to be perceived as mechanical implies, at the very least, that the listener is aware of the machine as an essential ingredient of the experience, and for most of the record-buying public this is simply not the case. The success of the recorded music industry is based on the illusion that recording is neutral, transparent, and does not interfere with or otherwise influence the transmission of a performance in any way.

Mechanical music is something else. It reproduces not the performance but the information. The underlying principle of mechanical reproduction of music is *elimination of the human interpreter.* Those who listen to music for vicarious emotion will think that eliminating the performer is completely missing the point. The music industry, after all, bases its success on the popular appeal of charismatic performers. But for a minority in the profession, a performer is only a means to an end, the end being the *thought process* encoded in an original score by a composer who has something significant to say. The distinctive contribution of the "authentic" early music movement since the seventies —a key to the unprecedented growth of early music in the classical music industry in the last quarter of the twentieth century—has involved a redirection of marketing and consumer attention away from the idea of the performer as star, in favor of fidelity to the original source and terms of performance. Early music restores the principle that performance is really about continuity of tradition, and that interpreters are the servants and not the creators of meaning.

Mechanical music has been around for a long time. The simplest forms are related to mechanical time-keeping. The clock in the steeple and the grandfather clock in the hall chime a simple music on the quarters and the hour. Touch-tone phone and digital watch alike are mechanical music devices. In each and every case digital information is stored and expressed in musical form. This information is absolute, ideal; you would not dream of "interpreting" it. If you want to get a touch-tone phone to dial up a number without touching the keys, you have to whistle the sequence of tones with complete accuracy. The musical box, the chiming clock, the fairground organ, and the player piano are all machines programmed to reproduce music on demand without the agency of human performers. Today the preferred instrument of mechanical music is the computer with sound card and bundled music software on which the amateur musician may experiment in the invention and realization of musical patterns.

Virtual reality

All mechanical musical instruments consist of a *storage* device and a *reproducer*. In evaluating such devices we need to consider
• What musical information is stored; and
• What musical information is reproduced.
The major categories of information that can be stored are pitch, time, timbre, dynamics, and expression. Pitch is the domain of

melody and harmony, and is stored in a machine as a scale or register of numerical values. In a chiming clock, pitch is stored as a sequence of bells or resonating metal rods, and in a digital watch or synthesizer pitch is stored as number sequences corresponding to voltages. These pitches are fixed. Pitch stability is crucially important if pitch-encoded information is to be preserved. Time information controls the location of pitches in sequence, and the rate at which stored information is scanned. Since time is a subjective experience, it follows that the impression of an unequal rhythm can be produced either from the strict performance of an irregular sequence or a distorted performance of a regular sequence. Because accurate reproduction by machine demands consistency in the speed of transmission, however, a mechanical musical instrument normally embodies the basic distinction between *clock* time or transmission time, which is absolute and uniform, and *musical* time which, being part of the content of experience music is designed to convey, has to be flexible. Absolute time being invariant, all subjective inequalities such as rhythms have to be measured and translated into exact spatial quantities and relationships. So encoding music for machine automatically involves a *quantifying of human experience*, which is a provocative exercise for any philosopher or scientist to be doing back in the seventeenth century when mechanical music began. It means that a listener comparing a mechanical musical performance with its equivalent in real life is in effect assessing how successfully human experience can be measured, a consideration today of interest to designers of robots, most of whom would be surprised, I am sure, to think of musical boxes as early robots.

Timbre or tone color becomes a factor when and if the mechanical sound is intended to resemble something other than itself. This is not the case with a musical box, which is not intended to sound like anything other than a musical box, but it is true of a grandfather clock or a cuckoo clock, which are designed to sound, respectively, like a bell-tower and a bird. A fairground mechanical organ imitates the music of a village band, while a pianola sounds like a piano *because it actually is a piano*.

Original sin

Most mechanical instruments are limited in dynamics, which is to say they don't try to reproduce loud or soft gradations of tone. In that respect, however, they are no different from conventional pipe organs, which, being unable to vary the air pressure, are obliged to find other ways of articulating loud and soft, typically by the performer varying the mixture of stops. Expression in a musical performance usually comes down to deviations of one kind or another, for instance, in intonation, loudness, or timing. The value a listener assigns to "human expression" can thus be expressed as *a value assigned to deviant behavior*, and the philosophical implications of deviancy being a defining issue for human character are really rather deep, even connecting with the scriptural doctrine of original sin ("we exist only inasmuch as we offend"). Expression not only deviates from exact musical values, it deviates differently for each and every individual (unless, of course, you want to make a living as an Elvis impersonator). In mechanical terms, deviation simply means a shift to a higher order of precision in data encoding, all encoding systems involving some compromise. Needless to say, mechanical instruments are designed not to deviate but to give back the same information time after time. The kind of deviancy available to mechanical instruments arises from old age and wear and tear, which is interesting aesthetically but not quite the same thing.[1]

Musical boxes

We described Bach's Brandenburg Concerto No. 2 first movement as music that goes by itself, and the clockwork style of the Baroque concerto has to do with a wider fascination with the watchmaker's art of creating self-powered, spring-loaded, timekeeping machines that store energy and release it in a controlled manner. Musical boxes of traditional design remain in production today as gift items expressing the craft of a bygone age. With a tone range normally limited to around twenty pitches in the high treble, a musical box is capable of playing both melody and harmony. Its sound is uniformly clear and precise, like a digital watch, and the music reproduced is normally a single item in considerably abbreviated form.

The musical information in a musical box is stored as a spatial array of pins projecting from the surface of a small revolving cylinder. As the cylinder is rotated by clockwork, the pins come into contact and displace metal tines of a comb structure that vibrate when plucked to produce delicate sounds of precise pitch. In layout, *and also visually,* the comb resembles human-powered serial pitch devices such as the panpipes, pipe organ, harp, piano, and xylophone. Like a clock or pocket watch, the power source or spring of a musical box is designed to produce a consistent energy supply, even though the cylinder in a musical box is considerably heavier, and therefore the amount of energy required to drive it is comparatively much greater than for the moving parts of a normal timepiece.

Early reproducers of cylinder design were grand and more versatile. The earliest performing astronomical clocks, erected in Salzburg, Austria, date from the late fourteenth century. In 1674 a carillon of forty-nine bells played from a programmable cylinder was installed in Malines cathedral, Belgium.[2] In 1715 the Parochial Church in Berlin inaugurated a massive Glockenspiel playing a selection of melodies on real bells in the church steeple from music stored on a cylinder five feet in diameter.[3] In miniature form, the conventional musical box plays only one piece of music and that item is repeated with every full turn of the cylinder. The data recorded on a cylinder is positional information only; every pin corresponds to the point in pitch and time order at which a note begins, and that is all—no dynamics, no duration, no expression of any kind.

The comb-like keyboard of a musical box often contains only as many notes as are required for the particular tune. You cannot change the cylinder in a normal souvenir musical box, though if you took it out of the box it would be difficult to know which way to put it back in again. For many years every summer an ice cream van used to pass by my home, playing a completely unrecognizable piece of musical box music that, in retrospect, I can only explain as the consequence of the cylinder having been taken out and put back in reverse, causing the music to play backwards and upside down. A musical box is a plucking instrument like the guitar or lute, and, like Dowland's lute its only power of sustain lies in the quality of vibration of each individual note. A feature of musical boxes is the apparent drifting away of the music as the spring finally winds down, a curious effect not of slowing down, but of time itself seeming to stretch away into infinity. A rare example of a modern composition for musical boxes is Stockhausen's *Tierkreis,* a collection of twelve melodies in separate boxes, each representing a different sign of the Zodiac.[4]

Follow the dots

If one were to take the cylinder out of a musical box and unwrap it so that the pin arrangement were laid out flat, the pattern could be seen to correspond to the distribution of printed notes on a musical stave in standard notation. In fact, these instruments could not have been invented until standard musical notation had come into existence. In order to store music on a cylinder in this manner, you have first to work out how pitch and time information can be recorded in two dimensions, and that means, in turn, having a conception of time and pitch as coordinates of a uniform space-time continuum. The notion of pitch and time as absolute and uniform is actually quite advanced, and it is not surprising that the first developments of musical boxes were contemporaneous with significant advances in map making. The bare coordinates of music here are pitch in the vertical and time in the horizontal domain. The musical information to be read extends the entire length of the cylinder, conforming to the pitch range from low to high. After one complete revolution, the music starts over again. The duration of the music depends on the size of the cylinder and the speed at which it revolves.

Although a musical box appears to be about music and nothing else, in reality it is about storing and reproducing process information *encoded as digital data*. The same method of encoding positional information can also be applied to non-musical tasks such as the reproduction of a drawing or of a signature by mechanical dolls. For example, instead of musical notes, the pin information may just as well correspond to successive positions of a marker on a plane surface on the same principle as those follow-the-dot puzzles we remember from childhood, and employing interpolation procedures equivalent to those now used in computer animation. In each case you have a programmed series of frames or positions that are reproduced in sequence to simulate an organic image or process.

That mechanical imagery affected musical expression in a wider sense can be heard, for example, in the slow movement of Mozart's Piano Concerto No. 23, not only in the emotional drama depicted in the relationship of solo piano and orchestra, but as vividly in the fortepiano's acoustic resemblance to a musical box by comparison with the living, breathing sounds of violins, clarinets, and flutes.

Haydn: Theme from "The Clock" Symphony
arranged for mechanical organ

Punched cards

Still a regular feature of fairground shows and circuses today, the fairground organ or calliope is a mechanical outdoor instrument on a larger scale that reads and performs music by means of compressed air blowing through coded perforations on a continuous paper roll or connected sequence of cards. The pneumatic reading and playing mechanism is an adaptation of Renaissance organ technology, and the perforated paper roll is a product of eighteenth-century advances in paper manufacture introduced from China. The information on a paper roll or card is positional data organized in a similar way to the pin cylinder of a musical box but in negative form, as a sequence of perforations. Unlike the musical box, perforations enable the *duration* of a note to be stored as well as its entry point, and the increased storage capacity of paper-based media permit a far greater range of musical sound effects. The fairground organ is intentionally more realistic than a musical box. It is, in fact, designed to sound like a village band in a natural outdoor environment, drawing on a wide repertoire of simulated brass, woodwind and even percussion. A cylinder musical box, by comparison, is a clockwork device for playing a single musical sample of limited duration, and is not designed to simulate or provoke the actions of a live performer. High on the list of significant improvements made to

mechanical instruments during the nineteeneth century, therefore, are 1. greater realism, and 2. improved storage capacity or
memory.

   Considerable development work on mechanical instruments
was already taking place in the late eighteenth and early nineteenth centuries. Both Haydn and his pupil Beethoven took an
informed interest in such musical devices and arranged pieces
from their own symphonic and other complex repertoire for the
mechanical creations, among others, of Friar Primitivus Niemecz,
a priest who also played cello in the Prince Nicolaus Esterházy
orchestra under Haydn's direction. From examples of transcriptions that survive, Haydn appears to have been particularly
interested in finding ways of making mechanically reproduced
music flow more naturally, for example, by inserting small-note
decorations into the melody in imitation of a concert-master or
soloist. The addition of extra notes leading up to a strong beat
may have been designed with the mechanical action in mind to
introduce a slight but controllable and arguably realistic *agogic*
hesitation to an otherwise clockwork regularity of performance.[5]

Jacquard's loom

Impressed at the improved information storage capacity of
mechanical organs in the Napoleonic era, French textile manufacturer Joseph Jacquard devised an industrial weaving process
using patterns encoded on punched cards to control the raising
or lowering of warp threads in sequence to create intricately
figured fabrics. Because these patterns were fixed in advance and
woven automatically, more elaborate patterns were feasible, the
speed of production was increased, and quality control was
naturally greatly enhanced. Putting skilled weavers out of work
may not have been a smart move politically in post-revolutionary
France, but it did indicate yet again that advances in music data
storage did impact significantly on other areas of industrial
manufacture and mechanical pattern reproduction.

Packaged entertainment

Further improvements in paper manufacture led to a range of
smaller instruments, such as the barrel organ and the player
piano, filtering on to street corners and into suburban chapels

and middle-class homes throughout the nineteenth century.[6] The rippling sound of the player piano inspired a new virtuoso aesthetic. Composers such as Chopin, Liszt, Moszkowski, and others developed a new Romantic style of pianism characterized by fabulous speed and lightness of touch in direct imitation of the distinctive fluidity of line and movement associated with the mechanical piano—features coincidentally more aligned to the lighter action of the eighteenth-century fortepiano than the newly fashionable, heavier action nineteenth-century pianoforte.[7] Toward the end of the nineteenth century, American composer Scott Joplin (best known for "The Entertainer," title music to the movie *The Sting*) popularized a deliberately laconic and mechanical piano-rag idiom that made a virtue of the expressionless style of a coin-operated or domestic pianola. (His published sheet music carries the advisory "Do not play this piece fast. It is never right to play ragtime fast—Composer.") Eubie Blake, a veteran of the Scott Joplin era, offers a unique glimpse of ragtime style of the turn of the twentieth century in recordings made in his ninth and tenth decades.[8] In turn of the century Paris the enigmatic French composer Erik Satie was independently pursuing an aesthetic of emotional restraint in piano compositions such as the *Trois Gymnopédies*, not only as a gesture of opposition to the emotional excesses of nineteeth-century Romanticism, but also, one suspects, in solidarity with the neo-classical purity of piano-roll music.[9]

Reproducing piano

German and American research into the dynamics of piano expression, using newly available electrical sensing and measuring techniques, led in 1900 to the manufacture of a refined mechanical *reproducing piano* that claimed to store and reproduce a full range of expressive variables in addition to the notes on the page. The piano, of course, is already a mechanical keyboard instrument designed for evenness of tone and touch and controlled entirely from the tips of the fingers, which means that the expressive dimension of a live performance on piano involves fewer variables than, say, a solo violin. Piano expression could be reduced to a combination of *timing* information, the distance between successive notes, and *velocity* information, the speed with which the individual key is struck. A piano, by definition, is pre-programmed for pitch, so intonation, which in other instruments is a significant ingredient of musical expression, is not part of the process. A piano cannot bend the note like a saxophone or violin, or indeed like a sitar, biwa, or koto. Since the piano is also a percussion instrument, a pianist is also unable to influence the tone after a note has been struck. So two major expressive elements of regular music, the ability to swell a note and the ability to bend a note, are not available to a piano player, thus greatly simplifying the challenge of encoding and reproducing a naturalistic performance using piano roll technology.

The unusual fidelity of the reproducing piano attracted concert pianists and composers alike who recognized the new instrument as a means of capturing and preserving an authentic live performance in full, expressive detail. At a time when the record industry was still in its infancy, the reproducing piano offered not only permanent storage and reproduction of a musical performance in full-frequency, but also freedom from the three-minute or so time constraint of a phonograph recording.

Composers such as Feruccio Busoni, Gustav Mahler, and Claude Debussy hastened to record authentic interpretations of their works, as did concert pianists anxious to make their mark. George Copeland was an American pianist who spent some time in Paris around 1911 and studied with Debussy during that time. His 1915 Duo-Art recording of Debussy's charming piano prelude "Clair de Lune" thus has a certain claim to authenticity, despite a few errors, and demonstrates the improved flexibility of dynamics and tempo available on the new instrument.[10] The exceptional fidelity of the reproducing piano is perhaps best illustrated by the large number of recordings now available of George Gershwin's performances on piano-roll, in particular his *Rhapsody in Blue*, an engaging fantasy for piano and orchestra that has been issued a number of times on compact disc in modern performances featuring the composer himself on piano roll as virtual soloist. Fifty years after Gershwin's death, an Australian orchestra issued a commemorative recording of the 1925 symphonic version of *Rhapsody in Blue* (orchestrated by Ferde Grofé) with "the composer" at the piano. A studio recording from 1927 featuring Gershwin live and Paul Whiteman and his band, now remastered on compact disc, can also be found in the catalogues; however, my preference is for the raunchier and faster original jazz band version of 1924, with the composer an ebullient soloist on piano roll.[11]

Multi-tracking

A number of composers realized that the piano roll was not limited to reproducing music for two hands. For some years Igor Stravinsky was contracted to produce piano-roll editions of his music for the Pleyela pianola company in Paris. He had no scruples against creating "multi-track" editions of his larger-scale ballet scores for piano roll. Such avant-garde music would have had very little Main Street sales potential, but there was a practical reason for creating piano-roll versions of the major ballets for use in rehearsals by ballet companies. Scores of the complexity of *The Firebird* and *The Rite of Spring* would otherwise require not one but two skilled *répétiteurs* (rehearsal pianists), whereas rehearsals using a pianola needed no preparation and made fewer demands on a solo operator. For many years lost and presumed destroyed, the original Pleyela piano rolls of *The Rite of Spring* eventually surfaced and were issued as a recording in 1991 along with a live orchestra performance based on the timing of the original mechanical version. The rolls, which had been cut according to the tempo and rhythmic indications of the original sheet music, revealed significant inconsistencies in tempo compared with the published orchestral score, most notably a greatly accelerated "Sacrificial Dance" that brings the ballet to a climax.[12] Stravinsky's music of his "Paris" period is composed in an idiomatically mechanical style that translates effectively into piano-roll idiom without loss of expressive character, so much so that it may even have inspired his famous remark (in *The Chronicle of My Life*) that "music, by its very nature, is essentially powerless to *express* anything at all."[13]

Another Stravinsky stage work on a Russian theme, *Les Noces* ("The Wedding") began life in 1917 as a glittering conception for grand orchestra and voices, celebrating old Russian peasant marriage customs. The work was subsequently revised in 1919 for chamber ensemble, including one player piano, before attaining its final form in 1923.[14] In subject and lavish scale the original version bears comparison with *The Rite of Spring*, but in the later versions, composed in the aftermath of the Russian Revolution, Stravinsky's orchestration is pared down by stages from the lush opulence of his earlier ballet style to an austere ensemble of four

pianos and percussion. Stravinsky intended these four pianos to
be mechanical instruments performing from piano rolls, emblem-
atically connecting Russian ancient culture with modernism and
the promise of new technology. Unfortunately, the four instruments could not be properly synchronized, which is why the
piano parts have been performed ever since by live players on
conventional pianos. Stravinsky's plans for the pianola in *Les
Noces* attracted the magpie attention of George Antheil and were
quickly incorporated, along with the siren and percussion of
Edgar Varèse's *Hyperprism*, in Antheil's noisily controversial
Ballet Mécanique of 1925.[15]

In the early 1990s, an enterprising Russian conductor named
Dmitri Pokrovsky produced a recording of traditional village
wedding songs from the Ukraine, sung in authentic peasant
style. Despite the composer's repeated denials during his lifetime, it had become clear after his death that Stravinsky had
borrowed extensively from the literature of ethnic Russian folk
music in composing *The Rite of Spring* and *Les Noces*; here, then,
was an opportunity to set the record straight. Lacking the
resources to hire pianists, Pokrovsky had the bright idea of
synthesizing the instrumental accompaniment for *Les Noces* on
computer, arguing with some logic that a mechanical reproduction of the piano parts was exactly what the composer had
originally intended. The result is a version of the work with a
distinctive and not unattractive edge and revealing a flavor not
unlike certain minimalist composers of more recent times.[16]
Certainly the listener is made aware of the level of precision
achievable by the use of mechanical instruments.

Others employed the player piano as a compositional aid.
Australian composer Percy Grainger—a much more interesting
and revolutionary figure than his output of folklorish or fairytale
arrangements might suggest—seized the opportunity while living in London during the 1910s to have piano rolls cut of a
series of experimental studies unplayable by a human performer,
incorporating such effects as sliding clusters. Some forty years
later, American composer Conlon Nancarrow dedicated himself
to the creation of piano-roll compositions that could reproduce
multi-track time layers of a complexity that were not only
unplayable by a human pianist, but also impossible for a composer to write down in conventional notation, such as rhythms in
the ratio of the square root of 2 to 1. Working directly on the

blank piano roll, Nancarrow measured out a music of complex, many-layered rhythms with dividers, a ruler, and colored pencils, often laid over a honky-tonk bass-line acting as a cantus firmus.[17] In the late 1950s Nancarrow's work in space-time notation attracted the attention of a generation of American avant-garde composers including John Cage, Morton Feldman, Earle Brown, and Elliott Carter, each of whom developed distinctively new notations based on the piano-roll principle of space being equivalent to time.[18]

Clocks and clouds

Mechanical music embodies many of the contradictions of Western music and its unnatural insistence on abstract standards of pitch, time, and touch (the keyboard). Human nature, however, has a way of infiltrating and eventually coming to terms with even the most mechanized of musical structures. There is a happy irony in the fact that the virtuoso piano music of Chopin, Lizst, Moszkowski, and others, which we now think of as the highest expression of human skill, should have been inspired by the high-speed action of a mechanical instrument, and that rather than being alienated by a music so distant from human emotion, the public found a way of being reconciled to what is really an eighteenth-century ideal of pure technique. The process of interaction continues. In the years following the release of Nancarrow's *Studies* for player piano pianists and musicians in Europe and elsewhere rose to the challenge of the composer's complex pattern-making, with the result that live performance editions for piano solo and chamber ensemble of many of the studies now exist, transcribed from the original piano-roll recordings.[19]

Notes

1. Eighteenth-century *Orgue de Salon* by Laprévote in the collection of Claude Marchal. *L'Art de la Musique Mécanique, Vol. 2* (Arion ARN 60406, 1997), track 1.

2. T. E. Crowley, *Discovering Mechanical Music* (Aylesbury: Shire Publications, 1975), 6.

3. Walter Bruch, *Vom Glockenspiel zum Tonband: Die Entwicklung von Tonträgern in Berlin* (Berlin: Presse- und Informationsamt des Landes Berlin, 1981), 7-11.

4. Stockhausen, *Tierkreis für 12 Spieluhren* (Zodiac for 12

musical boxes). (Stockhausen-Verlag SV 24, 1992).

5. Arthur W.J.G. Ord-Hume, *Joseph Haydn and the Mechanical Organ* (Cardiff: University College Cardiff Press, 1982), 97.

6. Hymn "Take it to the Lord in Prayer" played on a portable reed organ. In *L'Art de la Musique Mécanique, Vol. 2,* track 4.

7. Moritz Moszkowski, *Etude de Virtuosité* in A flat Op. 72 No. 11. Vladimir Horowitz (Sony S3K53461, 1993).

8. Eubie Blake, "Eubie's Classical Rag." *Wild about Eubie* (vinyl, Sony M34504, 1977).

9. Satie, *Gymnopédies.* Daniel Varsano (Sony SBK 48 283, 1992), tracks 1-3.

10. Debussy, *Suite Bergamasque* No. 3 "Clair de Lune." George Copeland, rec. Duo-Art, 1915 (Nimbus NI 8807, 1996), track 3.

11. George Gershwin, *Rhapsody in Blue,* 1924 jazz version. George Gershwin, rec. Duo-Art 1925, Columbia Jazz Band, cond. Michael Tilson Thomas (*Classic Gershwin,* CBS MK42516, 1987), track 1.

12. Igor Stravinsky, *The Rite of Spring,* original piano roll edition. (1) Boston Philharmonic, cond. Benjamin Zander; (2) Rex Lawson, pianola (IMP MCD 25, 1989).

13. Stravinsky, *Chronicles of My Life* (London: Victor Gollancz, 1936), 91.

14. Stravinsky, *Les Noces,* 1919 version with pianola. Orpheus Chamber Ensemble, cond. Robert Craft (vinyl, CBS 73439,1975).

15. George Antheil, *Bad Boy of Music* (London: Hurst and Blackett, 1945), 148-49.

16. Stravinsky, *Les Noces.* Version with synthesized instruments. Pokrovsky Ensemble, dir. Dmitri Pokrovsky (Elektra Nonesuch 9 79335-2, 1994), tracks 4-7.

17. Conlon Nancarrow, Study 3c for Ampico player piano. Conlon Nancarrow (Wergo WER 6168-2, 1988).

18. Elliott Carter, *Three Occasions for Orchestra.* South West German Radio Symphony Orchestra, cond. Michael Gielen (Arte Nova 74321 27773 2, 1995), tracks 4-6.

19. Nancarrow, Study 3c. Version for chamber ensemble. Ensemble Modern, dir. Ingo Metzmacher (RCA 09026-61180-2, 1993), track 6.

CHAPTER SIXTEEN

Mouth music

FORMANTS are cavity resonances in the voice, or in the tube or box enclosure of a musical instrument. Microphones and loudspeaker cabinets generate their own cavity resonances which are visible in analysis as peaks in the response curve. A violin box generates two resonances, the wood tone (the material) and the air tone (the cavity). Audio equipment manufacturers work hard to eliminate or conceal such resonances in the interests of *flat frequency response*. Cavity resonances are vital for music and speech, however, since only through recognizing and manipulating them are language and music able to function. By changing the inner airspace of the mouth a speaker is able to articulate different vowel sounds and diphthongs: [a] [e] [i] [o] [u] [ai] [oi] [uo] etc. These differences make the mouth a flexible and versatile instrument of communication. Natural differences in material and build mean that different woodwind and brass instruments sound differently, and the various pipes of a pipe organ—cylindrical, tapered, metal, wood—are constructed as they are to produce specific resonances and associated tone qualities.

Vo-de-o-do

To produce mouth music, the cavity resonances of the voice are employed without reference in the first instance to language. Instead, their purpose is to make the formants associated with a particular sequence of vowels sound as a melody. Mouth music is a very ancient musical tradition in parts of Europe and Asia. Recorded examples of mouth music from Tuva, Mongolia, and southern Siberia are in a pentatonic folk idiom that would readily be recognized by native singers of Scotland or Ireland. The sung (throat) pitch acts as a drone accompaniment, while the changes in vowel (mouth cavity) are heard as a whistling melody.[1] Since all human vocal structures are essentially the same, and since the overtone series responsible for differences in vowel resonances is also a constant, it is perhaps not surprising that mouth music from cultures as far apart as the Hebrides and Central Asia sound so alike.

The adoring suitor is in love with the music of his lady's voice. He is unconcerned with her words and will agree to anything as long as he may continue to bask in the sound of her speech, a transcendental experience that has its potential risks. It would be pleasant to say that coherent speech is always musical and that any given sequence of vowels that makes sense as part of a poem or speech also makes harmonic sense as melody, rhythm and harmony. Alas, this is not always the case. Poets understand the music of speech, and poems, in a sense, are attempts at reconciling the opposing demands of verbal meaning and musicality. Religious and magic incantations also contrive in their different ways to dissolve literal meaning in the sound of language.

Nonsense

Using the voice for purposes other than words is supposed to be nonsense, but speech without intelligence is actually *pure sense*. Laughter, the wordless keening of mourners at a funeral, groans of pain, gasping, or grunting of effort—all of these signals express *direct sensation* that is all the more authentic for the lack of any verbal intention or message. Beneath every song there is an instrumental backup or accompaniment; take away the song, consider only the backup, and you are dealing with the underlying pure sensation. For many younger listeners, music communicates primarily at the level of sensation and the verbal message is merely packaging. The same delight in meaningless sound that inspired Ella Fitzgerald and Louis Armstrong to improvise scat singing for the amusement of older generations, is cultivated today by the generation of hip-hop and its derivatives, forms of vocal music that employ words primarily for rhythm and texture and only incidentally for expressing a message.

MAKING THE CHEES MO'
BINDING
Skiddle up skat!
Skiddle up skat!
Oh, skiddle up, skiddle up,
Skat, skat! skat!

DOING THE "SCRONCH"
He, clapping, and singing
Ron kutta tun
Ron ka tung
She, feet firmly on the ground,
bending and swaying...[2]

Oral cultures understand that speech has tone and texture. Folk and popular music traditions like to play on the duality of language as text and sense. The *meaning* of a text is in its relation to time, both in the linear sense of words having to be in a certain order to express meaning, and also in the historical sense that words have their meanings conferred on them by tradition and use in an act affirming the continuity of human experience. The *sense* of a text, on the other hand, has to do with the physical experience of speaking. The medieval practice of reading aloud or ruminating celebrates the texture of language, expressed in the speech act, as a separate, distinct layer of meaning. The sound of speech is the incarnation of language.

Transcending words

What mouth music expresses as pure sensation, Gregorian and other ritual styles of vocal music express as a form of worship. Here a sacred text is the basic material, but the voice treatment of a text in plainchant sometimes suggests a different agenda, for instance, monitoring the tone quality of voice formants in their role as signifiers or embodiers of meaning. A sacred text, known from years of repetition, has embedded meaning. You don't have to think about it any more. It comes automatically. Any speech routine that becomes automatic can be observed from a different angle: as a gesture or pattern of physical effort, for example; or as a purely musical shape, as in the case of a mantric repetition of a prayer by an individual worshiper, or of a political slogan by a crowd of protesters: "Free-dom! Free-dom! Free-dom!"

A switch of attention from text to sound is most likely to occur in those parts of a chant or song where the musical line stays rooted to a particular pitch, and the only change that occurs is in the words being sung. In the monotone chanting of the Shinto priest (or a cattle auctioneer, or a horserace commentator, or a speak-your-weight machine), attention is ultimately drawn away from the melody line, which is a time-dependent indicator of meaning, to focus on the harmonic content or higher resonances that determine particular vowel sounds. Mouth music, like jew's harp music, reverses the process, consisting as it does of independent or meaningless sequences of pure vowels that make sense as audible melody.

Alternatively, a vocal line can be carried on one syllable like the melody on the syllable "Do" of "Dominus" in the "Antiphona ad Offertorium." In such instances, normal continuity of text and meaning is suspended and the listener's focus of attention diverted to alterations in the harmonic content of the vowel as it moves from one note to another. What is of interest here is the relative stability of the vowel resonance in comparison to the voiced pitch. A vowel resonance does not always have to be the same pitch; rather, it corresponds to the harmonic of a sung tone that is nearest in frequency to the cavity resonance of the mouth. In Tibetan Buddhist chant, the combined bass voices of massed monks generate a halo of high-frequency resonances that seem designed to convey the idea of a heavenly spiritual presence in company with the deep, earth-bound monotone of the human worshipers. This is *composed* music for voices singing in a reverberant temple environment. Unlike Western traditional music, Tibetan chant is less about singing a profoundly low tone exactly in tune than about cultivating a strict unison of pronunciation, since it is the vowel sound alone that determines the halo of high resonances. In a very artful way, the singers are able to make the two voice parts, the low sung tones and the higher resonance tones, appear to move in different directions, adding to the impression that the higher frequencies emanate from a separate voice of superhuman origin, resembling a response from the spirit world. A listener hears a change of pitch superimposed on a change of vowel: for example, a drop in tone coinciding with a shift from [a] to [e] is heard as a mysterious *upward* movement in resonance combined with a *descending* voice pitch.[3]

Magic and madness

Mantric utterances—the incessant repetition to oneself of particular words or fragments of text—are associated with religion and also with magic and madness. Each activity involves a sense of the power of words to change one's state of awareness. The mantra is supposed to elevate consciousness to a spiritual plane, while the magician's utterance ("Abracadabra!" "Shazam!") is an assertion of the higher reality of the invisible, parallel world of sound over the world of vision. In certain usages, advertising slogans or interjections for example, words carry a talismanic force that has little to do with conventional language use and meaning ("Yes!" "Eureka!"). We can understand the transformation of words from sense to gesture as a shift of mental focus from the meaning of words (language awareness) to an abstracted or objective perception of the sound and contour of a particular word or syllabic pattern (speech awareness). The practice of ceaseless repetition of a prayer or mantra is designed to induce the same mesmeric shift of perception of language from the mundane world of literal meanings to the transcendental domain of pure tone. The mantra is a feature of many ritual behaviors, and often therapeutic in intention, inducing a trancelike state designed to calm and release the mind from everyday stress. Very much in the spirit of the mantric tradition, Karlheinz Stockhausen in 1968 composed *Stimmung* for vocal sextet, a work exploiting traditional mouth music vowel resonances in a seventy-minute meditation on "magic names" drawn from world religions.[4]

Canticles of Ecstasy

Hildegard von Bingen may be the greatest woman composer who ever lived. Famous in her time as a scientist and scholar, she pursued a long and productive career among the closed religious communities of twelfth-century Rhineland Germany. For nearly eight centuries her vocal music, including a music drama, *Ordo Virtutum*, predating Monteverdi's pioneering operas by 450 years, lay undisturbed and forgotten.[5] The revival of interest in her music in the latter years of the twentieth century is both a tribute to the staying power and contemporary appeal of her musical vision, and unflattering testimony to the enduring masculine prejudice of classical music studies in Europe. As a medieval composer, she ranks with the best. Her status as a feminist icon today is grounded on music of an unprecedented emotional power that even today communicates a fresh, alternative, and distinctively female perception of a world hitherto known primarily through male-mediated experiences and rituals. Her music defends a woman's right to be heard, and the absolute right of woman's experience to be seriously acknowledged.[6]

The word *ecstasy* means a transcendental experience, usually sexual but also spiritual. In medieval Europe, both emotions were equally sacred. To modern ears, ecstasy signifies romantic love but in the era of Eloïse and Abelard, or of the ecstasy of St. Teresa, there was less of a distinction between spiritual and physical passion. What was important, above all, was making contact with a higher plane of existence. Just how this worked for Hildegard is not hard to understand if you compare the style of her music, a song like "O Vis Aeternitatis" for instance, with the male voice plainchant of "Antiphona ad Offertorium." Plainchant is anonymous, sung in public by male voices in a ritual that ultimately is about maintaining civil order. It represents the values of continuity and government in melody that neither rises too high nor sinks too low. The singers are civil servants. Like the judiciary today, they alone understand the arcane scripture on which their authority depends, but they also have a duty to interpret the law impartially, so plainchant is typically impersonal and a little aloof in character. Clarity, poise, precision, and evenness of tone are what you expect of a plainchant performance—not ecstasy.[7]

Now listen to Hildegard. This is the voice of an intelligent woman who has embraced a life of seclusion. As a woman, she has no civic role or public voice. Her music is intensely private and for sharing only within her female religious community. The object of her music is religious devotion, seeking after a transcendental experience. Like plainchant, her songs are based on religious texts, but unlike plainchant these words are not designed to instill good behavior among the local population but rather as a vehicle to push the envelope of human experience.

This did not commend her to the community of male bishops in her day, and a lingering distaste can still be detected among the masculine wing of the church even today:

> Critics remain divided as to the assessment of Hildegard's competence as a poetess and musician. Her colourful imagery and capricious melodies can appear inspired or unpolished according to taste. To some her songs appear repetitive and formulaic; to others they are coherent miniatures of genius. And while Hildegard's lack of formal training in Latin results in inconsistencies and poor construction, the lack of grammatical convention enables a torrent of original imagery to bypass traditional poetic shackles.[8]

"O vis aeternitatis" is a meditation on the incarnation. Incarnation means the Word made flesh, the disembodied idea re-embodied in human sensation. If plainchant is defined as a musical delivery system for public information, the words of Hildegard's canticle are simply a vehicle for exploring private emotion. The solo voice soars to the heights, descends, then soars again in a breathtaking expression of serene loneliness.

What kind of passion does this music express? Melancholy or private grief is not an explanation. If you listen and think about it, a stronger and more positive picture begins to emerge, just as for the Dowland song "Dear, If You Change" and the Mozart piano concerto slow movement. As for the Mozart piece, the extended range of the solo part from low to high tells you that this is not a music of depressed spirits, but music with the confidence to soar to the limits of human experience. Listen to the voice and its remarkable controlled evenness of tone as it rises and descends in a slow, seemingly endless undulation; that evenness and control signify stamina and strength of purpose. The song could only be sung by an individual of exceptional character and discipline, and then you also realize that this individual show of character is also a statement of personal commitment to an absolute ideal. That *is* impressive.

Mind games

Lewis Carroll and Edward Lear elevated literary nonsense to an art form. Along with composers of Romantic opera the poets of

"Jabberwocky" and "The Dong with the Luminous Nose" pursue a nineteenth-century preoccupation with invisible forces of nature, not only electricity and disease but also the unseen motivations of the human mind as revealed in patterns of speech and song. Peeling away the surface of language to reveal the thought patterns underneath is a form of inquiry into human psychology, and, unsurprisingly, nonsense as entertainment eventually segues into nonsense as science and psychic automatism in literature and art.

One of the first novelists to address the way the mind actually works was the eighteeth-century English novelist Laurence Sterne, whose "comic epic novel" *Tristram Shandy* anticipates the interior monologues of James Joyce and Samuel Beckett:

> Ptr..r..r..ing—twing—twang—prut—trut—'tis a cursed bad fiddle. —Do you know whether my fiddle's in tune or no?—trut ..prut..They should be fifths.—'Tis wickedly strung —tr...a.e. i.o.u. -twang. —The bridge is a mile too high, and the sound post absolutely down,—else—trut.. prut—hark! 'tis not so bad a tone.—Diddle diddle, diddle diddle, diddle diddle, dum. There is nothing in playing before good judges,—but there's a man there—no—not him with the bundle on his arm—the grave man in black.—S'death! not the gentleman with the sword on.—Sir, I had rather play a Capriccio to Calliope herself, than draw my bow across my fiddle before that very man.[9]

Studies of infant learning, of hearing, and of memory and the mind deliberately employ sequences of nonsense syllables or words to investigate how the brain learns and how it works. As long ago as 1755 Sterne's contemporary, playwright Samuel Foote, composed a piece of nonsense to test the memory of an actor acquaintance called Macklin who boasted he could remember anything after hearing it once:

> So she went into the garden to cut a cabbage leaf to make an apple pie, and along came a great she-bear, and popped its head into the shop. What? No soap? So he died, and she very imprudently married the barber; and there were present at the wedding the Picninnies, the Joblillies, the Garyulies, and the Great Panjandrum himself, with the little round button at top, and they all fell to playing

catch-as-catch-can, till the gun-powder ran out of the heels of their boots.[10]

For patients on the psychiatrist's couch, free association would become the key to unlock the doors of the subconscious. New technologies of acoustic recording and the telephone, by their very existence, provoked the kind of new research into the finer nuances of language celebrated, even parodied, by the character of Professor Higgins in George Bernard Shaw's *Pygmalion,* a play more familiar to many as the classic musical *My Fair Lady.*

The same bandwidth limitations of early acoustic technology that degraded consonants and emphasized vowels, also conspired to focus attention on the emotional subtext and ambiguities of spoken language, factors picked up, among others, by literary figures such as Gertrude Stein and James Joyce. The Surrealists, Dadaists, and Futurists all seized on the sounds of language as material for a new, non-representational vocal music tapping into the artist's subconscious. Thirty years before radio news dispatches from the front line, Filippo Marinetti's futurist novel *Zang Tumb Tuum* orchestrated text and typography to create a montage of voices and sound effects expressing the excitement and terror of battle in the Balkan trenches:

Down down at the bottom of the orchestra stirring up pools oxen buffaloes goads wagons pluff plaff rearing of horses flic flac tzing tzing shaak hilarious neighing iiiiiii stamping clanking 3 Bulgarian battalions on the march croooc-craaac (lento) *Shumi Maritza or Karvavena* TZANG—TUMB— TUUMB *toctoctoc* (rapidissimo) *crooc-craac (lento) officers' yells resounding like sheets of brass bang here crack there* BOOM *ching chak* (presto) *chacha-cha-cha-chak up down back forth all around above look out for your head chak good shot! Flames flames flames flames flames collapse of the forts behind the smoke Shukri Pasha talks to 27 forts over the telephone in Turkish in German Hallo! Ibrahim!! Rudolf! Hallo! Hallo, actors playlists echos prompters scenarios of smoke forests applause smell of hay mud dung my feet are frozen numb smell of saltpeter smell of putrefaction Timpani flutes clarinets everywhere low high birds chirping beatitudes shade cheep-cheep-cheep breezes verdure herds dong-dang-dong-ding-baaaa the lunatics are assaulting the musicians of the orchestra the latter soundly thrashed play on Great uproar don't cancel the concert . . .*[11]

In Paris that same year (1911), American author Gertrude Stein published *Tender Buttons*, a collection of meditative prose pieces intended to lull the reader (reading aloud, of course) into a trance through the mantric music of language. Stein studied with the philosopher William James, brother of novelist Henry James, was fascinated by the meaning of language as sound, and caught up in the new age of psychoanalysis and stream-of-consciousness literature. Stein defended her style of repetition by analogy with the movies. If you take a length of movie film, it looks like the same image repeated over and over again, but when you run it through the projector, the images reveal a dynamic process. So in "Portraits and Repetition" she declares, "any two moments of thinking it over is not repetition . . . As I said it was like a cinema picture made up of succession and each moment having its own emphasis that is its own difference and so there was the moving and the existence of each moment as it was in me."[12] She is equally persuasive as a reader, for example, her reading of "If I Told Him: A Completed Portrait of Picasso," as preserved on Caedmon.[13] In *Tender Buttons*, the underlying imagery is curiously domestic, evoking the comforts of food preparation, tidiness, and routine—a sense of the hidden beauty of ordinary things that one also associates with the found objects of Marcel Duchamp, whose female alter ego Rrhose Sélavy might well have been intended as an echo in turn of Stein's celebrated aphorism "a rose is a rose is a rose":

> Why is a pale white not paler than blue, why is a connection made by a stove, why is the example which is mentioned not shown to be the same, why is there no adjustment between the place and the separate attention. Why is there a choice in gamboling. Why is there no necessary dull stable, why is there a single piece of any color, why is there that sensible silence. . .
>
> South, south which is a wind is not rain, does silence choke speech or does it not. Lying in a conundrum, lying so makes the springs restless, lying so is a reduction, not lying so is arrangeable.
>
> Releasing the oldest auction that is the pleasing some still renewing.
>
> Giving it away, not giving it away, is there any difference.
>
> Giving it away. Not giving it away.[14]

Today's advertising copywriter owes a great deal to the pioneering wordplay of experimental writers of the earlier twentieth
century. Contrary to educated opinion, the function of language
is not primarily to make intelligible statements but to create
arresting abstract shapes, as a Japanese calligrapher creates
abstract shapes with letter forms. Our experience of language
begins with nonsense in the baby-talk with which parents
awaken the infant's sense of the textures and tonalities of speech;
it continues evolving through nonsense rhymes and games at
school, then into the slogans and chants of team competition and
group initiation rites. Scat singing in jazz is vocal nonsense
imitating real instruments; magical incantation and speaking in
tongues make contact with alternative realities. We do not pay
enough attention to nonsense as we do not pay enough attention
to noise.

Fümmsböwötääzää Uu, pögiff!

During the inter-war years, Austrian artist Kurt Schwitters experimented in typographic layout and speech sounds alike to
create an abstract and witty vocal music of which perhaps the
most famous example is his *Ur-Sonate* or Primeval Sonata.[15]
Drawn especially to consonants rather than vowels for their
noise and texture appeal, the generation of Schwitters, Raoul
Haussmann, and Franz Mon used abstract vocal music as a form
of protest against the debasement of language represented by the
megaphone politics of fascism. Radio and the movies helped to
reawaken twentieth-century ears to the music of the spoken
word. The 1930s travelogue also introduced Western movie audiences to exotic countries, their music and language, and thereby
to the understanding new to many people that for language to be
meaningful it does not always have to be understandable. The
1939-45 war effort drew attention to the role of intelligence
operations involving techniques of text and data encryption, and
at the end of the war many young artists adopted encoding
protocols in their art as a sign of the new age of cybernetics. The
European music avant-garde introduced a compositional technique known as *serialism* inspired by wartime code-making.
French writer Raymond Queneau and American William Burroughs experimented systematically with the deconstruction of a
text.[16,17] A generation earlier, dadaist poet Tristan Tzara declared

that one could make poetry by taking words at random out of a hat, a remark for which he was excommunicated from the surrealists by their leader André Breton. Ironically, Tzara's declaration was revived after 1945 by information scientist Johann von Neumann, and applied to research aimed at elucidating the roles of tone and texture as global features of language that remain potentially detectable even when a text is rendered down into a stream of disconnected words or syllables. The joint influence of these two unlikely bedfellows of cybernetics and dada is still very much alive in the subliminal communications world of advertising and the sound-bite.

Last word

Though he had little time for European serial methods, American composer John Cage was seriously preoccupied with the related idea of freeing language from its logical straitjacket. He evolved various methods of composing a text that in their own way are as carefully structured and objective as the music of any serialist. Cage invented a literary form called *mesostics*, words or phrases stacked in vertical array randomized from an original text and conforming to a secret code of which a *reader* is aware but an audience is not, since the code is a silent word or phrase that runs vertically through the column of text like a thread.

In 1988-89 Cage was honored by Harvard University with the Charles Eliot Norton professorship, an award entailing six lectures. Cage delivered his lectures in the form of six mesostics, taking as source material his essay "Composition in Retrospect" that is ordinarily rather heavy going.[18] The book and accompanying recording of Cage reading one of these lectures in a deliberately neutral tone suggest a style of ritual incantation in a spirit of priestly detachment. One wonders how these disconnected fragments were received by his largely academic audience. I can think of three possible reactions to an outwardly meaningless text. First, that Cage is making it up. Well, we can see and hear from the tone and repetitions of the text that that is not the case. Second, that the lectures should be regarded as music or magic incantation rather than meaningful speech and are therefore beyond the reach of reason and logic. That's possible, I suppose, but what, then, of the academics in the audience who make a career of defending the art and literature of surrealism, the mantric prose of Gertrude Stein, the cut-ups of William Burroughs, or the cyclic soliliquies of Samuel Beckett?[19] Would they also be defending a twentieth-century culture that is essentially meaningless? I don't think so.

Then there is the third option, which is the one I like. It says that Cage has a point, and his point is to deliver a message about human understanding. He is saying there is a difference between language, the delivery system which is necessarily continuous and structured, and the world of ideas, that only makes sense according to the way *the mind* actively deals with information. And the mind does not work logically. When you are thinking about something either actively or meditatively, what you are doing is letting the mind make its own connections from images and fragments retrieved from memory, and that process is governed not by grammar or sense or logic but by free association and inherent attraction. It is also true that when ideas are presented randomly and not in conventional formulae it is much more difficult for a listener to ignore them. By avoiding the structures of normal academic discourse, Cage is thus able to negotiate directly and powerfully with the minds of his audience.

Notes

1. Alexei Saryglar, "Fantasy on the Igil," throat music and *igil* (folk fiddle). *Tuva, Among the Spirits*, track 4. Also Jean Jenkins, *Vocal Music from Mongolia* (vinyl, Tangent TGS 126, 1977), side B, track 1.

2. "Jazz" picture captions from *Vanity Fair* c1926. In Lee Lorenz, *The Art of The New Yorker, 1925-1995* (New York: Alfred A. Knopf, 1995), 11.

3. "Padmasambhava Tsechu Sadhana: Invocation." Monks of the Tashi Jong community, Khampagar Monastery, rec. David Lewiston (*Tibetan Buddhism: The Ritual Orchestra and Chants*, Nonesuch 9 72071-2, 1995), track 1.

4. Stockhausen, *Stimmung* for six vocal soloists. Singcircle, dir. Gregory Rose (Hyperion CDA 66115, 1987).

5. Hildegard von Bingen, "Procession" from *Ordo Virtutum*. Oxford Camerata, dir. Jeremy Summerly. *Heavenly Revelations* (Naxos 8.550998, 1995), track 8.

6. Hildegard von Bingen, "O vis aeternitatis." Sequentia. *Canticles of Ecstasy* (DHM 05472 77320 2, 1994), track 1.

7. Offertorium "Domine Deus, in Simplicitate," Gregorian chant from the Proprium Missae in Epiphania Domini. Choralschola der Benediktinerabtei Münsterschwarzach, dir. Godehard Joppich. *Music*

of the Middle Ages (DG Klassikon 439424 2), track 6.

8. Nick Flower, liner notes for *Heavenly Revelations.*

9. Laurence Sterne, Book V, Chapter 15, *The Life and Opinions of Tristram Shandy, Gentleman* (9v. in 1v., London: Oxford University Press, 1951), 340.

10. Samuel Foote, 1755. Recalled from childhood memory by the author.

11. Filippo Marinetti, *Zang Tumb Tuum* (excerpt). Cited by Luigi Russolo in a letter to Balilla Pratella 11 March 1913. "The Art of Noises," tr. Stephen Somervell, in Nicolas Slonimsky, *Music Since 1900*, 3rd rev. ed. (New York: Coleman-Ross, 1949), 645. See also Caroline Tisdall and Angelo Bozzola, *Futurism* (London: Thames and Hudson, 1977), 93-9.

12. Gertrude Stein, "Portraits and Repetition." In *Gertrude Stein: Writings and Lectures* ed. Patricia Meyerowitz (London: Peter Owen, 1967), 105-07.

13. Gertrude Stein reading "If I Told Him: A Completed Portrait of Picasso." *The Caedmon Poetry Collection: A Century of Poets Reading Their Work* (New York: HarperCollins CD 2895(3), 2000), disc 2, track 12.

14. Stein, "Rooms" from *Tender Buttons*. In *Gertrude Stein: Writings and Lectures*, ed. Patricia Meyerowitz, 192-93.

15. Kurt Schwitters reciting the Scherzo to his *Ur-Sonate*, rec. 1932. In *Kurt Schwitters* CD-ROM (Hanover: Schlütersche 3-87706-771-9, 1996).

16. Raymond Queneau, *Exercices de Style* (Paris: Gallimard, 1947).

17. William Burroughs, "The Cut-Up Method of Brion Gysin" and "Fold-Ins." In *A William Burroughs Reader*, ed. John Calder (London: Pan Books, 1982), 268-82.

18. John Cage reading "Mesostic IV." From *I-VI: The 1988-89 Charles Eliot Norton Lectures* (book and audiocassettes; Cambridge, Mass: Harvard University Press, 1990).

19. Samuel Beckett, *Lessness*. In *Samuel Beckett: I Can't Go On, I'll Go On: A Selection from Samuel Beckett's work*, ed. Richard W. Seaver (New York: Grove Press, 1976), 555.

CHAPTER SEVENTEEN

Relativity

MUSIC deals in some pretty weird and pretty deep stuff. It's all the more impressive when you think that it does so without textbooks or mathematics. When people listen to music, there are thought processes going on and there is a message coming through. Sounds have meanings that the brain can recognize, and those meanings can be quizzed and evaluated in musical terms without reference to books. Some of the most interesting classical music is also the most abstract. The ideas in this music can take time to work through, which is why many string quartets and symphonies demand a lot of attention. This can be tiring if you don't know quite what you are supposed to be following. Just listening to a twelve-minute opening movement is not an easy task for a culture with a normal attention span for music of three-and-a-half minutes. The good news is that a piece of music tells you what it's about, and presents its argument in real time. Classical music is designed for listeners like you to understand. There are no instruction books; no special handshakes; no passwords are necessary. All the information is available for you to hear. It is just how you interpret it. That's not knowledge, just tactics. Tactics is easy.

Making a point subliminally, as in advertising, is still making a point. If an ad campaign is successful, then a message is getting across. If the music of certain classical composers is consistently more popular, then their message is getting across more successfully than the competition. What is this message and why is it successful? Some psychologists and scholars pretend that if there is a message in music it is purely emotional, not intelligent, but the response to a successful advertising campaign is emotional too, and we all know that advertising campaigns are intelligently planned. The fact that a response to music is expressed in emotional terms is no indication that the music itself is not organized in a rational way. Emotions don't have to be a lower form of human communication. If there are no words to convey an experience then emotions have to do. The feel-good factor and increased sales tell their story. Something is working here. And while you don't have to understand how it works in order to enjoy music, it can help if you do. If your child is going to benefit from "the Mozart effect," which claims that listening to classical music improves a child's intellectual performance, then you may want to know something about it. Or, at least, where to begin.

How fast is fast?

Abstract music deals in propositions of an objective kind. The clockwork style mechanism of J.S. Bach's Brandenburg Concerto No. 2 is an example of objectivity because this is not music for singing or dancing. It's the same for his piano Prelude No. 1 in C from *The Well-Tempered Clavier*: cool, objective, mechanical pattern-making in sound that does not generate any emotional response or rely on imitation of any regular human action. Such music arouses a listener's curiosity by its intricacy and consistency, in the manner that an example of intricate tiling can be satisfying and fascinating without having to resemble anything particular in human nature or nature itself. Music that is abstract has the potential to outlast music that imitates human behavior, for the very reason that it cannot be pinned down to recognizable images. The Muslim tile designers were forbidden to represent nature, perhaps for the related reason that abstract patterns are ideal, indisputable, and not subject to interpretation.

Bach's generation of classical composers is very preoccupied with time and motion studies. The typical instrumental form of Baroque music is the *partita* or suite, a sequence of short pieces based on different folk and courtly dance forms. These are not meant for dancing, but for reflecting on the relationship between the musical patterns associated with particular dances and the perceptions of time and motion associated with them. It is an entertainment, for sure, but it is also scientific in much the same way that a computer-animated movie is both entertaining and a demonstration of how human action can be precisely modeled and reproduced. So a suite is using the imagery of dance, that is, of actions on a human scale, to ask deeper questions about the nature and experience of time itself.

Fast and slow seem rather straightforward concepts to grasp. Fast is if you are running for a bus; slow is if you are dawdling along the street enjoying window-shopping. Fast action is manifested in a higher rate of heartbeat and more rapid footsteps, not to mention getting where you want to go in a shorter time. Fast is a trade-off of increased level of activity and energy expenditure for the sake of getting something done in a shorter time. By this token a high-leaping, fast-moving gavotte is one experience of time and a slow-moving pavan another experience. By modeling these and other forms of dance in standard musical notation, the composer is showing that the pattern, energy level, and perception can be quantified. You no longer have to do the dance in order to understand the difference. In fact, by making these dances available for objective review, the composer is freeing the discussion from being tied to individual human experience.

Foreplay

Of all the movements of a Baroque suite the one least like a conventional dance is the prelude. The word means "foreplay" and, believe it or not, a prelude is indeed a piece of music designed for the performer to get the feel of an instrument and a prevailing acoustic before launching into a sequence of country dances. The prelude marks a transition from tuning the instrument to performing. The Baroque keyboard instrument does not hold its tuning indefinitely, and, like a violin or guitar, ideally has to be tuned up for every new performance. After tuning up —a process intended to harmonize the internal structures of the instrument—there is normally a further "warming-up" process to exercise the fingers and also to bring the newly-tuned instrument into a state of readiness. The prelude is not a dance but rather a piece in improvisatory style that allows the performer to assess the acoustical balance of instrument and auditorium. By listening to the room response to single notes, chords, and passages of rapid fingering the performer can eventually decide on a neutral reference tempo that works for the particular combination of instrument and environment, against which the pace of faster and slower dance movements can be decided. The distinguishing feature of a prelude is that its tempo is decided in exactly the same way as medieval plainchant in a cathedral environment, by the length of time it takes for the sound to die away, not by a

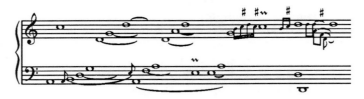

D'Anglebert: notation of an unmeasured prelude

mechanical tempo imposed on the performance, as it were, from outside. The prelude's relative freedom of timing is expressed graphically in the "unmeasured preludes" of seventeenth-century French harpsichord composers Louis Couperin and Jean-Henri D'Anglebert.[1] "Most unmeasured preludes are toccatas," declares one encyclopedia; that they are *tactile* is certainly true, but these pieces are actually less about showing off a performer's finger technique than about feeling one's way into the acoustic of a concert chamber. Many harpsichordists are uncomfortable with a notation such as this that does not fit into a regular visual pattern, but the spirit is familiar enough in the improvisatory style of flamenco guitar (where, of course, everything is done by ear and nothing is written down).

Equal and unequal steps

Movement in music is either one thing or the other. One is a music of even pulses, regular and uniform; the other is music of uneven pulses, long alternating with short. Even or uneven, that is the question. That is all. The difference between them is the same for music as for walking. If the musical flow is in even, regular pulses, it is like a person walking normally; when you are walking normally you are conserving energy and not focusing on the physical act of walking. Regular walking allows you to concentrate on where you are going, rather than what you are doing:

on the objective rather than the process. In music, a smooth-flowing accompaniment has the effect of directing attention to the line of a melody and where that melody is headed. If an underlying pulse is unequal, on the other hand, the effect resembles a person *limping* rather than walking normally. A person limps because of being in pain or being disabled. Either way the limp signifies that the act of walking requires *greater than normal effort*. People don't notice other people who walk normally when they pass them by in the street, but their attention is caught at once by a person who is limping. It's the same in music. If a piece of music has an uneven rhythm, it draws the listener's attention immediately. Uneven or limping rhythms generate a sense of greater effort, power, and energy.

We came across uneven or *dotted* rhythms in the opening "Deus in Adjutorium" of the Monteverdi *Vespers*, and again in "The Blue Danube" waltz of Johann Strauss II. Dotted rhythms are natural energizers for dance, drawing attention to the dance steps and to the effort of leaping, and adding to the dance's feeling of momentum. In Beethoven's Piano Concerto No. 4 second movement—which unusually for a slow movement is labeled *con moto* (implying "with a sense of motion")—the piano part is characterized by a calm even pace and the orchestra part by a fierce and impatient dotted rhythm.

Classical swing

As well as inventing a "no-time" notation for quasi-improvised keyboard preludes, the French devised a mechanism for measuring tempo. Being avid classical scholars, they borrowed from the Greeks the idea of a weighted string or pendulum, essentially the same as a length of fishing line with a sinker on the end. The longer the string, the slower it swings back and forth. Once you find the length of string that gives a swing rate of one per second (the French could do this, having chronometers to measure time), and then for one swing every two seconds, then for values in between corresponding to different speed sensations or dance tempi, you eventually arrive at a scale of string lengths that provides an objective measure of subjective time that is valid for the particular weight. The Greeks experimented with a form of executive toy, consisting of sets of pendulums hanging from the same bracket, and found that when a pendulum of a certain weight and periodicity (swing time) is set in motion, other pendulums of the same or related periodicities swing into motion as well. Those whose strings are related in length in simple ratios, such as 1 : 2, 2 : 3, 3 : 4, etc., are the ones that are able to move in synch with the drive pendulum and are therefore in harmony with its frequency of motion. Pendulums with string lengths or weight ratios that don't harmonize with the drive pendulum are able to draw very little energy from it and remain relatively motionless.

French overture

For orchestra suites, which involve large numbers of players, you can't start off with an unmeasured prelude. That would be silly. The players have to be coordinated. So instead composers came up with the French overture—music in the grand manner to express the idea of a symphony in an image of universal time, modeled on the movement of a pendulum. And this is precisely the point. Although it is the governing mechanism of a clock and thus emblematic of the inexorable majesty and constancy of time, a pendulum *does not move at a constant speed.* A typical example of an overture in the French style is the first movement of Handel's *Music for the Royal Fireworks.* A listener might easily imagine the music's slow and deliberate pace, with a slight pause between each step, as being the proper style for a royal procession and having no other hidden meaning. If we decide, however, that music in this style—which is certainly not a dance—is modeled on the movement of a great pendulum (not as big as Foucault's pendulum perhaps, but certainly larger than life), then it implies a powerfully symbolic connection between royalty and absolute time. Louis XIV of France, "the Sun King," is surely flattered to be reminded in musical fashion that he, the sun, has the power to control the motions of his subjects, the planets.

A pendulum moves back and forth to measure time objectively and with scientific accuracy. However, the motion of a pendulum is not continuous but *constantly varying* in a cycle of an acceleration, a retard, and a point of suspense before reversing direction. It follows that music modeled on pendulum motion, to represent the idea of divine or universal time, will not move in a regular uniform flow either, but in a series of measured steps, with a hint of a pause between each. And this is the

French style. A special kind of unequal or dotted rhythm was introduced to the French overture to express the peculiar swinging motion of the pendulum. This *notes inégales* or unequal notes style has been fiercely argued over by musical scholars, some saying you have to play the notes exactly as written, others in favor of a more organic, less literal interpretation. Since a present-day orchestra conductor waving a stick is also basing his gestures on the swing of a pendulum, and every conductor waves his stick in a slightly different way to achieve the same result, intuition is probably a more reliable guide, especially given the fact that the true motion gradient of a pendulum is impossible to realize exactly in the regular grid formation of standard notation.

Pogo sticks

The same point is made effectively if not very elegantly by a television ad showing a new model land cruiser easing its way past a group of well-dressed commuters traveling on pogo sticks up an unsurfaced country road. Wheels and tires give you a smooth carefree ride, it says; travel by pogo stick is rough and unsteady. Now wait a minute. If the piston-like motion of a pogo stick is the opposite of smooth wheel motion, *why is it* that a white dot painted on the tire of a car or a reflector attached to the rim of a bicycle wheel moves exactly like a pogo stick? Isn't it a paradox worthy of Zeno that a wheel rim as a whole moves in continuous motion, but every point on its rim is jumping like a pogo stick?[2] If this is a *motion* question, then the wheel is defined as a pogo stick raised to the power of infinity. If it is an *energy* question then the smooth drive arises from the efficient transfer of pogo stick energy (pistons) inside the engine to the wheels outside. Music for car ads is always smooth and gliding, even over the roughest terrain—even paced, effortless motion, focusing on getting there, melodic.

Of course in the time of Bach, Mozart, and Beethoven there weren't any cars, or even any bicycles, and the roads were rough. For these composers the imagery of slow and fast motion was either human, as in walking or dancing, or mechanical, as for a clock, or travel on horseback. A human being has two legs, but a horse has four. If your model of speed is human, then running is essentially the same action as walking, but on your toes and speeded up; but if you take a horse as your model, then an increase of speed is associated with a change in the pattern of motion, first from an even trot to a canter, which is an unequal rhythm, and finally a gallop, which is a rolling rhythm. For classical listeners, therefore, the difference between a regular and even pace and an unequal dotted rhythm is not only a difference in effort, but of effort *in association with* an impression of greater speed.

Eine kleine hop-la

Mozart's *Eine kleine Nachtmusik* is a familar popular classic. Its first movement starts off with a strong dotted rhythm. Does it sound energetic and fast? Well, yes it does. Mozart even calls it an *Allegro*, which signifies a lively pace. Would you use this music to advertise a luxury convertible? I don't think so. The power and energy of the unequal rhythm would go better as background music for father cranking the starting handle of the family Model T in an old silent movie. For Mozart's audience, however, the jerky rhythm is a perfect image of a sudden burst of energy. Pay attention to the strong beat of this fast, energetic movement.[3]

Thanks to the miracle of digital recording, a listener can switch back and forth between the first and second movements of the *Eine kleine Nachtmusik*. Listen briefly to the start of the second movement, "Romanza." As the name implies, it sounds rather languid and slow, much more like the music you would expect to hear advertising a luxury convertible. Flick back and forth between the first movement and the second, and pay attention to the beat in each case. The beat is *exactly the same* for the slow as for the fast movement.

What is going on? Why does one music come across as fast and the other as slow when both are exactly the same speed? *I'm* not asking the question. *Mozart* is asking the question. And the question is: If fast and slow are unrelated to speed, then what are they related to? And it turns out that the movement that seems to go at a faster tempo is also more *irregular* and more *unpredictable* than the so-called slow movement. Compare the two melody lines. In the first movement, the line is jagged and instrumental in character, like a bugle call, whereas the more voice-like melody in the "Romanza" moves in a gentle stepwise progression. In the first movement there are sudden changes of direction and style: zig-zagging up, then down, then dancing on the spot, then skidding away, full of uncontrolled nervous energy, like Charlie Chaplin skating on ice. The slow movement, by contrast, is elegant and poised, and knows how to move and how to wait, like a

Allegro and Andante movements in Mozart's
Eine kleine Nachtmusik.

fashion model showing off an elegant but very heavy costume. Finally, in the first movement there are sudden changes of mood and dynamics—now emphatic, now quiet, now speeding up, now slowing down. In complete contrast, the second movement is a model of even-toned serenity. What it tells you is that a listener's perception of fast and slow has as much to do with degrees of change and consistency of information in music as it has to do with the actual tempo.

Do they really mean that?

Good question. If a piece of classical music provokes intelligent thought, is it by accident or is it deliberate? Should we give the composer the benefit of the doubt? I think we should. And if we do, then composing music is not just instinctive but intelligent behavior, and it is also saying that music is capable of conveying deep ideas. The best evidence of intelligent music is classical first-movement form, also known as sonata form. In a standard piano sonata, string quartet, symphony for full orchestra, or concerto, there are normally three or four movements of which the first movement is a reasoned debate; the second is a slow movement study; the optional third study in cyclical motion in

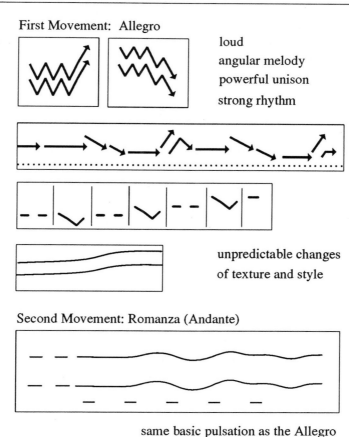

First Movement: Allegro

loud
angular melody
powerful unison
strong rhythm

unpredictable changes
of texture and style

Second Movement: Romanza (Andante)

same basic pulsation as the Allegro
smooth melody, stepwise undulations
regular pulse
singing style
flowing rhythm
consistent texture
predictable motion

Graphic analysis of tempi in relation to
contour in Mozart's *Eine kleine Nachtmusik*

triple time; and the last an exercise in fast motion. The model for first-movement form is a combination of logical proposition and chemistry experiment. The logical proposition model involves reconciling two opening statements that appear to have little in common; the chemistry experiment model (reflecting the establishment of agreed procedures for validating scientific experiments in the seventeenth century) has to do with defining a set of initial conditions, then altering the balance by introducing a new ingredient, then observing the resulting interaction, and finally drawing a conclusion based on a comparison of the set of conditions as they first appeared and those you end up with. Renaissance and Baroque composers tend to follow the logical model, the solution to which is inherent in the formation of the statements themselves (known as *subject* and *counter-subject*). After the mid-eighteenth century, however, classical first-movement form increasingly follows the chemistry experiment model. More dynamic and inherently less predictable in outcome, it relies much more on the *nature* of the thematic materials and their potential to interact.

Beethoven's fourth

Mälzel's mechanical metronome finally released music from its age-old dependence on human measures or perceptions of time such as the dance. Haydn was interested in reconciling mechanical time (the musical box, the mechanical organ) and naturalistic human gesture by introducing small-note ornamentation into a melody in the manner of the lead violinist or concert-master of a classical orchestra. The younger Beethoven, however, capitalizes on the essential *lack* of relationship between mechanical time and human time to create a music that operates on a basis of *psychological* rather than physical time. Psychological time is inertial time, meaning it is not measured by human external processes such as walking or running, but by mental activity such as problem-solving. In the *Eine kleine Nachtmusik* Mozart shows that the same pulse rate can give the impression of either fast or slow physical activity. In the Adagio introduction of his Symphony No. 4 Beethoven manipulates an audience's sense of time by systematically varying the density of information the brain is being asked to process.[4] Here there is nothing to sing, and no rhythm to dance to: this is music without any of the normal

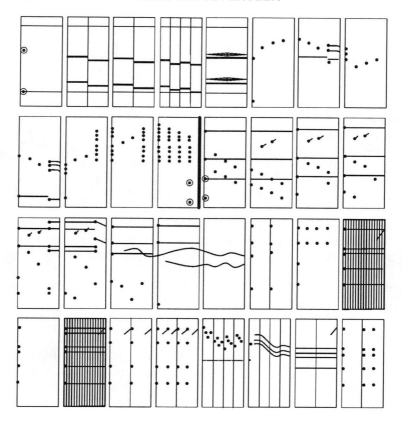

Graphic analysis of Beethoven: Symphony No. 4, introduction

sensory cues. It does not even imitate the motion of a pendulum, although everything that does happen conforms to a regular, chronometric beat that the audience subliminally recognizes.

What is psychological time? It is how the brain perceives an experience. While you are asleep, you have no sense of time. That's one extreme. When you are dealing with a crisis, you also have no sense of time, but your brain is working overtime. That's the other extreme. In between you have more or less to do. If less, maybe you're bored—that's another sense of time. If you are solving a problem or completing a task, maybe you're excited— that's another sense of time. All of these time senses are accommodated in the same framework of clock time or mechanical time. What Beethoven is doing in this example of abstract, first-movement form is testing the proposition of psychological time in a systematic way. Because it's systematic, we can be sure it's deliberate.

Critics following the score or listening to a recording (even my recommended recording, which is pretty good) will say "Aha! he's got it wrong. There is a definite *change of tempo* from the slow introduction to the faster Allegro." That's true, but it doesn't alter the fact that the music is *designed* around the proposition of a regular module of time, or modules if you consider the beat as the smaller unit and the measure as the larger. What happens in the reality of a musical performance is that acoustic considerations come into account to *modify* the absolute time values for the sake of audience perception, just as the columns of a Greek temple are *tapered* to give the illusion of *straightness*. Of course, a performance has to take live acoustics into account. The point is that any modification of timing should be undertaken in the spirit of maintaining the impression of an underlying, unvarying pulse.

Beethoven's major unit of time is the length of a bar or measure. With very few exceptions, all the notes in the introduction are treated digitally, meaning the sound is *either on or off* with nothing changing along the way. In the first measure flute and bassoon appear. Their sound does not change, so the listener has no way of knowing how long the measure will last. Then the entry of strings divides the measure into two, then two, then four. Two is okay, four beats is good, but then the pattern stops dead and instead of a beat all you hear is a change of loudness instead, a crescendo, then a diminuendo. That's two ways of creating a sense of time: one by subdividing the frame of reference, the other by a process of continuous change within the frame. All of this takes place within a context of constant tone, without any intervening breaks or silences.

Then Beethoven changes the pattern from constant tone to empty tone. A single line of points, four to a measure, go up, then down, then up, then down in a graceful curve. When the movement is upward, there is a lot of empty space between each note. It seems *slower* than the continuous four to a measure earlier. As the points on the curve descend, a bassoon provides a constant reference line, and this constant element contributes a sense of direction. This process is repeated at a lower pitch. Then

Beethoven thickens the texture by alternating single points and clusters of points. When there is only one short note at a time, the information content is meager and the brain can keep up very easily. In that situation the pace seems slow because you can deal with it without effort. However, as soon as you substitute clusters of points for single points, the information content becomes very much more complex. The brain has to register the different notes, their internal relations, and also their harmonic relationship with the previous cluster or note. That's a lot more for the brain to do, and so although the pulse remains exactly the same, the extra density of information means the brain finds it harder to keep up, and therefore *time seems to go faster*. So we have gone from *smooth* to *discontinuous*, from one to four divisions per bar, and from *simple* to *clustered* pulses of information.

At this point Beethoven repeats the process from the top. The second time around he develops toward the idea of a continuous melody, but by very cautious stages. First he combines long and short, legato and staccato elements, and begins to lean on the last beat of every measure to generate momentum. Then toward the very end, the music gets very quiet. A brief melody is discreetly introduced in the violins. No sooner has it arrived than the melody, and with it all harmony, comes to a halt. The entire orchestra is reduced to counting on a single note: *one, two, one-two-three-four*. The huge and unexpected climax that follows—a dense chord for full orchestra plus timpani, is the opposite extreme from the twilight emptiness of the very opening bar. And yet this intensely bright sound filling the entire pitch space is in its own way *equally timeless*. You have no idea from listening to the sound itself how long it is going to last. First the sound is on, then it's off, then it's on again, and then once again the pattern of *one, two, one-two-three-four*. Before you know what has happened the fast-movement Allegro is off and running. And you have no idea how he did it. At the very beginning of the Adagio, there is no sense of time. At the very end of the Adagio, you still have no sense of time, and yet the music has gone from no time to slow time and from slow time to fast time. The rest of the Allegro develops this scale of ambiguity of time relatively freely. The serious intellectual work is all done, equipping the listener with a vocabulary of terms by which to understand and articulate movement in different speeds. Like Picasso making a collage (for example, "Violon et Feuille de Musique," 1912),[5]

Beethoven weaves scraps of melody into the fabric of the Allegro, not to generate the usual subject and counter-subject dialogue, but as torn-off fragments that say: *Hey! Look at this. It fits. How about that.*

Odd versus even

Theme permutations in Schubert's "Great C Major" symphony

The first movement of Schubert's Symphony No. 9 (the "Great C Major") can be understood as a dialogue on two levels: between the two tonalities of C major and A minor, and between two rhythms, an even rhythm and an uneven or dotted rhythm. The main melody of the movement combines two simple, rising scales or modes, each of five notes: the C major scale *c d e f g*, and the cor-responding A minor scale *a b c d e*. The two scales pull the mood of the music in two directions, towards major (light, sunny) or towards the minor key (somber). In a way, the two correspond to a painted image of which one half is in the light, the other half in shadow. They are complementary, not in conflict.

In addition, Schubert alternates between an even rhythm and a lurching, dotted rhythm which is dynamic and restless. It adds

Permutations of a 3-note sequence

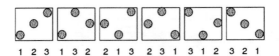

1 2 3   1 3 2   2 1 3   2 3 1   3 1 2   3 2 1

Permutations of even and uneven rhythms

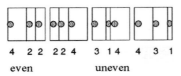

4   2 2   2 2 4   3 1 4   4   3   1
even   uneven

Cellular permutations in Schubert's
"Great C Major" symphony

energy, but also brings uncertainty into the musical action, and creates a sense of ongoing momentum. These were rhythmic conventions that the listener understood, and not ideas that Schubert dreamed up. The remarkable feature about this strong initial theme is that it seems so organic and well-knit, and yet when you look at how it is put together, it breaks down into a sequence of very basic rhythm units and simple ascending scales. If the opening theme were hard to accept as a product of calculation rather than inspiration, you would have to reconsider after hearing the clever permutations of the original theme, reassembling the melody units backwards and even upside down. These tricks don't happen by accident.[6]

It's a long movement, but essentially Schubert is conducting a dialogue between major and minor, and between even and odd rhythms. And that's it. Schubert lets the rhythm go for as long as it takes to make a point, and that takes time. He makes delicious gear-shifts from uneven to even beat patterns, and dramatic changes of mood from a stable rhythm to a sinister and threatening dotted rhythm. Equestrians will appreciate the meaning underlying these changes of pace from a trot to a canter, and from a canter to a gallop. Sometimes the dotted rhythm appears sinister and militaristic, as when the trombones (later horns) enter in the tenor range and rise in pitch and loudness with un-stoppable force to dominate the entire orchestra. At other times the dotted rhythms are lighter, suggesting the cantering rhythm of the hunt, or even a light and happy dance. The dynamics of dotted rhythms have many characters, it says. Equally, the even rhythms are able to check the musical pace and restore a sense of calm and stability. At the end of the movement it seems as if Schubert cannot decide which of the two rhythms shall prevail. For a time it seems as though the even beats are drawing ahead to a final cadence, but the dancing dotted rhythms will not be suppressed and the opening theme returns. At the very last moment, however, even beats supported by drums bring the dialogue of even and odd rhythms to an emphatic close.

Modern times

No study of musical time or timelessness would be complete without a reference to the undulating endless chord of "Farben" (Chord colors), the third of Austrian composer Arnold Schoen-berg's *Five Pieces for Orchestra* of 1909.[7] The mood of this study in tone colors is subdued, tranquil even, imagery that in its combi-nation of indistinct forms and pinpoint highlights is perhaps reminiscent of Whistler's painting *On Battersea Bridge*. The piece consists of an enigmatic chord sounding like a distant ship's siren ebbing and flowing through a moisture-laden atmosphere. The image of timelessness is a recurrent feature in Western music, from Pérotin to Wagner's overture to the opera *Das Rhein-gold*, to certain meditative pieces for harmonium by Sigfrid Karg-Elert, and of course Stockhausen's *Stimmung* of 1968.

A different kind of timelessness related to endless repetition is associated with the minimalist style of Steve Reich, John Ad-ams, Michael Nyman, Philip Glass, and others. Drawing on the ritual traditions of monotone chant—and indeed on the mantric style of Gertrude Stein and the graphisms of concrete poetry—minimalism also looks back to the mechanical precision of the early keyboard era of Frescobaldi and Bach, and further abroad to the musical traditions of Africa and Bali in the spirit of the generation of Colin McPhee, John Cage and Lou Harrison.[8] At its best, minimalist music induces a relaxation in the mind of the listener that allows subtly shifting resonances and emphases (accidental or contrived) to work at a subliminal level. In this respect, it offers an interesting foretaste of a therapeutic music that in a future era might be composed to alleviate specific disorders. At the other extreme, the full orchestra version of Stravinsky's great ballet *The Rite of Spring* of 1913 evokes the mechanical dynamism of futurism in the piston-like motion of the "Dances of the Young Girls" (in the original Nijinsky choreography "jumping up and down like knock-kneed Lolitas," as the composer later recalled), while in the culminating "Dance of the Sacred One" Stravinsky evokes a startlingly realistic sensation of clinical *arrhythmia*, a syncopated disruption of regular heartbeat rhythm, unique in classical music, an *absence* of pattern as powerful and evocative as a real heart attack.[9]

Notes

1. Louis Couperin, "Prélude à l'Imitation de M. Froberger." Laurence Cummings (*Louis Couperin Harpsichord Suites*, Naxos 8.550922, 1994), track 7.

2. For an image and an explanation see Irvin Rock, *Perception* (New York: Scientific American Library, 1984), 189-90.

3. Mozart, *Eine kleine Nachtmusik*. Philharmonia Orchestra, cond. Sir Colin Davis (Seraphim 7243 5 68533 2 4, 1995), tracks 1-4.

4. Beethoven, Symphony No. 4. Chamber Orchestra of Europe, cond. Nikolaus Harnoncourt (Teldec 2292-46452-2, 1991).

5. Pablo Picasso, "Violon et Feuille de Musique," collage, 1912. Musée Picasso, Paris.

6. Schubert, Symphony No. 9 "Great C Major." Belgian Radio Television Philharmonic Orchestra, cond. Alexander Rahbari (Naxos 8.550502, 1991), track 1.

7. Arnold Schoenberg, No. 3 "Farben" from *Five Pieces for Orchestra* Op. 16. Chicago Symphony Orchestra, cond. Rafael Kubelik (Mercury 289 434 397-2, 1998), track 7. There is a special aura to this reissue of a classic 1953 recording in mono.

8. John Cage and Lou Harrison, *Double Music*. Amadinda Percussion Group (Hungaroton HCD 31844, 1999), track 15.

9. Stravinsky, *Le Sacre du Printemps* (The Rite of Spring). Columbia Symphony Orchestra, cond. Igor Stravinsky (CBS MK 42433, 1988), tracks 17, 29.

CHAPTER EIGHTEEN

Inspiration

FILM MUSIC is about matching music to the moving image. What does this mean? It means, first, that you have a moving image to begin with. That's a precondition. *The moving image is the movie.* Second, there is no music already. Music is not already part of the moving image. That's mostly true. A movie drama is like real life in the sense that human behavior is not normally choreographed to music. There are very few situations in public life where music controls the action. We're not talking about concert events where the action is implicit and the participants are all sitting down, and we are also not talking about elevator music in a shopping mall that has little influence on the way people behave other than making the environment pleasant for shopping. You could say a wedding or a military parade are situations where there is a certain structure and pattern to human activity and that the music has a role in determining the timing and sequence of events. But these situations are rare. So the idea of putting music to a moving image implies, as a general rule, that the image comes first, that there is no music already accompanying the image, and that by putting music to the moving image you are enhancing it in some way for the viewer and making the underlying motive or pattern more immediately comprehensible.

A sight gag in an episode of the television cartoon series "The Simpsons" makes the point rather well. Late at night, Homer is seen running in terror down a dark and lonely forest road and the music track turns into Bernard Herrmann's "stab-bing" music for the famous scene between Anthony Perkins and Janet Leigh as the victim in Alfred Hitchcock's movie *Psycho*. Suddenly a bus pulls up carrying the Springfield Symphony Orchestra. They are playing the music on their way home from a concert. A violinist gets off, still playing the *squeak, squeak* theme while she walks up the drive. What was initially sinister back-ground music turns out to be part of the action, and as soon as the audience realizes it to be an incident in the action, the music becomes completely non-frightening.

Action and emotion

A handbook for cinema organists in the silent movie era declares:

> The music should reproduce, emphasise, insinuate or ref-lect the action of the photo-play. In this combination of action and music, the photo-play is analogous to the ballet; but whereas the term action implies choreographic effects, in the film the term also covers the emotions and even thoughts of the players.[1]

Whereas ballet is action choreographed to pre-existing music, film music is typically composed to accompany pre-existing edited action. So rather than the film-maker fitting movement to music, the film composer has to invent a music that not only suits the dramatic mood but also more or less matches and assists in uni-fying and coordinating actions that have been decided in advance and without the composer being consulted.

Shooting a movie breaks down dramatic actions into scenes, and splits scenes into individual shots, lines, and camera angles, each of which may be taken and retaken many times. To ensure dramatic continuity throughout a scene (as well as to mask the intrusive sound of the camera), music was often performed on set during the silent movie era and the convention continued into the era of sound. This music was not specially composed and did not appear in the subsequent movie. Its function was to allow the actors to stay focused on their characters and on the emotion of the scene throughout a long series of takes, to ensure consistency from start to finish when the scene was finally edited together, consistency that an audience would interpret as *unity of action* as well as *unity of time*. The use of so-called *mood music* to assist actors in remembering their roles is depicted tongue-in-cheek by Gene Kelly and Donald O'Connor playing set musicians for a fight scene early in the Stanley Donen movie *Singin' in the Rain*. For music to be employed in this way is extremely interesting and suggestive of music's function in much earlier societies as a memory aid and means of preserving the character or dignity of a ritual situation. In this respect, the numerous retakes of a movie scene implicitly correspond to the rites of the church or tribe, their purpose of constant repetition being in fact the same: to attain the equivalent of a perfect take in which everybody involved, leading and bit players alike, arrives at a total harmony of purpose.

Coordination

Paradoxically, the end result of marrying music and the moving image should be the same for the movie as for the ballet: it should look as though the visual actions are performed to the music. Since music does not in fact accompany our everyday actions, in both ballet and the movies music is implicitly acknowledged as an *unreal presence* that gives emotional and physical definition to actions that in themselves are often highly formal and stereotyped, and that would otherwise run the risk of appearing less realistic (or if not realistic, less believable) without a musical accompaniment. There is, of course, no orchestra hiding behind the curtain or around the corner. What the movies ask us to believe is that the expert choreography of a shoot-out or fist-fight is the credible behavior of characters with heroic presence of mind and supreme executive power. The music is simply a gloss enabling the movie audience to understand how a complicated and seemingly spontaneous sequence of actions is in fact anticipated and controlled by superior beings.

Poetry of motion

Moving picture entertainment began in the final decade of the nineteenth century as a side-show item in a public hall or marquee. The first black and white silent movies were short in duration and often unpremeditated in subject matter. The Lumière brothers made mini-movies of workers leaving a factory, a baby being fed from a spoon, and, famously, of a train arriving at a station. The subject matter was not the issue. What people wanted to see was *images of movement*. The movie taught viewers how to read the dynamics of real life in a completely new way. There were two reasons. One, technical, arises from the breakdown of continuous motion into a sequence of still frames; the other, perceptual, from the flat screen projection of cinema that allows an observer to keep *an entire field of view in focus*, which is not something you can do in real life. The attractions of the movie were *actuality* (these things actually happened), *realism* (this is the way ordinary people actually behave), and *motivation* (the instinctive dynamics of natural, unscripted human interaction). Looking at early newsreels, a viewer is struck by the banality of streetcars gliding around corners and pedestrians crossing the street, but the ordinariness is testament to the irresistible fascination of imagery of movement. If such a movie is speeded up, you see that people do not move in a rhythm, as experience would suggest, so much as glide. The human poetry of motion is enhanced rather than dampened by the distortions and speeded-up action of pioneer hand-operated cameras. Even today images of motion still play a powerful role in advertising: realistic movement superimposed on a fast-motion backdrop, the reverse action sequence that miraculously restores a house on fire, turns back the crash of a runaway car, or the slow motion replay that relives a winning goal or hole in one with enhanced intensity. Movement itself is still the dominating subtext of every movie, from the twitch of an eyebrow in a close-up, to the shark lurking in the distance, to the chase sequences that still connect the excitement of a *Terminator 2* with the adrenaline rush of a Mack Sennett comedy.

Real time

Movie music begins with the movies. Logical, but untrue. Early shows were noisy affairs, certainly. Movies were silent and paying customers did not know how to behave. Shows could also be unsafe. Movie stock was inflammable, the projector was fully exposed, a smoker in the audience could still light up a cigarette and throw away the match. It made sense to bring in a piano player to keep the audience members quiet and give them something to listen to while the movie was showing—or stop them rioting when the celluloid broke, which happened quite often.

From Augustus Voigt: *The Battle of Navarino*

The sound of a piano playing also helped to cover up the clatter of the projector. In the course of time, when piano-playing became an accepted part of the performance, the possibility arose of having music serve the interests of the visual action more directly. A quantum leap of the imagination is involved, how-ever, for a musician to go from *improvising* a medley of pieces for the purpose of keeping customers quiet, to *compiling* a sequence of pieces to match up with the scenes in a movie show. If the point of having music is to keep people quiet, then any music will do. For music to enhance the movie action you have to start thinking positively about the action and also change the circumstances of performance: put the noisy projection equipment out of earshot in an isolation chamber, provide comfortable seating and accommodation for the audience, and hire musicians who have experience in dramatic accompaniment.

Authenticating the visual

Composing, compiling, even improvising music to accompany a moving picture show changes the way you see it. Remember the soloist who could be making it up, and the chorus, who proves there is a script? In the movies the picture element corresponds to the soloist. You are watching the screen. You don't know what is going to come next. It's all happening on the instant. Music is to the visual image as the chorus is to the soloist. As soon as you hear music it tells you there is a script. *Music authenticates the visual.* It says: this is scripted.

The basic principle of having music as accompaniment and illustration of screen entertainment was established long before the arrival of moving pictures in the 1890s. Before the arrival of the telegraph around mid-century, news of battles abroad filtered through to the well-to-do public in artistically poeticized form. The example from Augustus Voigt's characteristic divertimento *The Battle of Navarino* is music composed to accompany a magic-lantern show dramatizing the famous 1827 naval engagement and captioned as a sequence of short scenes illustrating events in the battle.[2] Silent movies later adopted the same format of caption and image. (Erik Satie later parodied the genre with his whimsically captioned piano pieces *Embryons Deséchées.*)

With every new image on screen, Voigt's music for solo piano changes in character to suggest appropriate sound effects or emotions. Without the accompanying visual images this music makes little sense. It has no formal structure, no sense of continuity, no emotional integrity. The only claim to unity this music can uphold is that of conforming to a *timely sequence of events.* The sudden and random nature of real-life action is reflected in the unpredictable flow of the musical sequence. The popularity of such entertainments and their accompanying music tells us that audiences were prepared to accept the logic of a visual scenario as sufficient justification for an otherwise meaningless succession of musical gestures. The era of piano-accompanied silent action movies drew on these earlier skills of improvisation.

In larger movie houses the musical accompaniment was provided ideally by an orchestra, more usually by a small team of musicians, including a percussionist among whose special responsibilities was to provide the sound effects of gunfire (snare drum rimshot), horses' hooves (castanets) and so on. The era of the silent movie ended in 1927, but a few professional silent movie pianists were still in business fifty years later, when sisters Ena Baga and Florence de Jong were recorded improvising in 1973 to actual screenings of the Douglas Fairbanks classics *The Black Pirate* (1926)—a technicolor silent movie, believe it or not—and *The Thief of Baghdad* (1924) at London's Academy One cinema. Ena Baga explained:

> We don't necessarily have to see a film through first, and we certainly don't programme what we are going to play. The nature of the film sets the type of music we will use. *Black Pirate*, for instance, is Spanish, so we go armed with our memorised scores of Spanish music. Each change of scene and action then prompts a change of melody, tempo, or rendering. You watch the screen the whole time, and there has to be an instant translation and interpretation from eye to mind and fingers, with a certain amount of emotional response coming in along the way.[3]

The art of movie music improvisation arose at a time when the movies were a novelty attracting new audiences, especially in the provinces, whose acquaintance with music was limited to the music-hall or operetta. It was a time when copyright laws in their present form did not exist and all music was supposedly in the public domain. It made sense and was also simpler to illustrate a movie using themes and idioms that an audience would recognize rather than compose original music likely to add a further element of unfamiliarity to an already new movie experience. Recognition meant reassurance, which in turn added credibility to the unfamiliar visual drama unfolding on screen. Using familiar material also gave the improviser greater control in timing, since movies from the beginning were subject to changes of scene and mood to which the accompanist had to adapt as quickly as possible. Carl Stalling's music for the Warner Brothers 1952 cartoon *Feed the Kitty* contains a parody sequence of an idealized silent movie music montage that pays tribute to the music-hall clichés of the silent movie era.[4]

A professional composer brings a background of understanding to a movie drama, but an understanding based on the composer being completely in control of the development and timing of events. Throughout the nineteenth century, well in advance of the technology of the movies, Romantic composers were instinctively anticipating the demands of cinema entertainment by focusing more and more on two areas of key importance to the movie. One was psychological, the representation of states of emotion; the other was temporal, the representation of dramatic events in experiential time. Among the first original music scores for a movie is an accompaniment composed in 1908 by the French composer Camille Saint-Saëns for the silent historical movie *L'Assassinat du Duc de Guise*, music for small ensemble including a harmonium.[5] Although the story-line is violent, the film treatment is theatrical and silently verbose, striking attitudes in classical French style, with little to suggest the dynamics of later action movies.

The nineteenth century produced some of the best potential movie music well in advance of the real movies. Schubert's ballet music for *Rosamunde* of 1823 is a good example of action music having the ability to change instantly from one mood to another. Italian opera composer Gioachino Rossini displays a magical sense of controlled gesture for dramatic effect in his minimalist overture to *La Cenerentola* (Cinderella).[6] For dramatic atmosphere and special effects, the *Symphonie Fantastique* of Hector Berlioz offers another excellent model of musical design for the aspiring film composer. Songwriters Schubert and Schumann (he of the cello concerto) developed a highly economical language of emotional gesture. A towering figure in German opera, Richard Wagner conceived and realized a new art of spectacular music drama in a specially designed auditorium at Bayreuth. His operas bring together powerful voices, orchestra, stage and lighting effects in a total synthesis anticipating the epic cinema of fifty and a hundred years in the future. Despite the huge theatrical apparatus at Bayreuth, Wagner's is an idiom sensitive enough to amplify and convey the most delicate and nuanced expressions of human emotion in transition *in real time*.[7] At the very highest level, classical music was ready for the new realism and naturalistic action of cinema well before the technology arrived on the scene at the close of the nineteenth century.

Looking back, it seems perhaps strange that a stronger partnership between the movie industry and the serious composer did not eventuate. In part, this has to do with the rapid rise to dominance of the American movie industry and the absence of a powerful classical music tradition in the United States, but the critical reason is probably that the practical realities of moviemaking did not allow movies to develop the same collaborative relationship between scriptwriter, director, and composer that had obtained for opera and ballet in previous centuries.

As long as the visual image took priority, music was obliged to take a subordinate role. Nevertheless, the relative impact of music in the silent movie era can readily be gauged from the example of Russian director Sergei Eisenstein's 1925 masterpiece *Battleship Potemkin*, at the time a controversial film dramatization of politically-charged events of recent history. The movie opened to the accompaniment of a specially-composed orchestra score by Edmund Meisel, a score now lost. It is a sign of the singular influence of music over German audiences that government authorities in that country initially passed the movie for general release but *banned* Meisel's music on the ground that the combination of image and music might well be construed as "an exercise in political inflammation."[8]

Prototypes

Three examples of classical music that in very different ways anticipate the requirements of movie style are the symphonic poem *Till Eulenspiegel* by Richard Strauss, "Washington's Birthday" from *The "Holidays" Symphony* by Charles Ives, and the ballet *Petrushka* by Stravinsky. None of these composers ever actually composed music for film (although Disney negotiated with the reluctant composer to include an edited version of Stravinsky's *The Rite of Spring* in the movie *Fantasia*). Strauss's *Till Eulenspiegel* is concert music that abandons the traditional formalities of classical music in favor of an unprecedented narrative freedom of development. This is music that simply follows a story-line. Instead of the dance-related structures, symmetries, and repetitions of the traditional symphony, the symphonic poem develops a music of *action scenes* in which characters are identified by specific melodic and rhythmic signatures, a technique introduced by Wagner in his operas and later widely adopted by movie composers.[9] This is musical comedy of exaggerated style and Chaplinesque gesture, springing from the same folk traditions and middle-European culture as Disney's *Pinocchio*, while the anarchic spirit of Strauss's anti-hero lives on in the antics of such latter-day pranksters as Bugs Bunny and Daffy Duck.

Inclusive hearing

Among the many decisive technical innovations introduced to movie-making by the American pioneer D. W. Griffith was the elevated long shot allowing a panoramic view of complex large-scale events such as a battlefield engagement or public activity on a busy street. When Griffith's career as a maker of epic movies was launched in 1915 with *Birth of a Nation*, his contemporary Charles Ives had already composed a number of works for orchestra expressing a similar breadth of vision in musical terms. The movie long shot takes in a great deal of visual incident and allows the viewer a number of new freedoms. A Griffith long shot of a cattle stampede or a battle, for example, revealed the dynamics of collective behavior in ways never before seen or understood. In particular, it revealed that large-scale mass movements previously considered random are governed, in fact, by flow patterns similar in kind to turbulence in air or water. The long shot also introduced movie audiences to the idea of multiple incidents within a single frame, offering the viewer a choice of subject on which to focus rather than imposing a particular point or object of view. In "Washington's Birthday," completed in 1913, Ives creates a similar impression of multiple events in a musical image of a festive occasion in which different situations and actions are presented simultaneously even though experienced sequentially.[10] Like the movie composers of his generation, Ives is a compiler in that he borrows unashamedly from popular ballad material to illustrate a character or mood. His special genius lies, however, in maintaining a complex multi-layered music working on a number of different timescales simultaneously, in complete defiance of the rules of conventional notation and their inherent tendencies to simplify and integrate.

The latter half of "Washington's Birthday" is evocative of a movie scenario set in a grand mansion in a rural setting. It is evening. A church bell tolls in the distance. The music seems to track up to the front door and through into a parlor where family guests are dancing, but at the same time one can hear a different country dance music being played in another part of the house. There are some wonderful musical transitions, from the salon into the kitchen where a more boisterous square dance is playing with a solo jew's harp (a form of *mouth music*, incidentally) in the foreground. Then, after a little crisis or shriek, we seem to pass through the french doors into the peace and quiet of the back porch. In the background behind the door the muted jollity of the dance can still be heard (a miraculous touch for solo viola). Then, however, an air of nocturnal tranquility literally descends in the melody of a sentimental ballad, beautifully orchestrated, during which the sound of the church bell is heard once again in the distance. The Ives musical experience is richer aurally than most movies are visually, in fact, seeming to anticipate virtual reality simulations of the computer age.

Animation

Routine, everyday human activity is not like dancing. The old musicals were escapist fantasy. In reality, dancing—even Fred Astaire's tap dancing—is a formal affair. It takes longer to do something by dancing it than by just doing it, irrespective of whether the dance is fast or slow. There are good reasons for this. Dancing is pattern making and it's visual, and patterns take time not only to execute but also to register on the observer. Tap dancing has the added value of sound: the audience can hear as well as see the moves. The art of movement in movies is unlike ballet choreography, in that it aims to appear more naturalistic, but it does resemble choreography in having the function of telling a story. For an audience to be able to read a story from the actions of movie characters, the moves have to be planned and obvious, and that is not the way real life works. Normal actions tend to be discreet and economical because the point of normal behavior is to achieve results and not for the sake of being observed.

If the demands of legibility tend to slow down the pace of normal activity in a movie, editing is available to speed up the narrative sequence of events and restore the balance. An improvised piano accompaniment to a silent movie interprets the *action* within a scene as music of a style and tempo consistent with the dramatic situation. Achieving that sort of match is one thing; however, the *edits* from scene to scene that carry the action forward tend to provoke corresponding *changes* of musical character or tempo, and changes of musical character by definition are designed to draw attention to themselves and thus disrupt a natural continuity of events. French composer Claude Debussy was perhaps the first major classical composer to work out the logistics of a music appropriate to this new style of cinematic narrative. In his after-hours career as a music journalist, he once observed—in a characteristically ironic allusion to philosopher Arthur Schopenhauer—that music should aspire to the condition of the cinema. (Schopenhauer had said all the arts should or did aspire to the condition of music.)

> Is it not our duty [he wrote] to try and find the symphonic formulae best suited to the audacious discoveries of our modern times? The century of aeroplanes has a right to a music of its own. . . . Let us purify our music! Let us try to relieve its congestion, to find a less cluttered kind of music. There remains but one way of reviving the taste for symphonic music among our contemporaries: to apply to pure music the techniques of cinematography. It is the film—the Ariadne's thread—that will show us the way.[11]

Debussy's own music developed a transparency and dream-like quality perfectly adapted to the movies, even though he never composed a movie score. A typical example of his elusive, flickering imagery is "Gigues" from *Images* for orchestra, through which a tune based on the sea shanty "The Keel Row" provides a thread of continuity.[12] Unlike the solid, meat-and-potatoes fare of German music, the Parisian school favored a music of evanescent sensation and employed the symphony orchestra as a palette of tints and shades. The perfumed sensuousness of Ravel's *Daphnis et Chloé* greatly impressed Russian-born artist Wassily Kandinsky, whose vision of a new abstract art of music, light, and color entitled "The Yellow Sound" may well have influenced the abstract animation of the Bach *Toccata and Fugue* in Disney's *Fantasia* and anticipated the *son et lumière* movement of the sixties.[13,14] Another Russian, Alexander Scriabin, dreamed of a "color keyboard" on which modulated lighting effects could be realized as part of a larger symphonic creation in works such as *Prometheus—The Poem of Fire*.[15] Ironically, a color keyboard organ

was developed in 1936 by the English stage lighting manufacturers Strand Electric in association with the Compton Organ
Company, makers of electrical organs. Musical presentations
interpreted in colored light attracted a steady stream of distinguished visitors to the company's Covent Garden offices:

> The Duke of Kent was no exception, he had to take his dose
> of colour music. Only Mistinguette ever had the courage to
> back out. She lasted half-way through the Bach Toccata and
> Fugue in D minor and then got up from her seat. . . . Prob-
> ably the most famous visitor was Bernard Shaw. He, too,
> had a large dose of colour music including the whole of the
> first movements of Schubert's "Unfinished" Symphony and
> of Tchaikovsky's "Pathetique," respectively.[16]

One composer and film buff who anticipated the art of film choreography was Stravinsky: his 1911 ballet *Petrushka* expresses a
new style of musical animation, with very economical gestures
and open scoring. Whereas Debussy developed an art of musical
atmospheres, the young Russian composer and Debussy protégé
focuses in this particular work on a new poetry of *gesture*. A
huge symphony orchestra is treated very sparingly as a color
resource, keeping the power of numbers strictly in reserve for
sudden dramatic climaxes that vanish as abruptly as they appear.
The vividly descriptive actions of "Tableau II: Petrushka's
Room" are pared down often to one or two instruments at a time,
with the solo piano carrying most of the action. This new manner
of composing combines the overall time control of the symphonic
tradition with the improvisatory feel of the cadenza.[17] This music
really does sound as though it is being made up on the spur of
the moment. The aphorism "To the blind all things are sudden"
is one that not only perfectly describes the movies, but applies
equally well to Stravinsky's music, which solves the age-old
problem of *instant response* without the need for classical
preparation or musical argument. This same scene, singled out as
a critical influence on Scott Bradley, one of the great composers
of animated film music of the thirties and after, also includes the
mother of all cat and mouse chase sequences.[18] In Carl Stalling's
score for the 1956 Warner Brothers Road Runner cartoon *There
they go go go*, brilliantly orchestrated by Milt Franklyn, the
listener recognizes a superlative parody and genuinely masterful
tribute to Stravinsky's puppet character and his music.[19]

Unlike the epic historical dramas of the silent era, whose musical and dramatic roots lay in traditional opera, animated film arose as an art of *simulated motion* in a direct line of descent from the musical box and clockwork compositions of Bach, Mozart, Haydn, and Beethoven. Music was an essential ingredient. Walt Disney's "Silly Symphonies" banner said as much. Disney realized that animation was different precisely because the artist and composer could arrive at a complete fusion of animated action and music. Of course, the material was funny rather than serious dramatic art, but all the same, there was a serious point to prove in animation techniques that could use music to help create naturalistic illusions of body movement conveying a sense of body mass, object weight or weightlessness, and realistic spatial movement.

Disney's "Silly Symphonies" is one of several music-based animation series, including "Merrie Melodies" and the subversively infamous "Looney Tunes." Music was more important than spoken dialogue in those early years for not only artistic but also commercial reasons: if a modern viewer wonders why the voices of Donald Duck and Mickey Mouse are virtually unintelligible, it is not only because movie house sound systems in the early days were unreliable and music was easier to hear than spoken dialogue, but also for the sound economic reason that cartoons created for a world market needed to communicate in the universal language of music rather than in a specific tongue that only a fraction of the audience could understand. The classical music trend peaked after the war's end in 1945 with a number of classic prize-winning shorts featuring, among others, Tom and Jerry performing Liszt on the piano, and Bugs Bunny singing Wagner opera highlights opposite Elmer Fudd.

Synchronized sound

Cartoon animation is all about synchronized visuals and music, and only after the arrival of optical sound on film was it possible for the art to make real advances. In the last great years of the epic silent movie era, a number of significant name composers had been attracted to film music as a serious art medium, and with the arrival of the talkies, those same composers had to adjust to the much more precise demands of synchronized sound, not to mention the greatly diminished status enjoyed by composers in the movie world. During the silent movie era, exact coordination of music and action was not as critical an issue—capturing a prevailing emotion was what it was mostly about. The 1951 Warner Brothers movie *Singin' in the Rain*, in many ways an affectionate look back to 1927 and the transition from silents to talkies, draws attention to the mannerist style of Hollywood historical romance, its European rococo traditions and remoteness from reality. The genre was turned upside down by the arrival of Warner Brothers' *The Jazz Singer* featuring Al Jolson in blackface, a movie as remarkable today for political incorrectness as for technical innovation. The Vitaphone technique of synchronized disc and film was originally intended to allow the inclusion of musical numbers in an otherwise conventional silent movie. In the event American audiences were captivated by the sight and sound of an excited Al Jolson ad-libbing on camera. It was not merely that he talked and sang, but it was authentic, unscripted, in real time, and *his accent made him real*.[20] A skeptical industry was forced to embrace the talking picture, and the coming of optical sound within a mere two years brought higher quality sound along with totally reliable synchronization of visual narrative and sound-track.

Warner Brothers had risked everything for sound, and the investment paid off in a string of hard-edged movies exploiting the gritty realism of dialogue, music, and sound effects with enormous success. The urban gangster movie *The Public Enemy*, an early "hit" of 1931, features a sophisticated multi-layered soundtrack that in one notable murder scene, combines off-screen continuity music, staircase footsteps, empty room ambience, live on-scene piano, the sound of a punch on the jaw, gunshots on camera and typically slow-motion dialogue of contrasting voice roles including the abrasive accents of a young and charismatic Jimmy Cagney.[21]

Sound movies blurred the distinction between speech, music, and effects. Each layer of sound became an art form in itself. Loud noises such as gunshots proved especially difficult to deal with: it took a full *ten years* from 1929 to 1938 for film and radio sound effects experts, eventually in collaboration with the firearms industry, to design and build a pistol to fire blanks that sounded authentic but did not shatter the microphone.[22] The movie industry's continuing love affair with guns and loud explosions is perhaps as much about technical bravura as a taste for violence: *Do you hear that? We can do this.*

The advent of optical sound and music on film early in the 1930s renewed interest in the possibilities of fully synchronized music and action. Without click-tracks or electronic assistance composers looked for ways of achieving a tight fit of action and sound. In his chamber music score for Raymond Bernard's original 1934 movie dramatization of *Les Misérables,* composer Arthur Honegger alternates a metronome timescale of one beat per second (MM = 60) and three beats for every two seconds (MM = 90).[23] The one-second timing can be observed in listening to the music—for example, the scene "Evocation des forçats"—while keeping an eye on the digital second counter on the disc player as it blinks in time with the beat. Given a shooting script with climaxes at specified points within a scene timed to a fraction of a second, a composer can easily calculate on music paper exactly how many measures and beats need to be composed and precisely where changes in mood or emphasis have to be incorporated. Sticking to clock time is not, in fact, as restrictive a practice as it might appear on the surface. If evidence were needed, Beethoven's Symphony No. 4 is as good a proof as any that clock time is a measure with which human perceptions of time and associated emotions do not always conform. Among a wealth of examples of film scores composed to the "one beat per second" beat are classics such as the title music to *The Adventures of Robin Hood* (Erich Wolfgang Korngold, 1938), the chariot race from *Ben-Hur* (Miklos Rózsa, 1959), the stabbing scene from *Psycho* (Bernard Herrmann, 1960), even the love music to *Dances with Wolves* (John Barry, 1990).[24] Nobody would say that these pieces of music lack emotional variety.

Mickey-mousing

Throughout the thirties and forties European-trained Hollywood movie composers Max Steiner, Erich Korngold, and Franz Waxman did their best to adapt the Romantic style of Wagner and Richard Strauss to the timescale of movie action. Korngold, in particular, struggled to accommodate the terminology of European opera to a visual genre that was not only much faster and more kaleidoscopic but also one in which actions are deliberately contrived to happen more quickly on screen than in real life. Korngold's dedication to the idea of achieving the same total fusion of musical and visual action in a classic movie romance as in a classic cartoon (or Wagner opera) was admirable but futile. In animated movies and musical comedies, the story is told through exaggerated actions underscored by a synchronized score, but to do the same for a drama or action movie is to run the risk of travestying what should appear more natural, more made up on the spur of the moment. In under-scoring movie actions so meticulously Korngold made them appear not more real but simply exaggerated and ludicrous, provoking the unkind but accurate description of the technique as "Mickey-mousing":

> In *The Adventures of Robin Hood*, for example, he produced stirring martial accompaniments for the triumph of Robin Hood and the Merry Men marching through Sherwood Forest. Descending melodic sweeps in the winds and strings depict the men descending from trees.[25]

Composers reluctant to submit to the creative realities of the movie industry all the same experimented with film music techniques. Bravura examples include Maurice Ravel's opera *L'Enfant et les Sortilèges*, a children's fairy tale in which familiar objects spring to life,[26] and the Schoenberg cantata *A Survivor from Warsaw*, a musical mini-documentary with Ed Morrow style voice-over narrating a harrowing story of life and death in the wartime Polish ghettos—a more satisfactory exercise in the movie genre, one is bound to say, than the composer's earlier *Begleitmusik zu einer Lichtspielszene* (Accompaniment to a Cinematographic Scene), an essay in silent movie music that was virtually out of date at its time of composition in 1929.[27]

As late as 1962, the eighty-year-old Stravinsky, who had lived in Hollywood for over twenty years and been an avid filmgoer from the days of Little Tich, produced a major new work for television called *The Flood*, based on the tale of Noah.[28] Though composed in an advanced serial idiom for George Balanchine's New York City Ballet, his music for the scene "The Building of the Ark" is a perfect homage to animated film music in the noble tradition of Carl Stalling, Bugs Bunny, and friends.

Notes
1. George Tootell, *How to Play the Cinema Organ* (London: Paxton, n.d.), 74.

2. Augustus Voigt, *The Battle of Navarino: A Characteristic Divertimento for the Piano Forte. Composed and Dedicated to Adml. Sir Edwd. Codrington, Bart.* (London: Cramer, Addison & Beale, n.d.), 7.

3. Florence de Jong and Ena Baga interviewed by Arthur Jackson (*Piano Music for Silent Movies*, vinyl, Rediffusion Gold Star 15-13, 1974).

4. Carl Stalling, orch. Milt Franklyn, music for the Warner Brothers 1952 cartoon *Feed the Kitty. The Carl Stalling Project* (Warner Bros. 9 26027-2, 1990), track 6.

5. Camille Saint-Saëns, *L'Assassinat du Duc de Guise* Op. 128. Musique Oblique Ensemble (Harmonia Mundi HMT 790 1472, 1993).

6. Gioachino Rossini, Overture *La Cenerentola* (Cinderella). Zagreb Festival Orchestra, cond. Michael Halász (Naxos 8.550236, 1989), track 3.

7. Richard Wagner, "Der dort mich ruft" from *Die Walküre*, Act II. Bavarian Radio Symphony Orchestra, cond. Bernard Haitink (EMI Angel CDS 7 49534 2, 1989).

8. Mark Evans, *Soundtrack: The Music of the Movies* (New York: Hopkinson and Blake, 1975), 9.

9. Richard Strauss, *Till Eulenspiegel*, symphonic poem. Budapest Symphony Orchestra, Ken-Ichiro Kobayashi (Hungaroton HRC 082, 1988).

10. Charles Ives, "Washington's Birthday" from *The "Holidays" Symphony*. New York Philharmonic, cond. Leonard Bernstein (Sony SMK 60203, 1998), track 2.

11. François Lesure, ed., *Debussy on Music*, tr. Richard Langham Smith (London: Secker & Warburg, 1977), 297.

12. Debussy, "Gigues" from *Images*. Saarbrücken Radio Symphony Orchestra, cond. Hans Zender (CPO 999 476 2, 1997), track 1.

13. Maurice Ravel, Suite No. 2 from *Daphnis et Chloé*. Vienna Philharmonic, cond. Lorin Maazel (RCA Victor 09026-68600-2, 1997), track 2.

14. Ulrika-Maria Eller-Rüter, *Kandinsky: Bühnenkomposition und Dichtung als Realisation seines Synthese-Konzepts* (Hildesheim: Georg Olms Verlag, 1990), 65-120.

15. Alexander Scriabin, *Prometheus—The Poem of Fire*. Kirov Orchestra, cond. Valery Gergiev (Philips 289 446 715-2, 1998), track 16.

16. *Fifty Years in Stage Lighting: A History of Strand Electric*. Special issue of *Tabs* Vol. 22, No. 1 (London: The Strand Electric & Engineering Company, 1964), 68-9.

17. Stravinsky, *Petrushka*. Columbia Symphony Orchestra, cond. Igor Stravinsky (CBS MK 42433, 1988), track 5.

18. John Newsom, "A Sound Idea: Music for Animated Films." In *Wonderful Inventions: Motion Pictures, Broadcasting, and Recorded Sound at the Library of Congress*, ed. Iris Newsom (Washington, D.C.: Library of Congress, 1985), 68-69.

19. Carl Stalling, orch. Milt Franklyn, music for the 1956 Warner Brothers cartoon *There they go go go*. *The Carl Stalling Project* (Warner Bros. 9 26027-2, 1990), track 6.

20. Warner Brothers' *The Jazz Singer* 1927, trailer and Al Jolson dialogue "You Ain't Seen Nuthin' Yet." *The Golden Age of Hollywood Stars* (vinyl, United Artists Records, 1977), disc 1, side A, tracks 2, 3.

21. Sound track from Warner Brothers' *The Public Enemy* 1931. *The Golden Age of Hollywood Stars*, track 6. Also on video (New York: Turner Entertainment and Warner Home Video VHS 65032, 2000).

22. John S. Carlile, "The Story of a Shot," in *Production and Direction of Radio Programs* (New York: Prentice-Hall, 1946), 212-15.

23. Arthur Honegger, film score *Les Misérables* 1934. Slovak Radio Symphony Orchestra, cond. Adriano (Marco Polo 8.223181, 1989), track 3.

24. National Public Radio compilation "Music in Film" (Sony SMK 60991, 1990).

25. Mark Evans, *Soundtrack: Music of the Movies*, 24.

26. Ravel, *L'Enfant et les Sortilèges*. Montreal Symphony Orchestra, cond. Charles Dutoit (Decca 440 333-2DH, 1995).

27. Schoenberg, *A Survivor from Warsaw*; and *Accompaniment to a Cinematographic Scene* Op. 34. Simon Callow, London Symphony Orchestra, cond. Robert Craft (Koch 3-7263-2H1, 1995), tracks 6, 7.

28. Stravinsky, "The Building of the Ark" from *The Flood*. The Philharmonia cond. Robert Craft (MusicMasters 01612-67195-2, 1998), track 27.

CHAPTER NINETEEN

Memory

ALL MUSIC deals with memory as well as experience. Making music is a way of expressing contact and continuity with past experience both personal and cultural. The history of music is also a history of information management and delivery systems. Oral cultures rely on memory. The human memory is a system of limited capacity, so in order to qualify for memorization the information to be stored has to undergo a reduction process to eliminate superfluous detail, and an encryption process to make it easier to store and reproduce. Folk music and musical instruments express choices and limitations relating to the kind of information that can be stored, and therefore remembered. Those choices are not only determined by the physical structures of ear, brain, hand, or tongue, but also reflect cultural and environmental priorities. The ancient Chinese made instruments of bronze to demonstrate, among other things, that they had the materials and the technology to do so.

As technology advances, music changes. Road signs offer directions, but you need a map to tell you where you are; once you have a map and can read it, you have the freedom to plan other routes and set alternative goals. Plainchant notation developed from a simple sign language telling the voice which direction to turn, into a navigational chart offering both direction and global coordinates. Medieval four-line plainchant notation offers an overview of an entire melody sequence, allowing the reader to plan ahead and transforming a point-to-point journey dictated by time and tradition into a movement in space allowing for invented diversions along the way. The ability to visualize music gave rise to new possibilities of sequence and structure no longer restricted to the traditional practices and requirements of oral documentation. Renaissance standard notation replaced a variety of specialist tablatures and codes with a universal system of graphics for pitch and time that could be developed into mechanisms for the storage of musical information on cylinders and paper rolls, and that also stimulated the manufacture of keyboard instruments for music reproduction. The mythical battle between Apollo and Marsyas can be understood as a conflict of aesthetic and intellectual priorities over what information music should be storing and how it should be reproduced. Tablatures and violins emphasize the importance of emotional expression, whereas standard notation and keyboard instruments emphasize the primacy of objective documentation with the goal of preserving vital information intact.

Voice recording

Elvis lives? The personality of Elvis lives on in his recordings. And why? Because he made them in the first place. Listening to a music recording is, first and foremost, about reliving or recreating a past experience. Dedicated fans dress up in Elvis gear and impersonate the Elvis style of singing for much the same reasons as the singers of plainchant practice their art. It is a powerful way of identifying with the charisma of an individual who is no longer living, and it gives a form of meaning and purpose to people whose otherwise anonymous lives are enhanced by it. Collectors of old recordings feel a nostalgia for past artistic glories and a respect for and interest in the history and aesthetics of long-vanished performance practices.

Most of the music enjoyed today in the cities and suburbs of the world is music in recorded form, listened to casually on radio, in the car, on television, or on video. The technology of sound recording has, in one sense, *brought closure* to the art of music. To the extent that music originated in a desire to preserve significant information in acoustic form, the successful capture and storage of the actual atmospheric disturbance of an event is the ultimate act. The idea that Elvis lives is not necessarily a joke.

Florence Nightingale lives just as tangibly in her recorded message of 1890. In making the distinction between a voice and the
physical person the word *lives* acquires genuinely theological
overtones.

Phonautograph

The first audio recording machine was devised around 1850 by a
Belgian, Léon Scott, and adapted in 1859 by R. Koenig, a research
scientist and contemporary of the great Hermann Helmholtz, for
use as laboratory equipment for the study of voice tones. As the
name suggests, the original *phonautograph* was designed as a
"sound writing" device. It could record a trace corresponding to
the pressure changes of a sound, but it could not play back the
recording. The Scott-Koenig phonautograph combined the functions of microphone and pen tracer in a structure modeled on the
human ear. Pressure waves entering the open end of a barrel-
shaped vessel set in motion a diaphragm at the closed end emu-
lating the human eardrum. The diaphragm is fixed at the edges
and vibrates freely at the center like any drum skin. A bristle attached at the center and pivoting at an edge point converts the
difference in center-edge vibration into a corresponding sideways motion, tracing a white line through the waxy coating of a
paper wrapped around a cylindrical carriage rotated by hand.[1]

Storing the voice

Music had nothing to do with the *invention* of sound recording,
though virtually everything to do with its subsequent commercial development. Scott's phonautograph belongs to the same
area of medical and scientific interest as the doctor's stethoscope,
an instrument for monitoring the body's internal rhythms which
has not changed in almost two centuries. In the same way as
listening to a patient's *involuntary* heart rhythms assists in a diagnosis of *physical* well-being, such a device for tapping into the
voluntary wave-forms emitted by human subjects could conceivably have related to a corresponding nineteenth-century scientific goal of diagnosing a person's *mental* state from the sound of
his or her voice. A medical investigation is not interested in a
voice recording as an art work or performance. Ask yourself
about the purpose of a multi-track oscilloscope in an emergency

ward, monitoring the vital rhythms of a patient, and what the information is for. One is looking for evidence in the waveforms of *consistency*, of *pattern*, and above all, the possibility of *change*. The point of such equipment is through *observing* a patient's body rhythms to understand that patient's *state of being*, and Scott's original invention might well have been designed to serve a similar purpose.

To make a reasonable diagnosis you need a sample of a certain length. Vocal sounds are very rich and dense waveforms, and a sheet of waxed paper wrapped around a cylinder is good for samples of only a few seconds' duration. If you also want to play back an acoustic recording, it has to be recorded on a durable enough medium to withstand the tracking action of a stylus. All this was foreseen by French amateur inventor Charles Cros, who patented a proposal for etching in acid a soundtrack previously engraved on a wax-coated metal disc. The fragile, coated paper medium would eventually be replaced by a variety of materials, including tinfoil, rubber, wax compounds of varying hardness, and the shellac disc. The example of a lathe led to the development of carriage-mounted cylindrical recording devices for extending the duration of a recording from the single turn of a phonautograph to a spiral groove of several minutes' duration. The acoustically advanced *graphophone* of 1880, developed by telephone pioneer Alexander Graham Bell, his cousin Chichester Bell, and Charles Sumner Tainter, recorded directly on to a wax coated cylinder and incorporated a greatly refined pickup mechanism. His wife being deaf, Alexander Graham Bell had a special interest in human hearing and the problems of the hearing impaired, and one can interpret his interest in the graphophone in the first instance as a diagnostic tool for medical science as well as a hearing aid.[2]

Data compression

Thomas Alva Edison, the other great name in the early history of recording, came to his invention of a mechanism for capturing the human voice by a more serendipitous route. A canny businessman with a keen interest in battery-powered electric communications, in 1877 Edison was working on ways of improving the carrying capacity of telegraph lines for transmitting Morse code. The *dit-dah-dit* signals are simple, digital audio signals of constant intensity that transmitted more effectively by telephone wire over long distances in those days than voice signals of wide bandwidth and fluctuating strength. A telegraph signal is sent and received in real time, and demand is greatest at certain times of day. Edison reasoned that if the signal were pre-recorded on to a strip of paper (like a ticker tape) the information could be transmitted at a higher speed, increasing capacity; and it would also be possible to schedule onward transmission of pre-recorded communications more efficently, making better use of the full twenty-four hours in a day.

> Edison had invented a remarkable device for the automatic repetition of telegraph messages. It consisted of a simple apparatus by means of which the dots and dashes of the original message were recorded in a series of indentations on a long narrow strip of paper. This record could be fed into a sending machine and the message retransmitted without the service of an operator. In other words, Edison had made pictures on paper of the sounds communicated over the telegraph wires, thereby approaching the phonograph from another direction.
>
> "In manipulating this machine," Edison wrote in 1888, "I found that when the cylinder carrying the indented paper was turned with great swiftness it gave off a humming noise from the indentations—a musical, rhythmic sound, resembling that of human talk heard indistinctly."[3]

Edison was not thinking about music, and *he was not even thinking about the voice*. He was thinking about digital information storage, transmission and recovery, about data compression and time management. It was while listening to a sample of Morse code rewinding at high speed that Edison noticed that the accelerated pattern of *dits* and *dahs* sounded curiously like a speaking voice. That being so, it made sense to investigate whether a high-speed indentation recording could be made of a genuine voice signal. It would need a more robust medium than paper, for example, tinfoil, and a more compact form of storage, such as the cylinder. The editors of *Scientific American* were greatly impressed by a demonstration of a prototype machine bidding the assembled company good day and then reciting "Mary Had a Little Lamb." (One wonders if the gathering would have understood a recording of material less familiar in rhythmic and melodic inflection.)

The entrepreneur in him envisioned future markets for sound recorders and reproducers as speech and music storage devices for business and home entertainment. But, to his credit, Edison realized there was a great deal still to be done to make the tinfoil phonograph a viable commercial product. At the time he had other things to do, so he put the invention to one side while maintaining a close watch on the competition.

In all essential respects, the original phonautograph served as pattern and model for the recording machines that followed. The power of the voice signal drives the engraving stylus. The energy of a voice speaking or singing into a recording horn is concentrated as it converges on the diaphragm. The wider the aperture, the greater the initial area of wavefront and thus the amount of energy available to collect. Greater energy means a stronger signal moving the recording stylus, even though the waveform is less coherent at higher frequencies as a consequence of the addition of direct and horn-reflected signals. If a reproducing horn is of similar diameter to the recording horn, the emerging wavefront will be to scale with the original voice, and the audio image is likely to sound more realistic as a result. All the same, the listener's love affair with acoustic recording is a bit like Shakespeare's tale of Pyramus and Thisbe in *A Midsummer Night's Dream*, a romance conducted through a hole in the wall —or in today's terminology, through a wormhole in space and time.

Here and now

Telecommunications today relies on a number of competing technologies working together, but in the latter decades of the nineteenth century the telephone, the movies, and the phonograph served very different interests. The telephone converted audio signals into electrical information but did not store the information in recorded form. Nevertheless, the industry could capitalize on the combined assets of instant communication and online access to set up an early form of cable entertainment in a few major cities, many years before the arrival of wireless radio. The telephone *abolishes space* and allows for widely separated audience participation in the same event at the same time. Movies *abolish space and time*, converting organic, real-time events into a montage of still images that, despite being transportable and reproducable anywhere at any time, are no longer directly verifiable and may indeed be unreal, like a computer simulation today. Phonograph recording is different again. A sound recording *abolishes time*. As a *continuous trace* of an actual performance, not a fragmented succession of still frames, the image from a disc or cylinder gives a verifiable impression of the dynamics of human behavior in real time. When we watch an old movie, the message it conveys is one of *remoteness* in time, but when we listen to an old recording, the contrary impression prevails of *immediacy* in time. Movies are history, but audio is now.

Time travel

As a young man, Albert Einstein worked in a Swiss patent office, and it is not too far-fetched to imagine that his work brought him into contact with patent documents relating to acoustic recording and the relative merits of cylinder and disc as storage media. The information on a cylinder is constant in density, but the signal on an analog disc rotating at a constant speed gets more and more compressed as the stylus tracks toward the center. On the down side, cylinders are difficult to duplicate, whereas flat disc recordings are easy to manufacture in quantity. Today's compact disc combines the best of all possible worlds: digital information at constant density, easy to press, and reading the stored information at a constant speed.

Like early movies, the first audio recordings relied on manual rather than machine regulated drive. For that reason audiences could never be absolutely sure that the actions witnessed in a movie or the voices heard in a phonograph recording were being reproduced at their original speed, and by and large, they did not mind. A speech recording, such as Florence Nightingale's, can withstand irregularities in speed without noticeable impairment of clarity or quality. Once you start recording music, however, the situation changes. The steady tones of folk or classical music are apt to reveal all too clearly any speed inconsistencies arising from manual recording. For scientific and musical applications, it was clearly essential to play back the exact number of frames per second in a movie or number of turns per second of a phonograph recording in order to ensure that the reproduced speed was the same as the speed of recording, and that therefore the reproduced image was a true representation of the original. The same uncertainty applied to Scott's original phonautograph, which was rotated by hand. In the absence of a reliable mechanical drive for the instrument, Koenig had the bright idea of incorporating the constant tone of a tuning fork as a reference frequency—a reminder of the analogous role of the drone accompaniment in bagpipe or folk music (as, for example, in the song "Mandad' ei Comigo") as a tone regulator. As the industry expanded, W.H. Preece, representing the British Post and Telegraph Service, lobbied in favor of industry-wide standards of recording and playback speeds. For its part, the audio industry regarded *constant* speed recording as a higher priority than agreement on *absolute* speed. The movie industry remained undecided. For many decades film speeds fluctuated between twenty-two and twenty-five frames per second, commercial sound recording speeds likewise varying between a low of about sixty-five and a high of eighty or more revolutions per minute. Today's standardized professional audio equipment incorporates the digital equivalent of Koenig's tuning-fork as a matter of course.

The basic argument in favor of mechanically regulated timing for analog acoustic recording and reproduction is, all the same, highly intriguing because it involves a new thought process that distinguishes between the continuous time experience of an observer and the time experience of the thing observed. It argues, in effect, that *reality is in the speed of travel* and, by implication, that if the observer is moving at a different speed from the subject, the two experiences and their respective realities no longer coincide. H.G. Wells's novel *The Time Machine* plays with the notion of clock time and experiential time as separate processes. One can imagine young Einstein reflecting on the thought experiment that reality itself is a recording and that an individual experience of reality is conditioned by the speed at which it is "tracked"—or, how fast a person lives. Since we experience life in practice as a continuum and only in retrospect (in *memory*) as a loose configuration of fixed images, the relativistic analogy is closer to audio than to the movie.

Short, hot, round

In practice an acoustic recording and the reality of a concert performance were always going to sound very different. An acoustic

recording is driven by the power of the voice or instrument signal. That power is concentrated in the middle frequency range.
Because there is inertia in the system and resistance in the wax
recording medium the weaker high frequencies do not usually
survive. The acoustic image sounds veiled and distant as a result.
Lack of low frequencies is audible as lack of presence and weight
whereas lack of high frequencies is perceived as an absence of
sharpness, consonants, clarity, and intelligibility. The most successful recordings in the age of acoustic technology were made
by instruments and voices that conformed best to the frequency
limitations of the equipment. They were the soprano and tenor
voices and the solo violin and cello. The piano's weakly sustaining percussive tone did not record at all well, and members of
the orchestra situated at a distance from the horn, even less. Not
only were voices and violin especially powerful in tone in the
frequency bands where it mattered, they could also be brought
right into the center of the recording horn, and their ability both
to sustain and to modulate tone expressively were significant
added advantages.

Nineteenth-century opera, in effect, defined the repertoire
and the artistic standards of acoustic recording. To call grand
opera an art of screaming is to acknowledge the factors that
made classical recording a successful business in the twentieth
century, and opera singers rich and famous. Opera in the witty,
conversational style of Mozart and Rossini was not really suited
to the large commercial opera houses of the mid-Romantic era—
nor to their vast, middle-class audiences whose musical preferences tended toward the sentimental ballad. Even disregarding
considerations of nineteenth-century taste, one is still left with
the physical challenge to singers of producing signals capable of
being understood by frequently restless audiences seated at the
outer limits of "the gods"—the cheapest seats in the farthest
corner of the house, right under the heaven's angels painted on
the plaster ceiling. An opera house is not reverberant like a
cathedral and cannot sustain voices in the same way. Furthermore, opera by and large is a dynamic process and not a static
ritual. The large orchestra of Romantic opera has a job to do
supporting the voice, but at the same time is directly in competition with the singer. In 1829, at the age of thirty-seven,
Rossini gave up the struggle and abandoned opera for good.
Uniquely, Wagner had the resources and the foresight to address

the acoustical issues facing Romantic opera in the process of designing his own opera house at Bayreuth in Germany. Opera singing had become an art of *alarm*, and the practice of opera singing a competitive exercise in Olympian (if not Olympic) strength and stamina.

Pumped up and ready to go

The opera singer at the turn of the new century came ready equipped, therefore, with the physical skills and techniques required to make a successful career in acoustic recording, qualities that listeners continue to admire, for example, in the artistry of Italian tenor Enrico Caruso, who has remained a star ever since his recording debut in 1902 and most of whose recordings continue in circulation a century later. And, of course, opera singers also had a suitable repertoire of arias and items of appropriate length for an early disc or cylinder, which was just as important. Unless you already know the lyric you do not listen to an early Caruso recording for the words because the exact words are seldom clear, the sibilants largely unrecorded. But this is exactly how his voice would have sounded in a real-life opera performance to contemporary audiences sitting in the cheaper seats. A voice that communicates almost entirely in pure vowel sounds is best adapted to express what vowels are suited to express: not intelligence, which relies on consonants, but motivation or emotion.

> A voice would be shorn of the full richness of its harmonics. In some ways this could be flattering, giving that sort of ethereal effect we hear from a voice singing offstage in the theatre. . . . The loss was greater than any gain, however, for a voice remarkable for its purity of tone, as Melba's was, would be reduced to something pipelike when the harmonics were cut. . . . Tenors generally came off best; the characteristic tenor's sound is bright-edged, and the voice is well in the middle of the gramophone's frequency range.[4]

Recordings by opera singers were originally sold as *souvenirs* of a live performance, not as documents in their own right. They remain available nonetheless as powerful images of human personality and musicianship from a remote time.

Cleaning up the past

Remastering of historic recordings to compact disc raises aesthetic pros and cons similar to the debate over the cleaning and restoring of the Sistine Chapel frescoes. The argument that a layer of dirt on an old painting is part of its heritage—its sense of pastness—is met by counter-arguments that old dirt conceals the true picture, is unnecessary and potentially hazardous, and that the restoration process can also remove layers of earlier interventions, such as overpainted figleaves, that were not part of the artist's original conception. In recording terms, dirt corresponds above all to surface noise and, to a lesser degree, to the coloration of sound introduced by horn resonances. During the vinyl era, re-mastered archive recordings were often scrubbed clean by filtering away the entire higher frequency band. In the seventies an enterprising American engineer Thomas Stockham, pursuing a line of research initiated by Professor Dayton S. Miller in 1915,[5] produced a new series of remastered recordings of Caruso in which the added resonances of the recording horn are neutralized—in effect, taking out the "horn" quality and restoring a more natural balance, a literally "uncanny realism," to the voice.[6]

Nowadays excessive intervention in an archive recording is frowned upon, and the original hiss stays put because buried within it is a vestigial signal that nevertheless contributes to the artistic quality of an original recording. But since regular listeners don't appreciate hiss in a recording, the question is what to do to reduce the impact of high frequency noise without interfering irreversibly with the original recording.

In an admirable exercise in lateral thinking, the English manufacturer Nimbus came up with an answer by asking *why* surface noise is audible in the first place. The reasons why we hear surface noise in an old recording are 1. wear and tear, 2. abrasive impurities in the disc material, and 3. *the enhanced sensitivity of modern lightweight pickups and reproducers*. A record company can go back to archive recordings in original, unmarked condition, or can even make new pressings in vinyl from surviving stampers, and still have to deal with surface noise that remains as an artifact of the original mastering process. Nimbus decided to approach the problem by remastering not by direct transfer from an original disc but instead by recording *a performance* as it would have been enjoyed on a contemporary gramophone of the highest acoustic quality. In addition, the reproducing stylus used to play back these historic recordings is neither diamond nor an old-fashioned steel needle, both of which are very hard and unforgiving, but the more compliant cactus spine, a natural stylus of organic fibrous material that compresses into the record groove under the weight of the pickup head, and, in consequence, provides more accurate tracking and less incidental frictional noise. By recording from a gramophone playback Nimbus also managed to capture more bass response from cabinet resonance, and in their transfer of Caruso's 1904 recording of Leoncavallo's "La Mattinata" it is possible to hear the piano surge in loudness from time to time as the tenor moves to one side to prepare for a high note, allowing the piano behind him unimpeded access to the recording horn.[7]

Opera stars tend to acquire the physique of Sumo wrestlers because the work demands strength and stamina, and that means body mass. A singer of formidable bulk standing directly in front of the recording horn presents a significant obstruction in the way of any accompaniment, whether it be solo piano or full orchestra. Another Nimbus reissue of the Australian diva Dame Nellie Melba singing the aria "Caro Nome" from Giuseppi Verdi's opera *Rigoletto*, originally recorded by Victor in 1907 *with orchestra*, testifies to the remarkable fidelity of the best acoustic recordings of that era. To appreciate the quality of Melba's singing is to begin to understand the high standard of the recording process as well. Her voice is firm and clear, the melodic range from high to low impressive, the controlled evenness of tone very noticeable, and there are a number of climaxes of huge volume and incandescent radiance. Classical music is the most revealing of any weaknesses in any audio technology. Melba's sustained high notes have a laser-like power, and her trills (rapid alternations between adjacent pitches) are marvelously precise and even. These are long duration signals and there is no sign of wavering, either in the voice or in the turntable speed. A listener can detect a speed fluctuation in a recording as a change of voice pitch, and with a voice as clear and firm as Melba's the slightest fluctuation would be very obvious. Dame Nellie was built like a tank and terrorized everyone in sight except Fred Gaisberg, the father of classical recording and her producer for His Master's Voice (HMV); but her voice on disc has the explosive force and accuracy of a guided missile.[8]

Attention span

A major limitation of acoustic recordings is storage capacity—the time limit of three to four minutes of early cylinder and disc. After a century of technological advance, three to four minutes is still the normal length of a song, now elevated to a constant and routinely, if unofficially, defined as the attention span of the average listener. There is *no technological reason* why three to four minutes should continue to be the standard duration of a "single," although the economic viability of music radio is intimately linked to the number of songs that can be broadcast every hour, and thus to the standard length of a song. What is certainly true is that most classical music lies outside the time frame of early standard recording, and for that reason remained hopelessly uneconomic until the advent of the long-playing disc. Because of the limited duration of early cylinder and disc media, the classical music record industry through necessity created a market for excerpts and highlights. A century later, in the age of the compact disc, that market still exists.

It is altogether fascinating to consider the impact of acoustic recording on the classical composer. Somebody like Mahler, for example, following in the grand symphonic tradition of Schubert, Brahms, and Bruckner with its leisurely, contemplative style and overwhelming richness of color and effect, must have contemplated the future of music in the age of recording with disbelief. How do you compress the time-span of a Marcel Proust novel into a three-minute haiku? (Answer: you evolve into Samuel Beckett.) One is tempted to interpret the beautiful resignation of Mahler's "Adagietto for Strings" from the Symphony No. 5 of 1902 as a sense of musical values under imminent threat in a new age of information technology.[9]

Younger composers initially responded more positively to the challenge, producing epigrammatic new works for solo or small ensemble designed very much with the limitations of acoustic recording in mind. Erik Satie's *Trois Véritables Préludes Flasques (pour un chien)*, "limp preludes" for piano dedicated to Nipper, the HMV trademark listening dog, are an example.[10] Others include Stravinsky's *Three Japanese Lyrics* for high soprano and chamber ensemble,[11] or the *Six Bagatelles* for string quartet by Anton Webern.[12] Outstanding in the genre are two compositions by Schoenberg: the ecstatic miniature *Herzgewächse* for high soprano and acoustically miniaturized orchestra of celesta, harmonium, and harp,[13] and the 1912 monodrama *Pierrot Lunaire*, expressionist settings of twenty-one poems about madness by Albert Giraud, most of them under two minutes in length, for "reciter" and chamber ensemble, clearly designed with acoustic recording in mind and in the style of early spoken voice recordings.[14]

Popular classics

None of the above classical miniatures was ever issued on acoustic disc, which is a pity. The "early music" movement, if it has any grasp of the history of technology, ought to consider the possibility of authentic acoustic recording of such pieces, if only as an academic exercise to hear the effect of the veiled perspective of contemporary acoustic audio on a listener's perception of these rather special compositions. The reality of classical instrumental music in the acoustic era is represented by Austrian-born violinist Fritz Kreisler, who perhaps did more than any other single artist to promote public interest in non-vocal classical music in the formative years of the record industry. Kreisler was a genius without a repertoire of items of suitable length and character for recording on disc. He concealed the fact and promoted classical music by the clever deception of composing original works for violin and piano "in the style of" composers from the seventeenth and eighteenth centuries, titles such as *La Précieuse (in the style of Couperin)* and *Variations on a Theme of Corelli (in the style of Tartini)*. These engaging forgeries, charmingly written and elegantly performed, had an enormous influence on popular taste.[15]

Acoustic to electric

Data reduction as practiced in the acoustic record industry covered more than just bandwidth (lack of bottom and top) and time (the four-minute limit) as can be judged by the description of a landmark recording of English composer Sir Edward Elgar's major oratorio *The Dream of Gerontius* undertaken in the twenties for the Edison-Bell company:

> The first step was to obtain the permission of Novello's, the publishers of the score, to record the work. Now came

the cruel task of cutting a score in which every bar seemed indispensible and impossible to sacrifice. Besides the score, the limitations of pre-electrical recording also demanded a drastic curtailment of orchestra and strings. Instead of the usual complement of forty to fifty strings, I had to content myself with nine. A choir which in public performance consists of anything from sixty to three hundred voices had to be cut to eight. Another perplexity was the grand organ, anbe cut to eight. Another perplexity was the grand organ, an instrument which never yet had been recorded effectively; fortunately and surprisingly, the bass concertina made a convincing substitute.[16]

Some improvement in range and definition was being achieved despite the fundamental inefficiency of the system. The HMV company designed a recording studio like a loudspeaker cabinet with all-wood paneling to minimize the loss of ambient sound by absorption and send a clearer signal to the cutting head. But the improvement did not extend to dynamic range, and every performer in the orchestra had still to play as loud as possible.

> The recording studio was set in the uttermost interior [sic] of the building, completely shut off from daylight and outside noise. It was purely utilitarian: no soft lighting, no carpets or curtains brought warmth to the scene. The walls were of unpolished deal, the floor of hardwood and, in the absence of any absorbent of sound, my footsteps thundered on the bare boards, my voice boomed as if my head were in the resounding womb of some giant double-bass. . . .
>
> Mme Chemet and I were dealing with a *Berceuse* but [engineer] Arthur Clark, opening his kennel window, insisted on my playing *forte* all the time. I protested that it was impossible to bang out the notes of a lullaby; I should wake the baby. The result, in the test played back to us, was that I was unheard. I did not relish this. In the last reckoning I obeyed official recommendation and clattered my part of the lullaby like a charge of cavalry, to the approval of all.[17]

The breakthrough came with the arrival of electrical recording and amplification in 1925. Wireless telegraph research, stimulated by the events of the 1914-18 war and aimed at improving the transmission and detection of audio signals over greater distances, led to the development of modern microphone and amplifier technology. Instead of the acoustic energy of a performance having to drive the stylus, it could now be intercepted by a sensitive microphone and converted to an electrical waveform regulating the power output of an amplifier driving the cutting head. The difference is appreciable even for a solo performer such as Fritz Kreisler, if one compares the quite remarkable quality of violin and piano of his 1910 *Corelli Variations* acoustic recording with the noticeably more spacious sound of his *La Précieuse* recorded electrically in 1929.[18]

Peacetime communications

Musical entertainment, along with the news, defined the radio medium from its earliest beginnings in the nineteen-twenties. A communications medium has to communicate something, and the news is an unpredictable commodity, so music provided a useful filler between reports. Radio quickly acquired an image of social connection and coordination and at first interpreted its role of news medium as a responsibility to promote live concerts in preference to recorded music. This culture of immediacy gave rise to a new set of values for recordings of live concert events as distinct from studio productions. The same perception continues to flourish in the circulation of bootleg and live concert albums that carry the atmosphere and emotional mystique of an actual place and time, irrespective of performance or audio quality. A radio broadcast, needless to say, also had the advantage of not being restricted to items of short duration, and could therefore embrace an entire concert if the need arose.

Music, especially classical music, is the most critical of audio signals. A radio ham may listen to a voice news transmission and be content as long as the signal remains strong enough for the words to be understood. The listener to a classical music broadcast, on the other hand, is constantly aware of any shortcomings in the delivery chain. During the years of the Great Depression, competition between radio and the record industry grew ever more intense, and in many respects, classical music can be seen as the driving force behind the huge technical improvements in audio that took place during the glorious thirties. The record industry's answer to radio was to promote the 78 rpm disc as a collectible item offering higher musical content and more reliable audio quality in *permanent* form for *private* enjoyment, despite the fact that most consumers of electrical recording continued to listen to discs on non-electric record players.

Double take

In the nature of things, early recordings could not be edited or corrected. If a mistake was made, or a performance over-ran its allotted three or four minutes, the whole process would have to begin afresh. The single take dominated radio and recording alike until the arrival of commercial tape recording in the late nineteen-forties. The popularity of Maurice Ravel's *Boléro* for symphony orchestra—a work that began life as a ballet commission for Ida Rubenstein but is now nearly always performed as a concert item—may have to do as much with its reputation among musicians and studio engineers as for audience appeal. *Boléro* is a single movement for full orchestra without any breaks. It remains in exactly the same tempo for the best part of a quarter-hour, and makes a continuous crescendo from start to ear-splitting finish. One by one, soloists in every section except the strings are dangerously exposed, first to perform a melodic refrain in vaguely Middle-Eastern style in a very awkward register with very little opportunity to take a breath, and subsequently having to join in a rhythmic background *on one note* that requires the utmost concentration to achieve complete consistency of tone. If any soloist makes a mistake, there is very little option but for the whole orchestra to start again from the very beginning, so nerves are tense.[19]

In addition to anxiety over whether the orchestra will make it this time, a recording engineer faces the supreme technical challenge of finding the right level to capture the signal from start to finish without the risk of distortion. Should an edit be necessary, the level has to match as well as the beat, and the level is controlled by the players and not by the engineer. For the listener, the musical experience is of an irresistible force combining the inexorability of a march with the rolling momentum of a triple rhythm—the image of an armored division on the move, perhaps, rather than of foot soldiers.

Stereo

French inventor and telephone pioneer Clément Ader discovered

two-channel stereo by accident in 1881 as organizer of a demonstration live relay via telephone of a stage play at the Comédie
Française to a Paris exhibition site a few blocks away. After
initial complaints that the players had a habit of wandering off-
mike, Ader decided to connect left and right listening earpieces
to separate microphones (actually telephone mouthpieces), one
directed to the left and the other to the right side of the stage.
The unexpected consequence for many listeners was hearing a
stereo image in depth with realistic movement. The stereo effect
of these early telephone transmissions was surprisingly good,
but technology for recording two-channel sound on to cylinder
or disc did not then exist.[20] In 1931 and 1932 research teams in
the United States and England began to develop three-channel
stereo recording, though with optical sound and the movies in
mind. An RCA-Bell Labs team collaborated with Leopold Stokowski and the Philadelphia Orchestra to make test recordings
and transmissions in three-channel stereo that included excerpts
from Russian composer Alexander Scriabin's *Prometheus—The
Poem of Fire* and *Pictures from an Exhibition* by Modest Mussorgsky in the orchestration by Ravel.[21] EMI also made a number of
experimental stereo recordings including the final movement of
Mozart's Symphony No. 41 conducted by Sir Thomas Beecham,
using a different technique and microphones of advanced design
by Alan Blumlein.[22] At this time there was realistically no popular market for stereo music on disc. The industry was in a bad
way; the shellac 78 rpm disc was an unsuitable medium; the
depression was in full swing; and very few consumers actually
owned electric record players. Movies were another matter.
Optical sound offered high definition audio reproduction which,
together with the exaggerated scale of images on the cinema
screen, made left-right movie dialogue a suitable candidate for
stereo reproduction.

Essentially mono

Classical composers remained relatively unaffected by stereo
until 1956, by which time the long-playing vinyl disc introduced
by CBS in 1948 was firmly established. Despite its wide-screen
image, the traditional symphony orchestra is essentially a monaural music source intended for single-channel reproduction by
radio, television, or disc. The argument that stereo brings greater

realism to the listening experience is really rather thin. As long as the relationship of music and listener is static there is not much to be gained aesthetically or in any other sense by having part of the music come from one speaker and part from the other. However, if there is a genuine spatial dialogue going on in the music, that is another matter entirely. Among the small list of classical compositions that genuinely gain from multi-channel reproduction are the Monteverdi *Vespers*, the *canzone* of Gabrieli and Priuli, Thomas Tallis's extraordinary *Spem in Alium* for forty voices[23] and the *Lauda Jerusalem* by Antonio Vivaldi.

It is, all the same, intriguing to discover a small clutch of compositions dating from the thirties that are conceived in terms of experimental three-channel stereo. The most widely-known is the *Music for Strings, Percussion and Celesta* composed in 1936 by Hungarian composer Béla Bartók for double string orchestras situated left and right, flanking piano, harp, celesta, timpani and percussion in the center.[24] The deep melancholy of its third movement Adagio is often heard in documentary movies as accompaninent to imagery of the desolation of war. But Bartók's rapid left-right dialogue in the faster movements of this complex work is a bit too elusive for a live audience in concert to follow, and it is interesting to speculate whether it was conceived with recent developments in the movie industry in mind as a *movie performance in stereo*. A year later, in 1937, Bartók composed the *Sonata for Two Pianos and Percussion*, a second work using the same unusual three-channel formation.[25] There is a connection between this work and a 1932 ballet score, *L'Envol d'Icare* by the nineteen-year old Igor Markevitch, that much impressed Bartók in a reduced version for two pianos and percussion.[26] In 1938 the Czech Bohuslav Martinu composed a *Concerto for Double String Orchestra, Piano and Timpani* employing an almost identical layout to the Bartók *Music for Strings, Percussion and Celesta*.[27] Another piece of the puzzle is Swiss patron and conductor Paul Sacher, to whom both orchestral works are dedicated and who conceivably may have been entertaining the thought of creating his own stereo orchestral concert on film as a calculated rejoinder to Disney's cartoon classics.

Fidelity

Fidelity to what? asked Stravinsky on the subject of quality in a stereo recording. High fidelity in a recording has two distinct meanings. One is realism in the sense of *being there*: the integral spatial experience that Alan Blumlein had in mind for stereo in 1934, and that continues to drive purist audio engineers in their quest for stereo and surround-sound realism. In this connection one thinks of Decca in Britain and Mercury in the United States for minimalist stereo, and Nimbus, Hyperion, and other (mostly British) new labels as their heirs in the era of surround sound. The other meaning, the sense Stravinsky had in mind, is high fidelity as *high definition*: ideal rather than real.[28] If clarity is the goal, then more microphones can provide it, but that high definition sound is also quite unreal in spatial terms. If your end product is destined to be mono, then no matter, but if stereo is your goal, the audio panorama may end up sounding like a patchwork of arbitrary locations.

One of the great freedoms introduced by electrical audio technology was the freedom to position more than one microphone in the same performance to highlight a solo performer and give the engineer some control over the balance. During the radio era, the layout of a studio orchestra or big band on the platform was largely decided by the relative strengths of different instruments, the aim being to ensure a good balance at the microphone without the intervention of a balance engineer in a glass-fronted box. Popular music continued to play at more or less constant loudness in the tradition of acoustic recording. Much in the way that an electric guitar employs a distortion circuit to suggest loudness without necessarily making the sound any louder, so trumpets and trombones in the era of the big band employed a range of mutes as an expressive alternative to raising or lowering the sound level (a neat exception being the orchestrated fade at the end of the 1940 Duke Ellington/Billy Strayhorn classic "Take the 'A' train").[29] Band players standing up for a solo were not doing so just for effect. Standing up brought their instruments closer to the microphone, and sweeping them left to right produced a natural fade.

The microphone also inaugurated a new close-up style of intimate singing. Crooning, in the style mastered by Bing Crosby and Frank Sinatra, is a paradoxical combination of the intimacy of chamber music and the power of opera. The microphone made it easier for a laid-back vocalist in the late thirties to achieve the same dominance over an orchestra that a Dame Nellie Melba or a

Caruso were only able to achieve in their opera roles by sheer power, and in their acoustic recordings, by physical bulk.[30] The difference between a Caruso and a Sinatra recording is not one of realism, because both are equally unreal. It is a difference in the quality of intimacy that arises from the fact that, with the aid of a microphone and amplification system, an opera-sized audience *can hear the consonants just as clearly as the vowels.*

Global memory

The magic of audio inspired many of yesterday's avant-garde composers. Radiophonic allusions appear clearly in orchestral works by Italian Luciano Berio and German Karlheinz Stockhausen. The latter's *Trans* (pronounced "trance") of 1971 is a theatrical simulation of a short-wave radio transmission, complete with amplified sounds of switching channels. String and organ tones in cluster formation create a visual and acoustic "wall of sound" imitating radio interference that varies in opacity with the *clack-clack* of an amplified weaver's shuttle passing overhead from speakers to the left and right. Through these musical atmospherics, the audience strains to pick out the distant sound of big band, jazz-style harmonies from one or more wind bands concealed behind a sound-transparent curtain. From time to time, a rogue transmission bursts through the static—a viola suddenly erupting in a fiery improvisation, a shell-shocked violin, a demented cello, a sneering trumpet.[31]

Berio's postmodern *Sinfonia* of 1968 is an extended meditation on memory and the medium of recording, drawing on a range of literary (Samuel Beckett, James Joyce) and musical allusions. The third movement, "In ruhig fliessender Bewegung" (In a calm, flowing motion) employs the third movement of Gustav Mahler's Symphony No. 2 ("Resurrection") as a cantus firmus or musical stream of consciousness, the same image of eternal motion as Johann Strauss II's *The Blue Danube*, but adding layer upon layer to the flowing river of Mahler's theme. On the surface float fragments of a Bach solo violin partita, Beethoven's "Pastoral" Symphony, Stravinsky's *The Rite of Spring*, Ravel, Ligeti, Ives, and others in an uncanny prediction of the global electronic memory of today in which potentially all music is stored and permanently accessible. This is a work that asks the question whether there is anything new left for music to say, or whether all we can now do is recycle the past.[32]

Notes

1. Dayton C. Miller, *The Science of Musical Sounds* (New York: Macmillan, 1916), 71-73.

2. Roland Gelatt, *The Fabulous Phonograph 1877-1977.* 2nd rev. ed. (London: Cassell, 1977), 33-35.

3. Ray Stannard Baker, *The Boy's Book of Inventions: Stories of the Wonders of Modern Science* (London and New York: Harper & Brothers, 1903), 252-63.

4. J. B. Steane, *The Grand Tradition: Seventy Years of Singing on Record, 1900 to 1970* (London: Duckworth 1974), 7.

5. Miller, *The Science of Musical Sounds,* 156-74.

6. Enrico Caruso, *The Legendary Enrico Caruso: 21 Favorite Arias.* Stockham Soundstream computer process (RCA Red Seal 5911-2-RC, 1987).

7. Ruggiero Leoncavallo, "La Mattinata" rec. 1904. Enrico Caruso, Ruggiero Leoncavallo. *Caruso: The Early Recordings* (Nimbus Prima Voce NI 7900, 1999), track 1.

8. Giuseppe Verdi, "Caro nome" from *Rigoletto* rec. 1907. Dame Nellie Melba and orchestra. *Great Singers at La Scala* (Nimbus NI 7858, 1994), track 5.

9. Gustav Mahler, "Adagietto" from Symphony No. 5. Smithsonian Chamber Players, dir. Kenneth Slowik (DHM 054272 77343 2, 1995).

10. Satie, *Trois Véritables Préludes Flasques (pour un chien)* for piano. Aldo Ciccolini (EMI CZS 767282 2, 1991), disc 2, tracks 28-31.

11. Stravinsky, *Trois Poèmes de la Lyrique Japonaise.* Christiane Eda-Pierre, Ensemble du Domaine Musical, cond. Gilbert Amy (Adès 203512, 1991), track 4.

12. Anton Webern, *Six Bagatelles* for string quartet Op. 9. Artis Quartet (Sony SK 48059, 1992), tracks 9-14.

13. Schoenberg, *Herzgewächse* Op. 20. Christine Schäfer, Solistes de l'Ensemble Intercontemporain, cond. Pierre Boulez (DG 457 630-2, 1998), track 22.

14. Schoenberg, *Pierrot Lunaire* Op. 21. Christine Schäfer, Solistes de l'Ensemble Intercontemporain, cond. Pierre Boulez (DG 457 630-2, 1998), tracks 1-21.

15. Fritz Kreisler, *Variations on a Theme of Corelli (in the style of Tartini)* rec. 1910; and *La Précieuse (in the style of Couperin)* rec. 1929. *Kreisler plays Kreisler* (RCA Victor 09026-68448-2), tracks 6, 19.

16. Joe Batten, *Joe Batten's Book: The Story of Sound Recording* (London: Rockliff, 1956), 58-59.

17. Gerald Moore, *Am I Too Loud? Memoirs of an Accompanist* (London: Hamish Hamilton, 1962), 60-61.

18. *Kreisler plays Kreisler*, RCA Victor 09026-68448-2, 1997.

19. Ravel, *Boléro*. Vienna Philharmonic, cond. Lorin Maazel (RCA Victor 09026 068600-2, 1997), track 8.

20. Anthony Askew interviewed by Laurence Stapley. "Developments in Recorded Sound" No. 4 (British Library National Sound Archive C90/78/02, June 1985).

21. Modest Mussorgsky, orch. Ravel, *Pictures at an Exhibition*. 1931 experimental stereo recording. Philadelphia Orchestra, cond. Leopold Stokowski (vinyl, Bell Laboratories BTL 7901, 1977).

22. Mozart, Symphony No. 41 in C K551 "Jupiter" I. movement, (excerpt). 1934 experimental Blumlein stereo recording. London Philharmonic Orchestra, cond. Sir Thomas Beecham. *Centenary Edition: 100 Years of Great Music* (EMI promotional disc 7087 6 11859 2 8, 1997), track 15.

23. Thomas Tallis, *Spem in Alium* (40-voice motet). Tallis Scholars, cond. Peter Phillips. (Philips 289 462 862-2, 1999), disc 1, track 2.

24. Béla Bartók, *Music for Strings, Percussion and Celesta*. Orchestre Symphonique de Montréal, cond. Charles Dutoit (Decca 421 443-2, 1991).

25. Bartók, *Sonata for Two Pianos and Percussion*. Murray Perahia, Sir Georg Solti, David Corkhill, Evelyn Glennie (CBS MK 42625, 1988).

26. Igor Markevitch, *L'Envol d'Icare* for two pianos and percussion. C. Lyndon-Gee, K. Lessing; F. Lang, J. Gagelmann, R. Haeger (Largo 5127, 1993). Version for orchestra: Arnhem Philharmonic Orchestra, cond. Christopher Lyndon-Gee (Marco Polo 8.223666, 2000).

27. Bohuslav Martinu, *Concerto for Double String Orchestra, Piano and Timpani*. Czech Philharmonic Orchestra, cond. Jiri Belohlávek (Chandos CHAN 8950, 1991).

28. Igor Stravinsky and Robert Craft, *Memories and Commentaries* (London: Faber and Faber, 1960), 123-6.

29. Billy Strayhorn, "Take the 'A' train," rec. 1940. Duke Ellington and his Famous Orchestra. *The Duke Ellington Centennial Edition* (RCA Victor 09025 63458-2, 1999), track 8.

30. Kurt Weill, "September Song," rec. 1946. Frank Sinatra and orchestra, cond. Alex Stordahl. *Kurt Weill: from Berlin to Broadway vol. II* (Pearl GEMM CDS 9294, 1997), disc 2, track 5.

31. Stockhausen, *Trans* (Stockhausen Verlag SV 19, 1992).

32. Luciano Berio, "In ruhig fliessender Bewegung." III movement of *Sinfonia* for 8 voices and orchestra. Electric Phoenix, Orchestre de Paris, cond. Semyon Bychkov (Philips 446 094-2, 1994).

CHAPTER TWENTY

Outer space

THE CONTINUING popularity of sci-fi entertainments such as
Star Trek and *The X-Files* might seem to suggest that the idea of
making contact with hostile alien life-forms is a relatively modern
fixation. Wormholes in space, parallel universes, creatures from
another dimension, threats to the citizenry by national security
or to the national security by flying saucers are all part of an ongoing and essentially religious devotion to the idea of human
existence beyond the limits of ordinary perceptions. Any invention that extends human awareness beyond the normal reaches of
space and time is properly regarded with awe. Music is a part of
that natural drive toward self-perpetuation. In recording music
on disc or in notation, the individual creates a self-image that
will persist beyond the maker's lifetime, which for most people is
a natural though mysterious goal. Along with that impulse goes
a keen awareness of the possibility of other realms or states of
being. Earning a place in a hall of fame can be as potent an influence on the secular life as the prospect of heaven and hell on
the religious life. Just as the gods of ancient Greece embody unseen natural or emotional impulses that can reach down and
touch or control ordinary lives, so in George Pal's 1953 version of
H.G. Wells's *War of the Worlds* or George Lucas's more recent
Star Wars movies the threat from outer space is understood in
terms of human events being influenced by forces beyond the citizen's control. If it's not invaders from Mars, it's the government.

Among predominantly visual cultures fear of the unknown is fear of the unseen. In the *Alien* spaceship, light does not illuminate.

Connecting with the unseen

Music celebrates the invisible world of sound, and in so doing offers both a passport to outer space—the mysterious acoustic realm of ritual incantation—and an ability to conjure a response from unseen influences through the recorded sound of voices from a remote space or time. By 1900 the world was gearing up to a new era of the incorporeal and the transcendent, represented in the dazzling surfaces and yearning erotic impulses of Austrian artist Gustav Klimt and composer Arnold Schoenberg. Like many of his contemporaries, Russian artist Wassily Kandinsky was driven toward abstraction inspired as much by music as by a quasi-religious impulse to represent a transcendental reality. There are elements of microscope imagery and landscapes seen from the air in his watercolor improvisations, but their message is intentionally sensational rather than intellectual, more about the act of seeing than the merely visible. Like many composers from Alexander Scriabin to Olivier Messiaen, Kandinsky was a firm believer in synesthesia, the association of color and musical timbre.[1]

The most powerful influences on human development are always the least material: heat, which transmits energy; light, which makes visible; magnetism, which shows direction; sound, which transmits information. Perhaps the key event to signal a new era of transcendentalism in the twentieth century was Italian pioneer Guglielmo Marconi's 1901 wireless transmission across the Atlantic from Cornwall in England to the Newfoundland coast. To the lay person, wireless was more mysterious even than electricity. In their own ways, the telephone and the phonograph actually *demystified* the world of sound. In extending the range of human communication in space and time, these two great nineteenth-century innovations had succeeded in converting a mysterious dimension of human experience into a logical cause and effect process. With a telephone it was possible to talk to somebody far away and out of sight, but there had to be a person to talk, a person to listen, and a wire connecting the two along which the signal could pass. The phonograph introduced the possibility of storing the voice, but it also involved converting a previously invisible, evanescent, and intangible *sensation* into a visible incision on a solid and permanent body. Using a telephone or making a phonograph recording was no different in essence from playing a violin. The sound it produced was still the end product of human action (the cause) detected and amplified by passive technology (the effect). *Wireless telegraphy changed the paradigm,* eliminating the wire and restoring the mystery of audio communication. Nothing to see, nothing to touch. Communication between sender and receiver relied not on *doing* but *being,* not action but a state of readiness or *sympathetic vibration.* It was back to the age of Pythagoras, of the singing voice and a responsive harp, only this time the singing voice was an unknown disturbance in the ether and the listener the tuned receiver.

Switch on, tune in

Building on wartime advances in wireless technology, in the aftermath of the 1917 Russian Revolution electrical engineer Leon Termin introduced a musical instrument called the *theremin* that was the answer to every conceivable spiritualist fantasy of the Age of Electricity. The theremin reinvented music as a form of psychic resonance. Here was an instrument that produced voice-like melodies of an unearthly beauty *without direct physical contact.*[2] Generating music by silently waving your hands in the air was nothing new. A conductor or bandmaster waves a baton and the music plays. But an orchestra or band is a team of real life players who have their instructions and act on command, whereas a theremin is a device without visible musical hardware, that works without programmed software, is unable to see and rigorously isolated from human contact. This is a very strange combination of factors indeed, and no wonder a theremin concert was more like a séance than social entertainment.

Explaining how the theremin works does not in fact alter the mystery. *Outboard vertical and lateral aerials detect hand movements that influence the electrical capacitance of the system and control the frequency and amplitude of one of a pair of ultrasonic oscillators. The two oscillations added together generate a difference tone within the audible range.* There, what did I tell you? Unlike a conventional musical instrument which is relatively passive, the theremin is electrically charged and thus sensitive to changes in an electrical field. The hands being moist attract varying amounts of energy from the exposed elements in the system. The mysterious gestures of the theremin player simply vary the exposed skin area within range of the vertical aerial controlling pitch and the loop aerial to the side controlling amplitude. In effect, theremin and performer comprise a coupled oscillating system in which the soloist, instead of being the energy source, becomes a filter or modulator. H. G. Wells may have had the theremin in mind in his short story of a person discovering how to contact a parallel dimension by waving his hands in mysterious fashion in front of his face. The appeal of the theremin to contemporary audiences, however, was much more direct. First, the oscillator conjured up a stunningly realistic voice out of thin air; second, the voice sang with a wordless purity and power that conveyed the impression of a supernatural being.[3]

Ghost in the machine

Philosopher Gilbert Ryle refuted Descartes with a catch-phrase that, even if not inspired by the theremin, could have been designed for it. Children who play ghosts dress up in sheets, hide in the attic and wail a wordless "Woo-oo-oo!" out of the dark. The awesome, supernatural quality of the theremin relates to the fact that it *sounds* like a human voice, *is modulated* like a human voice, but can go *higher* than a human voice, *does not have to take a breath* like a human voice, *does not speak words* like a human voice, and *is more powerful* than a human voice. In her acoustic recordings, Dame Nellie Melba's voice is all of this too, including not speaking words (because you hear the vowels, but you don't hear the consonants). A wordless singing voice is the essence of the human spirit, just as the flute to the Japanese is the embodiment of the breath of life. The added *frisson* of a theremin performance lies in the creation of a disembodied voice out of a machine. That is Gilbert Ryle's ghost. Whatever your musical taste or mythology, whether you find it in the disembodied high soprano at the climax of Schoenberg's oratorio *Die Jakobsleiter* (Jacob's Ladder)[4] or the title music to the original *Star Trek* series, a wordless and invisible singing female voice mysteriously communicates a direct experience of transcendence. (Series director Gene Roddenberry, perhaps mindful of the mass hysteria that attended Orson Welles's all-too-realistic 1938 radio dramatization of *The War of the Worlds*, insisted that the theme music for *Star Trek* should at all costs *avoid* sounding alien or futuristic.) Hollywood's long association of sci-fi themes with fifties bubble-gum music—were our parents really *that* strange? —culminates in John Williams' bizarrely retro score for the deep space cabaret in *Return of the Jedi*.

Back to the future

Time travel is a paradoxical affair, as we know from the movies. The theremin was an instrument of the future, and yet when we listen to Clara Rockmore's exquisitely shaped melodies on compact disc the overwhelming image is of a musical and performance aesthetic of a bygone age, the silent movie era, including the gliding tones or sighing *portamenti* of orchestral strings so affectingly reproduced in the Mahler Adagietto recorded by the Smithsonian Players. Why the theremin was invented at all is still a mystery. Since the instrument made its debut around 1920, before electrical recording, perhaps it was originally intended as an instrument for making *acoustic* recordings. As a solo instrument, it would have offered a number of advantages. It could produce a convincing voice-like tone of exceptional quality in any range and at a lower cost than an opera singer like Dame Nellie Melba; it was completely tireless, and it was capable of producing a sustained signal stronger than the human voice and therefore better for recording. As we already know, the bandwidth of an acoustic recording even in 1920 did not extend into the high-frequency domain of consonants and language. Nor at that time was the classical record-buying public driven by a desire to experience the words of the song, but attracted rather to the sound of the voice.

Ironically, the same valve technology that made the theremin possible also led inevitably to electrical recording in the late twenties and the creation of a new high-definition aesthetic of audible consonants and enhanced clarity of diction. In 1928 RCA began manufacturing the theremin under the trade name Thereminovox, but the instrument was not a commercial success, falling victim, along with the reproducing piano, not only to economic depression but also to its now obsolete art nouveau aesthetic, not to mention the fact that it was so very difficult to master. The theremin ultimately found a more permanent niche, however, as an effects instrument in radio and the film industry, where it continues to evoke—as you might have guessed—the world of ghosts, horror, and little green men.

No ghost in the keyboard

Rapid progress in oscillator and valve circuitry led to the development of a range of alternative electric synthesizers. They included Friedrich Trautwein's trautonium, launched in 1928, for which Hindemith composed in 1930 a rather unexciting neoclassical *Konzertstück* for trautonium and string orchestra.[5] Adjudicating a contest between the theremin and the trautonium is surprisingly reminiscent of the mythical contest of Apollo and Marsyas between the violin and the pipes. The musical advantages of the theremin—its technical simplicity, beauty of tone, and voice-like expression—are met with the practical objections that the instrument is non-standard in pitch and very difficult to learn. By contrast, the trautonium incorporates a keyboard and fixed tuning to make it more user-friendly, but does not possess anything like the tonal beauty and essential musicality of the theremin. There is a law of synthesizers in general that says *any* electronic keyboard or music software, whether it be domestic or professional, hardware or software, transistor era or CD-ROM, analogue or digital—*all of them* suffer musically from their most practical design features. There is a reason. Science and technology created the keyboard and tempered tuning, *not for musical reasons,* but out of an intellectual preference for clarity and order, sustained by a marketing preference for simplicity in performance—and the industry continues to wonder why great music is not produced with advanced technology.

Point and click

A microphone is a sensing device that you place in the way of a musical performance or similar atmospheric disturbance. The microphone diaphragm covers an area of the size of a small coin, to all intents and purposes, *a point*. This extremely small sample of room atmospheric vibration is converted to a fluctuation of electrical energy. This minute fluctuation is magnified by means of an amplifier to create a signal powerful enough to make a recording or drive a loudspeaker. However, the surface area of a standard mid-range loudspeaker driver is many times larger than a microphone diaphragm. *There is a discrepancy here.* A sample vibration detected by equipment the size of a fingernail is reproduced as an atmospheric disturbance from a speaker cone perhaps as large as fifteen inches in diameter. Microphones are small because the smaller the diaphragm, the more coherent the signal; on the other hand, loudspeakers are large because the larger the surface area, the more roomy the sound. It's always a compromise. Real life sound is less coherent in any one location, but more coherent overall.

A synthesizer bypasses the microphone and generates its own electrical versions of the fluctuations a single microphone is designed to pick up. The engineers designing synthesizers theorize from their textbooks that the ideal signals for a musical effect are the purest waveforms. *This is not the case.* The trautonium incorporates simple circuitry to generate an approximation to the three basic waveforms: a sine wave, sounding something like a flute; a square wave, sounding something like a clarinet; and a sawtooth or ramp wave, sounding vaguely like a violin. These waveforms are the acoustic building blocks of the keyboard synthesizer industry. Electrical signals are continuous and uniform, essentially infinite in duration, but a traditional acoustic musical instrument—even a keyboard like a piano or harpsichord—produces sounds of greater complexity and limited duration. Notes start in a distinctive way, resonate within the body of the instrument, then die away as the energy dissipates. *The shaping of a musical note is part of its expressive meaning.* Electronic keyboards don't have that expressive capacity because they are not designed in the traditional way, and they are not designed in that way because they are designed by engineers. If you want expression, you have to look elsewhere. An electric guitar allows the performer to exploit a whole range of expressive distortions from gliding tones to vibrato to manipulating the basic waveform. But, of course, an electric guitar is harder to play, technically demanding, unreliable in pitch, etc. You can't win.

Original sin

Distortion, in fact, is the key. The radio message coming through is a distorted waveform. They call it modulation of a carrier frequency, but distortion is what it is. *The message is in the distortion.*

The lesson of synthesizer design is that the essential human quality in a musical signal is what we recognize as a capacity for *willful deviancy.* So much for public morals. Perhaps the missing ingredient in a work such as Milton Babbitt's *Philomel,* composed on the landmark Princeton-Columbia Mark II RCA synthesizer designed by Harry Olson in the fifties, is a capacity for original sin—because the instrument itself is simply too virtuous.[6]

Balance

Combining an electronic or amplified instrument with an orchestra changes everything. An orchestra controls its own output. Each player understands what is involved in playing loud and soft, and the conductor is available on the podium to ensure a good overall balance of tone. A full choir and orchestra can produce a hugely powerful signal. For sheer range from loud to soft it would be hard to beat the forty-minute "Dies Irae" from the *Messa da Requiem* by nineteenth-century Italian composer Giuseppe Verdi.[7] There is an enormous difference between natural acoustic power in a musical performance, and *amplified* power. It is not just the power of numbers but also the quality of the musical signal. In a live acoustic performance, a huge sound is generated by a wide spread of human energy pushing a vast amount of air, whereas in a heavy metal concert with speaker towers what you are hearing is relatively little human energy vastly amplified pushing out of a comparatively small area. To even approximate the broad-band impact of a Verdi *Requiem,* a rock band audience has to endure dangerously intense signals at far greater energy levels emitted from very much smaller sources (the speaker columns). The fewer players and direct feed or close microphones of a rock band also mean that the amplified signal is *more coherent,* especially at high frequencies, and for that reason potentially more dangerous to hearing. You don't run that risk at an orchestra concert because the signal is generated over a wide area and there is consequently much less coherence in the high frequency bandwidth.

A classical composition incorporating a keyboard synthesizer in an orchestra has to deal with two imponderables: first the *nature* of the electrical signal in relation to a body of acoustic instruments, and second the imbalance between an amplified instrument and the orchestra, an imbalance both of *scale* and *size.*

The difference in nature arises because a synthesizer and its performer are not a coupled system like a normal instrument and performer, for which reason the synthesizer does not respond to the human performer in the normal way. The imponderable of scale arises because the synthesizer signal is internally generated and relatively meager and has to be amplified to become audible in the first place. Size is another, because neither the performer nor the conductor has the same measure of control over the dynamics of an electrical instrument as over one that is unamplified. In addition, loud and soft in a regular instrument like a horn or cello are associated with differences in tone quality. Changes in tone color and brightness occur because the instrument resists, and a higher energy input generates a sharper waveform. You don't get that dynamic response by turning a knob, so a synthesizer doesn't necessarily sound loud even when playing at high volume.

Ondes of the Baskervilles

French inventor Maurice Martenot's *ondes martenot* ("Martenot waves"), also launched in 1928, is a particularly successful hybrid instrument that combines the expressive features of the theremin and the practical advantages of a keyboard. A touch-sensitive ribbon, similar to a slide guitar in operation, allows the performer to create a soaring melody line just like the theremin, but for discrete and rhythmic passages integrating with an orchestra a regular keyboard is also available. Robust and reliable in operation, the ondes martenot has become a permanent feature of the classical music scene in the main because a significant body of music has been written for it since the thirties, most of the best of it by French composer Olivier Messiaen.

Messiaen is an interesting figure and a true transcendentalist whose compositions combine the visionary and the arcane. As an prisoner of war in a German internment camp, he composed the remarkable *Quartet for the End of Time*, in which eerie cello harmonics hovering over the ticking clockwork mechanism of piano, violin, and clarinet evoke the ghostly wail of the ondes martenot.[8] Messiaen's monumental ten-movement *Turangalîla Symphony* completed in 1948 celebrates the end of the 1939-45 war in a hymn of religious and erotic ecstasy in which ondes martenot and piano play leading roles.[9]

Intelligence matters

America's obsession with invaders from outer space, triggered by the 1938 *War of the Worlds* broadcast, was re-energized after the war following rumors of alien spacecraft being shot down near a secret military base in Roswell, New Mexico. The atomic age itself awakened new fears of unseen forces, including radiation and communism, that found expression in Hollywood in a flurry of B movies dealing with unseen enemies, unexplained new technologies, covert action, and mind control. In Hollywood the old electronic instruments were brought into service as symbols of alien invasion. The sensational impact of Messiaen's *Turangalîla Symphony* provided Hollywood composers with a musical imagery appropriate to Cold War angst, directly influencing among others Dmitri Tiomkin's 1951 expert but overloaded score for the movie *The Thing (From Another World)*, incorporating flexatone (banshee wail), wind machine, two pianos, three harps, electronic organ, pipe organ, and one very discreet ondes martenot.[10]

For a new generation of composers, the war experience ushered in a new era of machine intelligence, data encryption, and information theory. Whether they realized it or not, avant-garde composers in both America and continental Europe found themselves involved in activity related directly to voice communication, work that was ultimately to lead to real-time voice synthesizer technology, but which had begun as a semi-covert initiative to develop automated speech recognition technology for Cold War intelligence. The strangest and most alien musical creations of the fifties, works such as Boulez's eerily atomistic *Le Marteau sans Maître* (based on a poem by René Char that might be translated as "The automatic hammer"),[11] Stockhausen's electronic cantata *Gesang der Jünglinge* (Song of the Youths),[12] and Luciano Berio's *Thema: Omaggio a Joyce* (Theme: Homage to [James] Joyce).[13] All share a common intellectual agenda of examining how the human voice conveys meaning and how that meaning is related to individual speech elements and configurations. We are right back in medieval times.

Approaching the same topic from another direction, John Cage addresses the conceptual bases of music and speech communication from a position of machine-like objectivity. In order to grasp how an acoustic signal is programmed to produce an intelligent response, we first have to ascertain what an acoustic signal is in the abstract. The best way to do that is to create performance events that are totally uncontaminated by intention. Cage's preoccupation with chance procedures in works such as the *Music of Changes* for piano can be seen as bringing a composer's knowledge of musical signification (and how to avoid it) to bear on the problem, for information science, of coming to terms with the underlying mental processes relating to the perception of significance itself.[14]

Monsters from the Id

The years 1956-58 proved unusually productive musically. Among the more scientifically interesting exercises in musical speech synthesis dating from this period are two miniature studies created in the electronic music studio of Cologne Radio. Taking a leaf out of Bell Labs' visible speech transcoder, composer Herbert Eimert and studio technician Heinz Schütz produced two short studies "Musik und Sprache" (Music and speech) and "Zu Ehren von Igor Stravinsky" (In honor of Igor Stravinsky), in which sample spoken phrases (the respective titles) are deconstructed and reassembled by a data reduction process involving analog filtering and compression that reduce the speech samples to impulse patterns on a frequency grid.[15] In Nicholas Nayfack's 1956 sci-fi epic *Forbidden Planet*,[16] not only is the subject of alien life addressed more maturely (human beings rather than aliens being the invaders) but also effectively conveyed in radically abstract "electronic tonalities" by Louis and Bebe Barron, representing a lost race of intelligent life-forms transformed into pure energy. This remarkable tape composition, produced with very little special equipment, employs gliding tones, radiophonic distortion, and loop echo for a musical imagery that makes no concession to tradition and contains few of the usual clichés representing fear of the unknown.[17] It is also remarkable for dynamic range, with some of the loudest sounds in cinema music prior to the ear-splitting electronic scream of the mystery plinth in Kubrick's *2001*. Dynamic presence is also a feature of Edgar Varèse's tape composition *Poème Electronique*, a spacious, polyphonic montage of pre-recorded and synthesized audio images created for the Le Corbusier-designed Philips Pavilion at the Brussels World Fair of 1958.[18]

The link between music and alien life took a bizarre twist in Stephen Spielberg's 1977 movie *Close Encounters of the Third Kind*, in which a visiting flying saucer communicates with a reception party on Earth using a tone language based on the teaching method of the Hungarian composer Zoltán Kodály. An oddly similar theme is taken up by Stockhausen in the 1976 cantata *Sirius*, depicting the arrival and departure by flying saucer of four solo musicians to the accompaniment of an eight-channel surround-sound "spin-dizzy" effect producing peculiar strobe-like acoustic illusions.[19]

Twilight of the gods

As a young man, Pierre Boulez played ondes martenot for Jean-Louis Barrault's Théâtre du Petit Marigny in Paris alongside Maurice Jarre, the film composer and father of electronic pop idol Jean-Michel Jarre,[20] later producing two influential early studies in *musique concrète* at Pierre Schaeffer's Paris Radio studios in 1952.[21] Boulez finally realized a lifelong ambition to create a music of exactly controllable electronic timbres with the inauguration in 1977 of Studio IRCAM, located under the piazza fronting the Pompidou Center in Paris, a facility with customized, parallel processing computer technology for real-time waveform interpolation and synthesis. In Boulez's showcase composition *Répons* (1981-84), humanity's age-old dream of making contact with the supernatural reappears in a music designed to radiate outward like a sound wave from a traditional orchestra seated in a center circle to six keyboard instruments forming an outer perimeter. These soloists have a priestly function of interceding, as it were, with the higher intelligence of the 4X computer, communicating with it in mantric lines of notes or arpeggios, corresponding to data strings.[20] Their signals occur at unscripted intervals and are intercepted by the computer that transforms the interval material *in real time* into new melodic and tone-color configurations. The remarkable apotheosis of this extraordinary work takes the form of a conversation of the gods, or at least of artificial intelligences, whose glistening, metallic tones pass back and forth over the heads of the audience.

Notes

1. Sergei Eisenstein, "Colour and Meaning." In *The Film Sense*, tr. Jay Leyda (London: Faber and Faber, 1948), 92-122.

2. Steven M. Martin, *Theremin: An Electronic Odyssey* (MGM, VHS 1000667, 1993).

3. Clara Rockmore, *The Art of the Theremin* (Delos DE 1014, 1987).

4. Schoenberg, *Die Jakobsleiter* (fragment). Mady Mesplé, BBC Symphony Orchestra, cond. Pierre Boulez (Sony SMK 48 462, 1993), track 11.

5. Paul Hindemith, *Konzertstück* for trautonium and string orchestra. Oskar Sala, Munich Chamber Orchestra, cond. Hans Stadlmair (Erdenklang 81032, 1998).

6. Milton Babbitt, *Philomel* for soprano, recorded soprano, and synthesized sound. Bethany Beardslee (New World 80466-2, n.d.).

7. Giuseppe Verdi, *Messa da Requiem*. Leontyne Price, Rosalind Elias, Jussi Björling, Giorgio Tozzi, Singverein der Gesellschaft der Musikfreunde Wien, Vienna Philharmonic, cond. Fritz Reiner (Decca 444 833-2, 1995), disc 1, track 2.

8. Olivier Messiaen, "Liturgie de Cristal" from *Quatuor pour la Fin du Temps*. Fabio di Castola, Ricardo Castro, Emilie Haudenschild, Emeric Kostyak (Accord 201772, 1990), track 1.

9. Messiaen, *Turangalîla Symphony*. François Weigel, Thomas Bloch, Polish National Radio Symphony Orchestra, cond. Antoni Wit (Naxos 8.554478-79, 2000).

10. Dmitri Tiomkin, Suite: *The Thing (from Another World)*. National Philharmonic Orchestra, cond. Charles Gerhardt. *Spectacular World of Classic Film Scores* (RCA Victor 2792-2-RG, 1977), track 18.

11. Pierre Boulez, *Le Marteau sans Maître*. Yvonne Minton, Ensemble Musique Vivante, cond. Pierre Boulez (tape, CBS DCT-40173, 1985).

12. Stockhausen, *Gesang der Jünglinge* (Stockhausen Verlag SV 3, 1991).

13. Berio, *Thema: Omaggio a Joyce* (RCA Victor 09026-68302-2, 1998), track 1.

14. Cage, *Music of Changes* Books 1-4 for piano. Herbert Henck (Wergo 60099-50, 1989).

15. Herbert Eimert and Heinz Schütz, "Musik und Sprache;" "Zu Ehren von Igor Stravinsky." Electronic music. *Einführing in die Elektronischen Musik* (vinyl, Wergo 60006, n.d.).

16. Nicholas Nayfack, *Forbidden Planet* (VHS, MGM/UA M202345, 1991).

17. Louis and Bebe Barron, Title music to *Forbidden Planet*. In *OHM: The Early Gurus of Electronic Music 1948-1980* (Ellipsis Arts CD

3670, 2000), disc 1, track 8.

18. Varèse, *Poème Electronique*, 1998 transfer. *The Complete Works* (London 289 460 208-2, 1998), disc 1, track 3.

19. Stockhausen, *Sirius* (Stockhausen Verlag SV 26, 1992).

20. Jean-Louis Barrault, *Memories for Tomorrow: Memoirs of Jean-Louis Barrault*, tr. Jonathan Griffin (London: Thames and Hudson, 1974), 161.

21. Boulez, "Eventuellement. . ." in *Rélévés d'Apprenti*, ed. Paul Thévenin (Paris: Editions du Seuil, 1966), 177-80.

22. Boulez, *Répons*. Ensemble InterContemporain, cond. Pierre Boulez, electronic music asst. Andrew Gerzso (DG 289 457 605-2, 1998).

Feedback

"PLEASANTLY surprised is probably the best way to describe my experience at the symphony concert the other night. It wasn't what I had expected. Classical music has never set me on my ear and I do admit I was expecting more of the same, although once the evening began it became clear that I was in for more than I could imagine. The experience was expanding! I was not ready for it. In my experience with live music it has been the unexplainable of the live performance that has made the experience great or just okay. The unexplainable is that tight fit of the players that gives the music life and makes the listener a part of the experience, and amazingly enough it was the one thing I was not expecting. It still seems too stale when I listen to classical recordings yet when I remember that night it was anything but stale. You could see the interaction of the violins with the cellos, as well as the lead and the supporting roles. When I watched the lead cellist to the left of the piano play it was stupendous the emotion he carried, and the part the supporting cast played seemed to fill the empty space that he couldn't fill. The music was together, it painted a complete picture, you could see the lead and feel what he was trying to accomplish, and at the same time you could see the way the support dropped back to allow him his freedom while all the while being there ready for him to come back to the fold. It was wonderful!"
—*Mike Warren*

"There is a poetic music of life that surrounds us in all the sounds we hear, if only we would listen correctly. How does one communicate this with others without speech? Tapping a pencil in a notebook in class tells everyone around us that we are bored. Humming or whistling lets everyone know that we are happy or scared. Clearing a throat tells someone you are present. We are all musicians, if only we would admit it to ourselves as we do to the world!"
—*Nicole Guerriri*

"Growing up in my family required me to visit church every Sunday. I didn't mind much but I found myself either bored or frustrated from all of the kneeling, standing, and sitting. It was not until my mother joined the choir that I had a new appreciation for the hour and a half long service.

"Most churches that I know come with high or vaulted ceilings, perhaps to make you feel small and become aware that there is a higher power up there, but also because of the way in which the sounds reflect in the space (acoustics).

"Seated up and behind the parish we don't get to see the choir and musicians but their singing resonates throughout the space, like a higher power, maybe like an angel singing from above. Oh it is a glorious noise! The entire church is filled with the beautiful music.

"Occasionally, about five choir members will leave the loft and go behind the wall directly behind the altar for a different singing effect. Again, no-one can see them. We can only hear the voices reverberating off the long and narrow hallway that is open at the top. These are the *behind the wall* singers and they chant Latin phrases or sing small hymns. When they are behind the wall the sound is quite different and the mood is a bit darker. The songs are somber and don't have as many notes or high notes in them. Since only about five are back there it is softer but yet rings just as loud as the whole choir because of the reflecting off the walls."
—*Kitty Blanton*

"Before I entered this music appreciation class I never thought about all the different elements and meanings that are in music. It was just something to listen to. In a very short time I learned how much music really intertwines with everything. The most interesting to me is how we use elements of music like pitch and tone in our language, speaking and communicating through sound. It is possible to have a conversation with someone based only on noises and sound patterns.

"It's very interesting to me that topics discussed in lectures are relatively simple ideas and concepts. For example, punctuation marks can change the meaning of a word, just like changing the sound patterns can change the whole feeling of a song. I just never thought of it like that before.

"I'm getting the notion that the better understanding I have of music and its elements, the more I can incorporate it into my work, because music and sound are a part of everything whether it's clearly recognized or very subtle."
—*Kasey Bartlett*

"Music can emulate any emotion, express thousands of ideas, and can be understood by people of all cultures. It is truly a universal language. Instrumental music has the ability to capture and express moods and feelings, and inspire images and old memories. Lyrical pieces amplify the inflections and emphasis of certain words and phrases, making their meaning more clear than if they had been spoken. The flourish of each word and phrase is a celebration of its meaning. Each drawn out note gives time to ponder each idea. Beside the emotion that can be expressed, song also creates a consistency in the expression of a text. The tone of each person's voice is different, but there is conformity in a note."
—*Jennifer Mangini*

"Music uses symbols positioned on a type of grid. It communicates several types of information. In many ways it resembles map coordinates and alphabets. It also has a connection to mathematics. The computer has taken symbolic and mathematical language to unbelievable levels of speed and understanding. Early musical information was recorded with a type of picture. Lines were drawn that represented strings or keys of an instrument. From left to right a mark was put on a line to indicate which string was to be played. Symbols were developed to represent specific notes. Musical sounds could then be written, printed, shared, and reproduced for everyone to enjoy. This grid-like structure of musical notation reminds me of a map grid. By

using a latitude and longitude number a specific point can be located by the intersection. That's very simplified but enough to show the relationship to music. The symbols tell you a position or key to play. When an instrument is played many locations must be found very quickly. A graphic bar allows compound understanding. The symbol has meaning and its location on the bar has more meaning."
—*Ron Patrick*

"What struck me the most was that a musician can create a piece of music out of the imagination, an art work that can be played and enjoyed just as it was originally intended. It's not like a painting: you can't recreate a painting brush stroke by brush stroke. Music can be fully experienced through the re-creation of itself. The experience is formed by the players, conductor, and the audience, all on different levels. The artist's intention is realized through the different layers of appreciation. It is also amazing how the physical motion of the girl's dancing was created with different instruments and notes. There is no language barrier: it's a universal interpretation of Stravinsky's image of the Russian ritual of spring. His imagination created a sound that explained the image in his head. How many artists can turn a picture into sound in such an elegant way? The image is clear in my mind: as the music takes shape it slowly adds definition. It was as if he was painting on a canvas in my mind. A painter cannot create that reaction. Sound can be absorbed into the soul and can be felt with vibration. It touches the senses and invades the mind. What other kind of art can touch us so deeply? The masters of art don't compare to the masters of sound. The person who forms sound creates an environment that affects emotions and thoughts. It is very complex and strange."
—*Selena Smith*

"There is an irregular rhythm to the piece [Stravinsky's *Petrushka*], but it seems to be different from the irregular rhythm of Beethoven or Schubert in its lack of repetition. Each time the music returns to the bass drums, trumpets etc. from the quieter segments, there is a difference to the sound. A Beethoven piece seems to return to the same sound when revisiting a section within the music. Yet again, this aspect of variation lends itself to the dramatic ideas in a ballet or cartoon. Instead of dulling itself

down with a sound that the listener would immediately
recognize, the music *differs from itself* and creates a more interesting and stimulating moment that allows the action to surge
forward."
—*Jon Edwards*

"Stanley Kubrick chose *The Beautiful Blue Danube* by Johann
Strauss II for its long drawn out flow. The long swings of the
music follow the long slow arcs of orbiting bodies in space. The
Viennese must have been a very affluent people to crave such
rich and full sounding music, but not extremely good dancers.
There is no subtlety to the beat; it is driven home by strong accents in the harmony.

"Human dance is regular, just like that of objects in space.
Both flow in predictable ways, but human dance is set to a comparatively quick rhythm, as opposed to space travel which is in
long drawn out arcs and unchanging circles. Dance is quick and
full of variety to keep the dancers interested. Space works on a
rhythm that is longer than a person's lifetime.

"Rivers are metaphors of time and its relentless movement
forward. Stanley Kubrick may have chosen this waltz because his
movie looks to the future, just like the image of the river. The
movie is more than just a futurist movie; it is about evolution
which moves on the same timescale as the river, both working
over eons to shape the planet.

"The real difference between dance and movement in space
is that human motivation is fleeting, while the course of a
planet's movement is constant for millions of years. Humans
sway back and forth, and spin each other round, while the 'space
ballet' is slow and timeless. The beat of the dance of the cosmos
spans thousands of lifetimes, while we operate on the tapping of
a foot."
—*Justin Horvath*

"Mozart's slow movement [from the Piano Concerto in A, K488]
was in my opinion a dialogue of friends. The piano's conversation seemed to be a bit melancholy and I could see how one
could compare it to the melancholy sound of Dowland's piece
'Dear, if you change.' However, after deliberation of 'Dear, if you
change' the listener understands that the song is a love song and
not really melancholy at all, which leads me to believe that

Mozart's piece is not necessarily melancholy either. The orchestra in this piece is the voice of reason and consolation. I get the impression that the piano is a child and the orchestra its guardian. The fact that the orchestra is not overbearing and overwhelming leads one to believe that as a guardian it allows the child to explore and grow, only stepping in occasionally to offer support, guidance, and reassurance. The piano solos—beautiful by the way—seem to be a wee bit precarious, somewhat like a child learning to ride a bike, not wanting the parent to let go of the seat but at the same time yearning 'to be big' and self-sufficient (at least in terms of bike-riding) to gain the pride of its parent. Another factor which supports this parent/child relationship is the fact that the piano becomes stronger and *more fearless* when accompanied by the orchestra, just as a child with the love, support, and acceptance of its parents becomes stronger and more fearless with age. I believe that this music is a dialogue of mother and child, perhaps even a coming of age for the child when he must let go and find his own way in the world. Uncertainty in the part of the piano followed by reassurance from the orchestra."
—*Katie Gago*

"Could Erik Satie's *Gymnopédies* be played by a mechanical piano? Yes, but it would lose what intensity and emotion or dramatic moments it has, even though they are rather discreet to begin with. Time and pitch are rather uncomplicated. The rhythm is static or unchanging but there are moments when the volume rises and the music intensifies as if the pianist is trying to stay completely unemotional but slips up every once in a while."
—*Sarah Komelasky*

"In the park early this morning there is barely any sound. The soft white noise of the big fountain drifts across to the bench where I am sitting. I stare across the park, listening to the sound of the birds chattering back and forth: first one from my right, then another from the far left. Their chirps set off a chorus of birds. With a fluttering of leaves and wings over my head a bluebird lands above me. He does not seem happy though: his sharp squawks and chirps sound of danger or anger. As I look up I see him fight with a squirrel. The squirrel seems to be invading the bird's space. He chatters back and forth for a minute or two.

Then the advancing bird becomes louder and louder as he gets closer to the squirrel. The squirrel, now quiet, slowly backs away and then runs, scraping and clawing his way to the next tree as fast as he can. The bird, seeming content, begins to sing back and forth with the other birds again before flying off. The air begins to quiet down again and all that is left is the calming white noise of the fountain in the background hissing away, with little drops splashing into the water."
—*Everett Freyberg*

"My trip down River Street was almost exactly the same as my other trips. The thing that was different was that I actually started listening to the sounds around me. I know that I have heard them every other time that I have gone down there but this time was different. I really listened to all of the sounds around me. . . . I decided to take the steps down to River Street instead of walking down the car ramp or taking the elevator. As soon as my head fell beneath the wall that separates Bay Street from River Street I could hear a big difference. When I finally got to River Street I could barely hear the traffic on Bay Street at all.

"I really enjoyed the sound of people talking. I found out that if you look at a couple and pay attention to them you can hear their conversation. What I really enjoyed was *not* paying attention to any particular person and just listening to it all. I found it worked best if you looked toward the water or sat on a bench and admired the architecture of the street. When I did this all I could hear was words: it was like throwing confetti up in the air, your eyes cannot pay attention to any specific piece but just the whole cloud of paper. It was like a cloud of words and syllables floating all over the street."
—*Ian Heffernan*

"In Picasso's *Man with a Blue Guitar* an older gentleman is sitting cross-legged against a wall. Perhaps he is sitting on a sidewalk in front of an abandoned building. Although this man is obviously alone and possibly lonely, he seems to have no ambition to search for love. His hair and beard have turned gray and his skin is wrinkled and thin. . . . I feel Picasso used the older man with gray hair and torn clothing to depict human nature and the course of life and death. The man appears to be ashamed of himself and his life in some ways, yet he also seems content with

his place in life, and realizes he can never change nature's course. I feel the guitar was placed upright in the old man's hands to symbolize life and growth. . . . The man symbolizes just the opposite. In my opinion, he is a reminder of how the body begins to change physically with age, and in the end decays. Much like music that lives on forever, the man and his guitar do share one common interest, and that is a soul. . . . The man could easily sell his guitar for food and clothing, but he sits alone barefoot and thin with his guitar in his hands. This in itself is a very powerful statement about music and man. Human beings desire music and the emotions felt through it. I think the man would do without material things in order to hear music, whether it be sad and depressing tunes or fast uplifting beats. The point is, he chooses to hear."
—*Amanda Stanley*

On John Cage's remark that the purpose of music is "to quiet the mind thus making it susceptible to divine influences":[1]

"What this means to me is: let the music that the listener is hearing take control of the mind. Try not to think what was played but let it think for you. The music will open your mind to an entire new horizon and take you places you have never been or dreamed about. The divine influence of music is the soul, and letting go of our mind lets the soul take control. So let your mind go and the rest (body and soul) will follow. In short, in my own words 'Do not think or interpret, let the music do the thinking and let the influence of the music take your soul and interpret for you.'"
—*David Hernandez*

"I believe that John Cage is trying to discuss the emotional freedom felt when listening to great music. Cage is also probably talking about how the mind responds to music as well. Through listening to calming music one can actually put to ease the troubles of the day and therefore one's mind becomes open to thoughts and possible 'divine influence' that one might not have been aware of before. Through listening the mind can also work toward finding meaning and beauty in the actual piece of music. I agree with this and feel that through this class I have become more aware of how wonderful a tool music can be as a release to

creativity in all forms of media."
—*Karen Heston*

"Music isn't just a form of entertainment, it is pure inspiration and it is therapeutic. I think Cage is saying that music opens your eyes a little wider, and lets you see everything you would not normally see."
—*Merry Shuart*

Notes

1. Reprinted in *John Cage: Documentary Monographs in Modern Art,* ed. Richard Kostelanetz (London: Allen Lane, 1971), 77.

Bibliography

ANTHEIL, George. *Bad Boy of Music*. London: Hurst and Blackett, 1945.

ASKEW, Anthony. Interview by Laurence Stapley. "Developments in Recorded Sound," No. 4, C90/78/02. London: British Library National Sound Archive, June 1985.

BAKER, Ray Stannard Baker. *The Boy's Book of Inventions: Stories of the Wonders of Modern Science*. London: Harper & Brothers, 1903.

BALLANTYNE, Deborah. *Handbook of Audiological Techniques*. London: Butterworth-Heinemann, 1990.

BARRAULT, Jean-Louis. *Memories for Tomorrow: Memoirs of Jean-Louis Barrault*, tr. Jonathan Griffin. London: Thames and Hudson, 1974.

BATTEN. *Joe Batten's Book: The Story of Sound Recording*. London: Rockliff, 1956.

BAZELON, Irwin. *Knowing the Score: Notes on Film Music*. New York: Van Nostrand Reinhold, 1975.

BECKETT, Samuel. *I Can't Go On, I'll Go On: A Selection from Samuel Beckett's Work*, ed. Richard W. Seaver. New York: Grove Press, 1976.

BEGLEY, Sharon. "Into the Heart of Darkness." *Newsweek* (27 November 2000), 70-74.

BENTHALL, Jonathan. *Science and Technology in Art Today*. London, Thames and Hudson, 1972.

BERIO, Luciano. *Two Interviews: With Rossana Dalmonte and Balínt András Varga*, tr. ed. David Osmond-Smith. London: Marion Boyars, 1985.

BOLINGER, Dwight, ed. *Intonation: Selected Readings*. Harmondsworth, England: Penguin Books, 1972.

351

BOULEZ, Pierre. *Conversations with Célestin Deliège*. London: Eulenberg, 1975.

_____. *Orientations: Collected writings*, ed. Jean-Jacques Nattiez, tr. Martin Cooper. London: Faber and Faber, 1986.

_____. *Rélévés d'Apprenti*, ed. Paul Thévenin. Paris: Editions du Seuil, 1966.

BUCHNER, Alexander. *Musical Instruments: An Illustrated History*, tr. Borek Vancura. London: Octopus Books, 1973.

BRUCH, Walter. *Vom Glockenspiel zum Tonband: Die Entwicklung von Tonträgern in Berlin*. Berlin: Presse- und Informationsamt des Landes Berlin, 1981.

BURROUGHS, William. *A William Burroughs Reader*, ed. John Calder. London: Pan Books, 1982.

CAGE, John. *I-VI: The 1988-89 Charles Eliot Norton Lectures*. Cambridge, Mass.: Harvard University Press, 1990.

_____. *John Cage: Documentary Monographs in Modern Art*, ed. Richard Kostelanetz. London: Allen Lane, 1971.

CARLILE, John S. *Production and Direction of Radio Programs*. New York: Prentice-Hall, 1946.

CATCHPOLE, Clive K. *Vocal Communication in Birds*. Vol. 115 of *Studies in Biology*. London: Edward Arnold, 1979.

CHASINS, Abram. *Leopold Stokowski: A Profile*. London: Robert Hale, 1979.

CLANCHY, Michael T. *From Memory to Written Record: England 1066-1307*, 2nd ed. Oxford: Basil Blackwell, 1993.

COPELAND, Peter. *Sound Recording*. London: British Library, 1991.

CRITCHLEY, Macdonald, and R.A. Henson, ed. *Music and the Brain: Studies in the Neurology of Music*. London: Heinemann, 1977.

CROWLEY, T.E. *Discovering Mechanical Music*. Aylesbury: Shire Publications, 1975.

CULHANE, John: *Walt Disney's Fantasia*. New York: Abrams, 1983.

DAYTON, Leigh. "Rock Art Evokes Beastly Echoes of the Past." *New Scientist*, 28 November 1992.

DEBUSSY, Claude. *Debussy on Music*, ed. François Lesure, tr. Richard Langham Smith. London: Secker & Warburg, 1977.

DEUTSCH, Diana. "Memory and attention in music." In *Music and the Brain: Studies in the Neurology of Music*, ed. Macdonald Critchley and R.A. Henson, 95-130. London: Heinemann, 1977.

EAMES, Charles, and Ray Eames. *A Computer Perspective*. Cambridge, Mass.: Harvard University Press, 1973.

EISENSTEIN, Sergei. *The Film Sense*, tr. Jay Leyda. London: Faber and Faber, 1948.

ELLER-RUTER, Ulrika-Maria. *Kandinsky: Bühnenkomposition und Dichtung als Realisation seines Synthese-Konzepts*. Hildesheim, Germany: Georg Olms Verlag, 1990.

EMBER, Ildikó. *Music in Painting: Music as Symbol in European Renaissance and Baroque Painting*. 2nd ed. tr. Mary and András Boros-Kazai. Budapest, Hungary: Corvina, 1989.

ERNST, David. *The Evolution of Electronic Music*. New York: Schirmer, 1977.

EVANS, Mark. *Soundtrack: Music of the Movies*. New York: Hopkinson and Blake, 1975.

FLETCHER, Neville H., and Thomas D. Rossing. *The Physics of Musical Instruments*. New York: Springer-Verlag, 1991.

FORSYTH, Michael. *Buildings for Music: The Architect, the Musician, and the Listener from the Seventeenth Century to the Present Day*. Cambridge: Cambridge University Press, 1985.

GELATT, Roland. *The Fabulous Phonograph, 1877-1977*. 2nd rev. ed. London: Cassell, 1977.

GREAT BRITAIN. Department of Scientific and Industrial Research. Building Research Board. *Sound Insulation and Acoustics*. Post-War Building Studies no. 14. London: His Majesty's Stationery Office, 1944.

HELMHOLTZ, Hermann. *On the Sensations of Tone: As a Psychological Basis for the Theory of Music*. 2nd rev. ed. tr. Alexander J. Ellis, 1885. Reprint, New York: Dover Publications, 1954.

HOOVER, Cynthia A., ed. *Music Machines—American Style*. Washington D.C.: Smithsonian Institution, 1971.

HUTCHINS, Carleen Maley, ed. *The Physics of Music*. San Francisco: W.H. Freeman, 1978.

HUYS, Bernard. *De Grégoire le Grand à Stockhausen: Douze Siècles de Notation Musicale*. Brussels: Bibliothèque Albert Ier, 1966.

JONES, Daniel. *The Pronunciation of English*. Cambridge: Cambridge University Press, 1956.

KANDINSKY, Wassily. *Concerning the Spiritual in Art*, tr. M.T.H. Sadler. 1914. Reprint, New York: Dover Publications, 1977.

_____. *Point and Line to Plane*, tr. Howard Dearstyne and Hilla Rebay. 1947. Reprint, New York: Dover Publications, 1979.

KNUDSEN, Vern O. "Architectural acoustics." In *The Physics of Music*, ed. Carleen Maley Hutchins, 79-83. San Francisco: W.H. Freeman, 1978.

LANZA, Joseph. *Elevator Music: A Surreal History of Muzak, Easy-Listening, and Other Moodsong*. New York: St Martin's Press, 1994.

LONDON, Kurt. *Film Music*, tr. Eric S. Bensinger. London: Faber and Faber, 1936.

MANVELL, Roger, and John Huntley, *The Technique of Film Music*. London: Focal Press, 1957.

McLAUGHLIN, Terence. *Music and Communication*. London: Faber and Faber, 1970.

McLUHAN, H. Marshall. *Understanding Media: The Extensions of Man*. London: Routledge & Kegan Paul, 1964.

MACONIE, Robin. *The Concept of Music*. Oxford: Clarendon Press, 1990.

_____. *The Science of Music*. Oxford: Clarendon Press, 1997.

_____. *The Works of Karlheinz Stockhausen*. 2nd rev. ed. Oxford: Clarendon Press, 1990.

MACTAGGART, Peter, and Ann Mactaggart, eds. *Musical Instruments in the 1851 Exhibition*. Welwyn, England: Mac & Me, 1986.

MILLER, Dayton C. *The Science of Musical Sounds*. New York: Macmillan, 1916.

MOORE, Gerald. *Am I Too Loud? Memoirs of an Accompanist*. London: Hamish Hamilton, 1962.

MYERS, Rollo. *Erik Satie*. London: Dennis Dobson, 1948.

NEWSOM, Iris ed. *Wonderful Inventions: Motion Pictures, Broadcasting, and Recorded Sound at the Library of Congress*. Washington, D.C.: Library of Congress, 1985.

NEWSOM, John, "A Sound Idea: Music for Animated Films." In *Wonderful Inventions* ed. Iris Newsom, 68-69. Washington, D.C.: Library of Congress, 1985.

OLSON, Harry F. *Music, Physics, and Engineering*. 2nd rev. ed. New York: Dover Publications, 1967.

ORD-HUME, Arthur W.J.G. *Joseph Haydn and the Mechanical Organ*. Cardiff: University College Cardiff Press, 1982.

OUELLETTE, Fernand. *Edgard Varèse: A Musical Biography*. London: Calder and Boyars, 1968.

PARSONS, Denys. *Directory of Tunes and Musical Themes*. Cambridge: Spencer Brown, 1975.

PARTCH, Harry *Genesis of a Music*. 2nd rev. ed. New York: Da Capo Press, 1974.

PRIEBERG, Fred K. *Musica ex Machina*. Berlin: Ullstein Verlag, 1960.

PIERCE, John R. *The Science of Musical Sound*. New York: Scientific American Books, 1983.

QUENEAU, Raymond. *Exercices de Style*. Paris: Gallimard, 1947.

READ, Herbert. *Art and Industry: Principles of Industrial Design*. 4th rev. ed. London: Faber and Faber, 1956.

REIS, Claire R. *Composers, Conductors, and Critics*. New York: Oxford University Press, 1955.

RICHTER, Hans. *Dada: Art and Anti-art*, rev. ed. tr. David Britt. London: Thames and Hudson, 1995.

ROCK, Irvin. *Perception*. New York: Scientific American Library, 1984.

RODGER, Ian. *Radio Drama*. London: Macmillan, 1982.

ROOLEY, Anthony, ed. *The Penguin Book of Early Music*. Harmondsworth, England: Penguin Books, 1980.

ROTHSTEIN, Edward. *Emblems of Mind: The Inner Life of Music and Mathematics*. New York: Avon, 1996.

SABANEEV, Leonid. *Music for the Films*. tr. S.W. Pring. London: Pitman, 1935.

SAUSSURE, Ferdinand de. *Course in General Linguistics*, rev. ed., ed. Charles Bally, Albert Sechehaye, Albert Reidlinger, tr. Wade Baskin. Glasgow: Fontana, 1974.

SCHLAIN, Leonard. *Art & Physics: Parallel visions in space, time and light.* New York: Quill, 1991.

SLONIMSKY, Nicolas. *Music Since 1900*. 3rd rev. ed. New York: Coleman-Ross, 1949.

STEANE, J. B. *The Grand Tradition: Seventy Years of Singing on Record, 1900 to 1970*. London: Duckworth, 1974.

STEIN, Gertrude. *Gertrude Stein: Writings and Lectures*. ed. Patricia Meyerowitz. London: Peter Owen, 1967.

STEINER, Fred. "Music for *Star Trek*: Scoring a Television Show in the Sixties." In *Wonderful Inventions*, ed. Iris Newsom, 286-301. Washington, D.C. : Library of Congress, 1985.

STERNE, Laurence. *The Life and Opinions of Tristram Shandy, Gentleman.* London: Oxford University Press, 1951.

STORR, Anthony. *Music and the Mind*. London: HarperCollins, 1992.

STRAVINSKY, Igor. *Chronicles of My Life*. London: Victor Gollancz, 1936.

STRAVINSKY Igor, and Robert Craft. *Memories and Commentaries*. London: Faber and Faber, 1960.

TAYLOR, Charles. *Sounds of Music*. London: British Broadcasting Corporation, 1976.

TISDALL, Caroline, and Angelo Bozzola. *Futurism*. London: Thames and Hudson, 1979.

TOOTELL, George. *How to Play the Cinema Organ*. London: Paxton, n.d.

VAN BEUNINGEN, Charles. *The Complete Drawings of Hieronymus Bosch*. London: Academy Editions, 1973.

VITRUVIUS (Marcus Vitruvius Pollio). *Ten Books of Architecture*, tr. Morris Hicky Morgan, 1914. Reprint, New York: Dover Publications, 1960.

WINCKEL, Fritz. *Music, Sound and Sensation: A Modern Exposition*, tr. Thomas Binckley. New York: Dover Publications, 1967.

Discography

ANTHEIL, George. *Ballet Mécanique*, 1925 version. Rex Lawson, The New Palais Royale Orchestra and Percussion Ensemble, cond. Maurice Peress. MusicMasters 01612 67094-2, 1994.

Art de la Musique Mécanique, Vol. 2. Arion ARN 60406, 1997.

BABBITT, Milton. *Philomel* for soprano, recorded soprano, and synthesized sound. Bethany Beardslee. New World 80466-2, n.d.

BACH, Johann Sebastian. Brandenburg Concerto No. 2, BWV 1047. Academy of St. Martins in the Fields, cond. Sir Neville Marriner. EMI 7243 5 69877 2 2, 1987.

_____. *A Musical Offering*, BWV 1079. Capella Istropolitana, dir. Christian Benda. Naxos 8.553286, 1998.

_____. Partita No. 3 for solo violin. Yehudi Menuhin. Rec. 1936. (EMI CHS7 6035-2, 1989.

_____. Prelude No. 1 in C from *The Well-Tempered Clavier*. Vladimir Feltsman. MusicMasters 01612 67105-2, 1998.

BARRON, Louis, and Bebe Barron. Electronic music for the movie *Forbidden Planet* (excerpt). In *OHM: The Early Gurus of Electronic Music 1948-1980*. Ellipsis Arts CD 3670, 2000.

BARTOK, Béla. *Music for Strings, Percussion and Celesta*. Orchestre Symphonique de Montréal, cond. Charles Dutoit. Decca 421 443-2, 1991.

_____. *Sonata for Two Pianos and Percussion*. Murray Perahia, Sir Georg Solti, David Corkhill, Evelyn Glennie. CBS MK 42625, 1988.

BEETHOVEN, Ludwig van. Overture "Coriolan", Op. 62. Slovak Philharmonic Orchestra, cond. Stephen Gunzenhauser. Naxos 8.550

072, 1987.

_____. Piano Concerto No. 4 in G, Op. 58. Anthony Newman, Philo-
musica Antiqua of New York (on period instruments), cond.
Stephen Simon. Newport NCD 60081, 1991.

_____. Symphony No. 4. Chamber Orchestra of Europe, cond. Niko-
laus Harnoncourt. Teldec 2292-46452-2, 1991.

_____. Symphony No. 6, "Pastoral." Berlin Philharmonic, cond. André
Cluytens. Seraphim CDL 7243 5 69017 2 8, 1995.

_____. Violin Concerto in D, Op. 61. Schlomo Mintz, Philharmonia
Orchestra, cond. Giuseppe Sinopoli. DG 463 064 2GH, 1988.

BERIO, Luciano. Thema: Omaggio a Joyce. Electronic music. RCA Victor
09026-68302-2, 1998.

_____. "In ruhig flessender Bewegung." III. Movement of Sinfonia for
eight voices and orchestra. Electric Phoenix, Orchestre de Paris,
cond. Semyon Bychkov. Philips 446 094-2, 1994.

BERLIOZ, Hector. "Scène aux Champs" from Symphonie Fantastique.
London Classical Players, cond. Roger Norrington. EMI CDC7
49541-2, 1989.

BLAKE, Eubie. "Eubie's Classical Rag." In Wild about Eubie. Vinyl,
Sony M34504, 1977.

BOCCHERINI, Luigi. Cello Concerto No. 9 in B flat. Steven Isserlis, Ostrobothnian Chamber Orchestra, cond. Juda Kangas. Virgin VC7
59015 2, 1992.

BOULEZ, Pierre. Le Marteau sans Maître. Yvonne Minton, Ensemble
Musique Vivante, cond. Pierre Boulez. Tape, CBS DCT-40173,
1985.

_____. Répons. Ensemble InterContemporain, cond. Pierre Boulez,
electronic music asst. Andrew Gerzso. DG 289 457 605-2, 1998.

BRAHMS, Johannes. Hungarian Dance No. 5 for piano four-hands.
Silke-Thors Matthies, Christian Köhn. In Brahms Four Hand Piano
Music, Vol. 2. Naxos 8.553140, 1997.

_____. Hungarian Dance No. 5, orch. Ernst Schmeling. London Sym-
phony Orchestra, cond. Neeme Järvi. Chandos CHAN 8885, 1991.

CAGE, John. Daughters of the Lonesome Isle for prepared piano. Boris
Berman. In John Cage: Music for Prepared Piano, Vol. 2. Naxos
8.559070, 2000.

_____. Music of Changes, Books 1-4 for piano. Herbert Henck. Wergo
60099-50, 1989.

_____. The Seasons. American Composers Orchestra, cond. Dennis
Russell Davies. ECM 1696 465 140-2, 2000.

_____. John Cage reading "Mesostic IV." From I-VI: the 1988-89
Charles Eliot Norton Lectures. Audiocassette. Cambridge, Mass:
Harvard University Press, 0 674 44007 2, 1990.

CAGE, John, and Lou Harrison. *Double Music*. In *John Cage: Music for percussion, Vol. 1*. Amadinda Percussion Group. Hungaroton HCD 31844, 1999.

Carmina Burana: Passion Play. Mittel-alter-Ensemble der Schola Cantorum Basiliensis, cond. Thomas Binkley. DHM 05472 77689 2, n.d.

CARTER, Elliott. *Three Occasions for Orchestra*. South West German Radio Symphony Orchestra, cond. Michael Gielen. Arte Nova 74321 27773 2, 1995.

CARUSO, Enrico. *The Legendary Enrico Caruso: 21 Favorite Arias*. Stockham Soundstream computer process. RCA Red Seal 5911-2-RC, 1987.

Chants de la Cathédrale de Benevento. Ensemble Organum, cond. Marcel Pérès. Harmonia Mundi HMC 901476, 1993.

CODAX, Martin. "Mandad' ei Comigo." Sinfonye, cond. Stevie Wishart. In *Bella Domna*. Hyperion CDA 66283, 1988.

CORELLI, Archangelo. Concerto Grosso in D. Clarion Music Society (on period instruments), cond. Newell Jenkins. In *Hidden masters of the Baroque, Vol. 1*. Newport Classics NCD 60075, 1988.

COUPERIN, Louis. "Prélude à l'imitation de M. Froberger." Laurence Cummings. In *Louis Couperin Harpsichord Suites*. Naxos 8.550922, 1994.

CRUMB, George. *Ancient Voices of Children* for soprano and chamber ensemble. Jan DeGaetani, Contemporary Clamber Ensemble, cond. Arthur Weisberg. Elektra Nonesuch 79149 2, 1970.

DEBUSSY, Claude. "Gigues" from *Images pour Orchestre*. Saarbrücken Radio Symphony Orchestra, cond. Hans Zender CPO 999 476 2, 1997.

_____. *Prélude à l'Après-midi d'un Faune*. Belgian Radio Television Philharmonic Orchestra, cond. Alexander Rahbari. Naxos 8.550 262, 1989.

_____. *Suite Bergamasque* No. 3 "Clair de Lune" for piano. George Copeland rec. Duo-Art, 1915. Nimbus NI 8807, 1996.

DODGE, Charles. *Earth's Magnetic Field: Realizations in Computed Electronic Sound*. Bruce R. Boller, Carl Frederick, Stephen G. Ungar, scientific associates. Vinyl, Nonesuch Records H-71250, 1970.

DOWLAND, John. "Dear, If You Change." *First Book of Songs*, No. 7, 1597. Emma Kirkby, Anthony Rooley. In *The English Orpheus*. Virgin Classics 0777 7595212 4, 1989.

_____. "Semper Dowland, Semper Dolens" for viols and lute, from *Lachrimae*, 1604. Fretwork. In *Goe Nightly Cares*. Virgin VC7 91117-2, 1990.

EIMERT, Herbert, and Robert Beyer. "Musik und Sprache," "Zu Ehren von Igor Stravinsky." In *Einführing in die Elektronischen Musik*.

Vinyl, Wergo 60006, n.d.

FELDMAN, Morton. *Coptic Light*. Deutsches Symphonie-Orchester Berlin, cond. Michael Morgan. CPO 999 189-2, 1997.

GABRIELI, Giovanni. *Canzon del Duodecimi Toni a 10*. The Wallace Collection, cond. Simon Wright. In *Gabrieli & St Mark's: Venetian Brass Music*. Nimbus Records NI 5236, 1990.

GERSHWIN, George. *Rhapsody in Blue*, 1924 version for jazz orchestra. George Gershwin, rec. Duo-Art 1925, Columbia Jazz Band, cond. Michael Tilson Thomas. In *Classic Gershwin*. CBS MK42516, 1987.

Music of the Middle Ages: Gregorian Chant, Music of the Gothic Era. Choralschola der Benediktinerabtei Münster-schwarzach, dir. Godehard Joppich. DG Klassikon 439424 2, 1982.

HANDEL, Georg Frideric. *Music for the Royal Fireworks* (original version for wind orchestra 1749). The English Concert, dir. Trevor Pinnock. DG Archiv 453 451-2, 1997.

_____. *The Water Music* (original 1717 version). English Chamber Orchestra, cond. Johannes Somary. Vanguard Classics SVC-47, 1996.

HAYDN, Franz Joseph. Symphony No. 45, "Farewell." The Hanover Band, cond. Roy Goodman. Hyperion CDA 66522, 1991.

_____. Symphony No. 94 in G, "Surprise". The Hanover Band, cond. Roy Goodman. Hyperion CDA 66532, 1991.

_____. Symphony No. 101 in D, "The Clock." La Petite Bande cond. Sigiswald Kuijken. DHM 05472 772451 2, 1994.

_____. "O vis aeternitatis." Sequentia. In *Canticles of Ecstasy*, DHM 05472 77320 2, 1994.

HILDEGARD VON BINGEN. "Procession" from *Ordo Virtutum*. Oxford Camerata, dir. Jeremy Summerly. In *Heavenly Revelations*. Naxos 8.550998, 1995.

HINDEMITH, Paul. *Konzertstück* for Trautonium and string orchestra. Oskar Sala, Münchner Kammerorchester, cond. Hans Stadlmair. Erdenklang 81032, 1998.

HONEGGER, Arthur. *Les Misérables*. Slovak Radio Symphony Orchestra, cond. Adriano. Marco Polo 8.223181, 1989.

IVES, Charles. "Washington's Birthday," from *The "Holidays" Symphony*. New York Philharmonic, cond. Leonard Bernstein. Sony SMK 60203, 1998.

Japan: Traditional Vocal and Instrumental music. Ensemble Nipponia, dir. Minoru Miki. Elektra Nonesuch 9 72072-2, 1976.

Kagura: Japanese Shinto Ritual Music, rec. János Kárpáti. Hungaroton SPLX 18193, 1988.

KHACHATURIAN, Aram. Violin Concerto in D minor. Hu Kun, Royal Philharmonic Orchestra, cond. Yehudi Menuhin. Nimbus NI 5277, 1988.

KREISLER, Fritz. *Variations on a Theme of Corelli (in the Style of Tartini)*, rec. 1910, and *La Précieuse (in the Style of Couperin)* rec. 1929. In *Kreisler plays Kreisler*, RCA Victor 09026-68448-2, 1997.

LASSO, Orlando di. "Hark! Hark! the Echo" (O la o Che Bon Eccho), *Libro de Villanelle, Moresce, et Altre Canzoni* (1581) No. 14. Glasgow Orpheus Choir, cond. Hugh Roberton. Vinyl, His Master's Voice DLP 1020, n.d.

LEONCAVALLO, Ruggiero. "La Mattinata," rec. 1904. Enrico Caruso, Ruggiero Leoncavallo. In *Caruso: The Early Recordings*. Nimbus Prima Voce NI 7900, 1999.

LIGETI, György. *Atmosphères*. New York Philharmonic, cond. Leonard Bernstein. In *Music of Our Time*. Sony SMK 61845, 1999.

MAHLER, Gustav. Adagietto, from Symphony No. 5. Smithsonian Chamber Players, dir. Kenneth Slowik. DHM 054272 77343 2, 1995.

MARCELLO, Alessandro. Concerto for Oboe and Strings in D minor. Ferenc Erkel Chamber Orchestra, dir. József Kiss. Naxos 8. 550 475, 1993.

MARCELLO, Benedetto. Concerto for Trumpet and Strings, Op. 2, No. 11. Miroslav Kejmar, Capella Istropolitana, cond. Petr Skvor. Naxos 8.550243, n.d.

MARKEVITCH, Igor. *L'Envol d'Icare* for two pianos and percussion. C. Lyndon-Gee, K. Lessing; F. Lang, J. Gagelmann, R. Haeger. Largo 5127, 1993. *Version for orchestra*: Arnhem Philharmonic Orchestra, cond. Christopher Lyndon-Gee. Marco Polo 8.223666, 2000.

MARTINU, Bohuslav. *Concerto for Double String Orchestra, Piano and Timpani*. Czech Philharmonic Orchestra, cond. Jiri Belohlávek. Chandos CHAN 8950, 1991.

McPHEE, Colin. *Balinese Ceremonial Music* for piano four-hands. Peter Hill, Douglas Young. In *East-West Encounters*. Vinyl, Cameo Classics GO CLP9018 (D), 1982.

MESSIAEN, Olivier. *Quatuor pour la Fin du Temps*. Fabio di Castola, Ricardo Castro, Emilie Haudenschild, Emeric Kostyak. Accord 201772, 1990.

_____. *Turangalîla Symphony*. François Weigel, Thomas Bloch, Polish National Radio Symphony Orchestra, cond. Antoni Wit. Naxos 8.554478-79, 2000.

MONTEVERDI, Claudio. *Vespers*. Boston Baroque (on period instruments), cond. Martin Pearlman. Telarc 2CD-80453, 1997.

MORRISON, Herbert. "Hindenberg [sic] Disaster." Rec. May 6, 1937. In *20th Century Time Capsule*, ed. Glenn Korman. Buddha Records 7446599633 2, 1999.

_____. "1937 Hindenburg Air Disaster" (same as above). In *The*

Century in Sound, ed. Richard Fairman. The British Library NSA CD8, 1999.

MOSZKOWSKI, Moritz. *Etude de Virtuosité*, Op. 72, No. 11. Vladimir Horowitz. Sony S3K53461, 1993.

MOZART, Leopold. Concerto for Trumpet in D. Wynton Marsalis, English Chamber Orchestra, cond. Raymond Leppard. Sony SK557497, 1995.

MOZART, Wolfgang Amadeus. *Eine kleine Nachtmusik*. Philharmonia Orchestra, cond. Sir Colin Davis. Seraphim 7243 5 68533 2 4, 1990.

_____. German Dance No. 3. Philharmonia Orchestra, cond. Sir Colin Davis. Seraphim 7243 5 68533 2 4, 1990.

_____. *Notturno in D major*, K286. London Symphony Orchestra, cond. Peter Maag. Decca Legends 289 466 500-2, 2000.

_____. Piano Concerto No. 23 in A, K488. St. Luke's Orchestra, dir. Julius Rudel. MusicMasters 01612 671649 2, 1998.

_____. Symphony No. 40 in G minor, K550. The Cleveland Orchestra, cond. George Szell. Sony SBK 46333, 1990.

_____. Symphony No. 41 in C, "Jupiter," K551. I. Movement (excerpt). 1934 experimental Blumlein stereo recording. London Philharmonic Orchestra, cond. Sir Thomas Beecham. In *Centenary Edition: 100 Years of Great Music*. EMI promotional disc 7087 6 11859 2 8, 1997.

Music in Film. National Public Radio. Sony SMK 60991, 1990.

MUSSORGSKY, Modest. *Pictures at an Exhibition*, orch. Ravel. 1931 experimental stereo recording. Philadelphia Orchestra, cond. Leopold Stokowski. Vinyl, Bell Labs BTL 7901, 1977.

NANCARROW, Conlon. *Study 3c* for player piano. Conlon Nancarrow. Wergo WER 6168-2, 1988.

_____. *Study 3c*. Version for chamber ensemble. Ensemble Modern, dir. Ingo Metzmacher. RCA 09026-61180-2, 1993.

NIGHTINGALE, Florence rec. 1890. In *The Wonder of the Age: Mister Edison's New Talking Phonograph*, ed. Kevin Daly. Vinyl, Argo ZPR 122-3, 1977.

PARTCH, Harry. *And on the Seventh Day Petals Fell in Petaluma*. Gate 5 Ensemble, dir. Harry Partch. CRI CD 752, 1997.

PENDERECKI, Krzysztof. *Anaklasis*. London Symphony Orchestra, cond. Krzysztof Penderecki. EMI CDMS 65077-2, 1994.

PEROTIN. "Viderunt Omnes." Early Music Consort of London, cond. David Munrow. In *Music of the Middle Ages*. DG Klassikon 439424 2, 1993.

Piano Music for Silent Movies. Florence de Jong, Ena Baga. Vinyl, Rediffusion Gold Star 15-13, 1974.

PRIULI, Giovanni. *Canzona Prima a 12*. Early Music Consort of London (on period instruments), dir. David Munrow. In *Monteverdi's Companions*. Virgin Veritas 7243 5 61288 2 8, 1996.

PURCELL, Henry. *Fantazias for Viols*. Rose Consort of Viols. Naxos 8.553957, 1995.

RAVEL, Maurice. *Boléro*. Vienna Philharmonic, cond. Lorin Maazel. RCA Victor 09026 068600-2, 1997.

_____. Suite No. 2 from *Daphnis et Chloé*. Vienna Philharmonic, cond. Lorin Maazel. RCA Victor 09026-68600-2, 1997.

_____. *L'Enfant et les Sortilèges*. Montreal Symphony Orchestra, cond. Charles Dutoit. Decca 440 333-2DH, 1995.

ROCKMORE, Clara. *Art of the Theremin*. Delos DE 1014, 1987.

ROSSINI, Gioachino. Overture, *La Cenerentola*. Zagreb Festival Orchestra, cond. Michael Halász. Naxos 8.550236, 1989.

SAINT-SAENS, Camille. *L'Assassinat du Duc de Guise*, Op. 128. Musique Oblique Ensemble. Harmonia Mundi HMT 790 1472, 1993.

_____. "Organ" Symphony, No. 3 in C minor, Op. 78. Philippe Lefebvre, Orchestre National de France, cond. Seiji Ozawa. Seraphim 7243 5 73430 2 2, 1997.

SATIE, Erik. *Gymnopédies* for piano. Daniel Varsano. Sony SBK 48 283, 1992.

_____. *Parade: Ballet Réaliste*. Orchestre Symphonique et Lyrique de Nancy, cond. Jérôme Kaltenbach. Naxos 8.554279, 1997.

_____. *Trois Véritables Préludes Flasques (pour un chien)* for piano. Aldo Ciccolini. EMI CZS 767282 2, 1991.

SCELSI, Giacinto. *Konx-Om-Pax*, Cracow Radio Television Orchestra, cond. Jürg Wyttenbach. Accord 200402, 1988.

SCHOENBERG, Arnold. *Accompaniment to a Cinematographic Scene*, Op. 34. London Symphony Orchestra, cond. Robert Craft. Koch 3-7263-2H1, 1995.

_____. No. 3 "Farben" (Colors) from *Five Pieces for Orchestra*, Op. 16. Chicago Symphony Orchestra, cond. Rafael Kubelik. Mercury 289 434 397-2, 1998.

_____. *Herzgewächse*, Op. 20. Christine Schäfer, Solistes de l'Ensemble Intercontemporain, cond. Pierre Boulez. DG 457 630-2, 1998.

_____. *Die Jakobsleiter*. Mady Mesplé, BBC Symphony Orchestra, cond. Pierre Boulez. Sony SMK 48 462, 1993.

_____. *Pierrot Lunaire*, Op. 21. Christine Schäfer, Solistes de l'Ensemble Intercontemporain, cond. Pierre Boulez. DG 457 630-2, 1998.

_____. *A Survivor from Warsaw*. Simon Callow, London Symphony Orchestra, cond. Robert Craft. Koch 3-7263-2H1, 1995.

SCHUBERT, Franz. *Rosamunde*, Suite No. 2. Orchestra of the Age of Enlightenment, cond. Sir Charles Mackerras. Virgin VC7 91515-2,

1992.

_____. Symphony No. 9, "The Great C Major." Belgian Radio Television Philharmonic Orchestra cond. Alexander Rahbari. Naxos 8.550502, 1991.

SCHUMANN, Robert. Cello Concerto in A, Op. 129. Jürnjakob Timm, Gewandhausorchester Leipzig, cond. Kurt Masur. Curb D2-78028, 1995.

SCHWITTERS, Kurt. *Ur-Sonate*, rec. 1932. In *Kurt Schwitters*. CD-ROM, Schlütersche 3-87706-771-9, 1996.

SCRIABIN, Alexander. *Prometheus—The Poem of Fire*. Kirov Orchestra, cond. Valery Gergiev. Philips 289 446 715-2, 1998.

STALLING, Carl, orch. Milt Franklyn. *Feed the Kitty*. In *The Carl Stalling Project*. Warner Bros 926027-2, 1990.

_____. *There they go go go*. In *The Carl Stalling Project*. Warner Bros 926027-2, 1990.

STEIN, Gertrude. "If I Told Him: a Completed Portrait of Picasso." In *The Caedmon Poetry Collection: A Century of Poets Reading Their Work*. New York: HarperCollins CD 2895(3), 2000.

STOCKHAUSEN. *Gesang der Jünglinge*. Electronic music. Stockhausen Verlag SV 3, 1991.

_____. *Sirius* for soloists and electronic music. Annette Meriweather, Boris Carmeli, Markus Stockhausen, Suzanne Stephens. Stockhausen Verlag SV 26, 1992.

_____. *Stimmung* for six vocal soloists. Singcircle, dir. Gregory Rose. Hyperion CDA 66115, 1987.

_____. *Tierkreis für 12 Spieluhren* (Zodiac for 12 musical boxes). Stockhausen Verlag SV 24, 1992.

_____. *Trans*. (1) Southwest German Radio Symphony Orchestra, cond. Ernest Bour; (2) Saarbrücken Radio Symphony Orchestra, cond. Hans Zender. Stockhausen Verlag SV19, 1992.

STRAUSS, Johann II. *The Beautiful Blue Danube*. Vienna Philharmonic, cond. Herbert von Karajan. DG 439 104 2GDO, 1968.

STRAUSS, Richard. *Also Sprach Zarathustra*. London Philharmonic Orchestra, cond. Klaus Tennstedt. Seraphim 7243 5 73560 2 2, 1999.

_____. *Till Eulenspiel*, symphonic poem. Budapest Symphony Orchestra, Ken-Ichiro Kobayashi. Hungaroton White Label HRC 082, 1988.

_____. *Eine Alpensinfonie*, Op. 64. Staatskapelle Dresden, cond. Karl Böhm DG 447 454-2 GOR, 1996.

STRAVINSKY, Igor. "The Building of the Ark" from *The Flood*. The Philharmonia, cond. Robert Craft. MusicMasters 01612-67195-2, 1998.

_____. *Les Noces*. 1919 version with pianola. Orpheus Chamber Ensemble, cond. Robert Craft (vinyl, CBS 73439,1975).

___. *Les Noces*, arr. synthesizer. Pokrovsky Ensemble, dir. Dmitri Pokrovsky. Elektra Nonesuch 9 79335-2, 1994.

_____. *The Rite of Spring*, ed. after the Pleyela piano roll. (1) Boston Philharmonic, cond. Benjamin Zander; (2) Rex Lawson, pianola. IMP MCD 25, 1989.

_____. *Le Sacre du Printemps* (The Rite of Spring). Columbia Symphony Orchestra, cond. Igor Stravinsky. CBS MK 42433, 1988.

_____. *Petrushka*. Columbia Symphony Orchestra, cond. Igor Stravinsky. CBS MK 42433, 1988.

_____. *Trois Poèmes de la Lyrique Japonaise*. Christiane Eda-Pierre, Ensemble du Domaine Musical, cond. Gilbert Amy. Adès 203512, 1991.

STRAYHORN, Billy. "Take the 'A' train," rec. 1940. Duke Ellington and his Famous Orchestra. In *The Duke Ellington Centennial Edition*. RCA Victor 09025 63458-2, 1999.

Susan Baker's Fiddles and Follies. Susan Baker. Vinyl, Argo ZK86, 1979.

TALLIS, Thomas. *Spem in Alium*. Tallis Scholars, cond. Peter Phillips. Philips 289 462 862-2, 1999.

TCHAIKOVSKY, Peter Ilyich. Ballet suite *The Nutcracker*, arr. Stokowski. Philadelphia Orchestra, cond. Leopold Stokowski. In *Fantasia: Remastered Original Soundtrack Edition*, Walt Disney D STCS 452 D, 1990.

Tibetan Buddhism: The Ritual Orchestra and Chants. Monks of the Tashi Jong community, Khampagar Monastery, rec. David Lewiston. Nonesuch 9 72071-2, 1995.

TIOMKIN, Dmitri. Suite from the movie *The Thing (from Another World)*. National Philharmonic Orchestra, cond. Charles Gerhardt. In *The Spectacular World of Classic Film Scores*. RCA Victor 2792-2-RG, 1977.

Tuva, Among the Spirits: Sound, Music and Nature in Sakha and Tuva, prod. Ted Levin and Joel Gordon. Smithsonian Folkways SFW 40452, 1999.

VARESE, Edgar. *Hyperprism* (with siren), rev. Richard Saks. Asko Ensemble, cond. Riccardo Chailly. In *Varese: the Complete Works*. London 289 460 208-2, 1998.

_____. *Poème Electronique*. Tape music. In *Varèse: the Complete Works*. London 289 460 208-2, 1998.

VERDI, Giuseppe. "Caro nome" from *Rigoletto*, rec. 1907. Dame Nellie Melba and orchestra. In *Great Singers at La Scala*. Nimbus NI 7858, 1994.

_____. *Messa da Requiem*. Leontyne Price, Rosalind Elias, Jussi Björling, Giorgio Tozzi, Singverein der Gesellschaft der Musikfreunde Wien, Vienna Philharmonic, cond. Fritz Reiner. Decca

444 833-2, 1995.

VIVALDI, Antonio. Concerto No. 1 in E, "Spring," RV 269, from *The Four Seasons*. Felix Ayo, dir. I Musici. Philips 438 344-2, 1999.

_____. *Lauda Jerusalem* RV609. Margaret Marshall, Ann Murray, John Alldis Choir, English Chamber Orchestra, cond. Vittorio Negri. Philips 420 648-2PM, 1988.

_____. Lute Concerto in D, RV93, arr. Bream. Julian Bream, Monteverdi Orchestra, cond. John Eliot Gardiner. In *Concertos and Sonatas for Lute*, RCA 09026 61588-2, 1993.

Vocal Music from Mongolia, rec. Jean Jenkins. Vinyl, Tangent TGS 126, 1977.

WAGNER, Richard. "Der dort mich ruft" from *Die Walküre*, Act II. Bavarian Radio Symphony Orchestra, cond. Bernard Haitink. EMI Angel CDS 7 49534 2, 1989.

_____. "Entry of the Gods into Valhalla" from *Das Rheingold*. Cleveland Orchestra, cond. George Szell. Sony SBK 48175, 1992.

Warner Brothers: The Golden Age of Hollywood Stars. Vinyl, United Artists USD 311, 1977.

WEBERN, Anton. *Six Bagatelles* for string quartet, Op. 9. Artis Quartet. Sony SK 48059, 1992.

WEILL, Kurt. "September Song" rec. 1946. Frank Sinatra and orchestra cond. Alex Stordahl. In *Kurt Weill: From Berlin to Broadway, Vol. II*. Pearl GEMM CDS 9294, 1997.

WYATT, Sir Thomas. "Blame Not My Lute." The Consort of Musicke, dir. Anthony Rooley. *Musicke of Sundrie Kindes*. Vinyl, Editions de l'Oiseau-Lyre 12BB 203-6, 1975.

Index

Adams, John, 281
Ader, Clément, 319-20
Aeldred of Rievaulx, 159
Amati, Nicoló, 41
Ambrosian chant, 89
amphitheater, acoustics of, 86-8
Angelico, Fra, 124-5, 140
Antheil, George, 232, 235, 247, 250
Aristoxenus, 88
Armstrong, Louis, 252
Astaire, Fred, 293

Babbitt, Milton, 334, 339
Bach, Johann Sebastian, viii, 101, 115, 117, 160, 161, 162, 163-5, 169, 173, 174, 180, 189, 266, 271, 294-6, 323
Bacharach, Burt, 84
Bacon, Roger, 92
Baga, Ena, 289, 300
Baker, Susan, 218
Balanchine, George, 299
Balinese music, 232-3
Balla, Giacomo, 230
Barrault, Jean-Louis, 338-9
Barron, Louis and Bebe, 337, 339
Bartók, Béla, 166, 321, 325
Barry, John, 298
Barrymore, John, 85
Bataille, Gabriel, 144, 154
Beckett, Samuel, 94, 258, 262, 264, 315, 323
Beecham, Sir Thomas, 184, 186, 320
Beethoven, Ludwig van, 166, 173, 196-200, 230, 235, 243, 269, 271, 275-8, 282, 296, 344
Bell, Alexander Graham, 306
Bell, Chichester, 306
Bell Laboratories, 63, 66, 320, 337
Berio, Luciano, 323, 325, 336, 339
Berkeley, Busby, 218
Berlioz, Hector, 166, 217, 230, 235, 290
Bernard, Raymond, 298
Bingen, see Hildegard von Bingen
Blake, Eubie, 245, 249
Blumlein, Alan, 320, 321
Boccherini, Luigi, 193-4, 199

Bosch, Hieronymus, 2, 16, 131-3, 140, 227
Boulez, Pierre, 153, 336, 338, 339
Bradley, Scott, 295
Brahms, Johannes, 212-13, 219, 315
Breton, André, 261
Brown, Earle, 179, 248
Bruckner, Anton, 173, 315
Brueghel, Peter, the Elder, 120
Burne-Jones, Edward, 213
Burroughs, William, 261, 262, 264
Busnois, Antoine, 227
Busoni, Feruccio, 246
Byron, Lord, 214

Cage, John, 179, 188, 232-3, 248, 262-3, 264, 282, 336-7, 339, 348-9
Cagney, James, 297
Canaletto, Antonio, 24
Caravaggio, Michelangelo, 136-7, 140
Carmina Burana, 103, 116-17
Carpenter, John, 232
Carroll, Lewis, 257
Carter, Elliott, 248, 250
Caruso, Enrico, 312-14, 322, 324
cathedral acoustic, 104-6
cave acoustics, 69
cavity resonance, 47, 251-2
Chagnon, Napoleon, 7
Chaplin, Charlie, 272-3, 292
Char, René, 336
Chopin, Frederic, 244, 249
Clanchy, Michael, 92, 99
Cocteau, Jean, 230
Codax, Martin, 96-7, 99, 156, 227, 310
color organ, 294-5
Copeland, George, 246, 249
Corelli, Archangelo, 113-14, 117, 163, 191, 192
Couperin, Louis, 268, 282

Cristofori, 183
Cros, Charles, 306
Crosby, Bing, 322
Crumb, George, 76, 82
Ctesibius, 157-8, 159

Dalí, Salvador, 63
D'Anglebert, Jean-Henri, 268
da Vinci, see Leonardo da Vinci
de Jong, Florence, 289, 300
Debussy, Claude, 176, 188, 246, 249, 294, 295, 300
Disney, Walt, 216-17, 291, 321
Donen, Stanley, 229, 235, 285
Doppler shift, 23
Dowland, John, 101, 141-3, 144, 145-7, 149-53, 177, 257, 345
Duchamp, Marcel, 230, 260
Dürer, Albrecht, 142-3, 154

ear mechanism, 31-2, 35-6
Edison, Thomas Alva, 59. 306-08
Edo Lullaby, 176-9, 188
Eimert, Herbert, 337, 339
Einstein, Albert, 309-10
Eisenstein, Sergei, 291, 338
Elgar, Sir Edward, 316-17
Ellington, Duke, 187, 322
Esterházy, Prince Nicolaus, 243

Fairbanks, Douglas, Senior, 289
Fantasia (1941), 216-18, 219, 291
Feldman, Morton, 179, 234, 248
fireworks, 227-8
Fitzgerald, Ella, 252
Fleming, Renée, 10
foley, 228-9
Foote, Samuel, 258
Forbidden Planet, 337, 339
formants, 47, 251-2
Franklyn, Milt, 295, 300, 301
frequency range, audible, 32, 34, 38, 105
Frescobaldi, Girolamo, 281
fugue, 115

Gabrieli, Giovanni, 106, 111-13, 114, 116, 117, 146, 160, 167, 189, 320
Gainsborough, Thomas, 198
Gaisberg, Fred, 314
Galileo Galilei, 160
Garbo, Greta, 195
George I, King of England, 23
Gershwin, George, 246, 249-50
Ghent Altarpiece, 125-8, 140, 159
Giotto di Bondone, 121-3, 125, 140
Giraud, Albert, 315
Glass, Philip, 281
Gluck, Christoph Willibald von, 184
Goethe, Johann Wolfgang von, 198
Grainger, Percy, 248
Greek drama, 58, 80
Griffith, D.W., 292
Grofé, Ferde, 246
Grünewald, Matthias, 134-5, 140
Guarneri, Andrea, 41
Gutenberg, Johann, 133

halo, 120-1
Handel, Georg Frideric, 23-6, 115. 166, 227-8, 235, 270
Harmon, Leon D., 63-4, 65, 66
Harrison, Lou, 281, 282
Haussmann, Raoul, 261
Hawking, Stephen, 76
Haydn, Joseph, 101, 166, 169-70, 174, 184, 192, 197, 208, 229, 243-4, 296
Helmholtz, Hermann, 305
Herrmann, Bernard, 284, 298
Hildegard von Bingen, 255-7, 263

Hindemith, Paul, 135, 332
Hindenburg Disaster newscast, 57-8, 68
Hitchcock, Alfred, 284

Honegger, Arthur, 298, 301
Huygens, Christiaan, 86

infrasound, 51-2
Ingres, Jean-Auguste, 137
intonation, 56
IRCAM, 338
Isenheim Altarpiece, 134, 140
Ives, Charles, 217, 291-3, 300, 323

Jacquard, Joseph, 244
James, Henry, 260
James, William, 260
Japan, 73-4, 175-9
Jazz Singer, The, 297, 301
John of Salisbury, 92-3
Jolson, Al, 297, 301
Jones, Daniel, 56, 68
Joplin, Scott, 245
Joyce, James, 258-9, 323, 336

Kaigal-ool, Tuva singer, 71, 81
Kandinsky, Wassily, 3, 16, 231-2, 235, 294, 300, 328
karaoke, 173
Karg-Elert, Sigfrid, 281
Kells, Book of, 94
Kelly, Gene, 229, 235, 285
Khachaturian, Aram, 200, 201
Kington, Miles, 173
Klimt, Gustav, 328
Knudsen, Vern O., 87-8, 99
Kodály, Zoltán, 338
Koenig, R., 305, 310
Korngold, Erich Wolfgang, 298-9
Kreisler, Fritz, 316-18, 324
Kubrick, Stanley, 106, 213-16, 219, 234-5, 337, 345

larynx, 36, 45-6
Lasso, Orlando di, 110, 117
Le Corbusier, 337
Lear, Edward, 257
Léger, Fernand, 232

Leigh, Janet, 284
Leonardo da Vinci, 88
Léoncavallo, Ruggiero, 314, 324
Ligeti, György, 233-4, 236, 323
Lincoln, Abraham, 63, 75
lips, 36-7
Liszt, Franz, 184, 244, 249, 296
Louis XIV, King of France, 270
Lucas, George, 327
Lumière brothers, 285-6

Mälzel, Johann Nepomuk, 275
Mahler, Gustav, 173, 246, 315,
    324, 331
Man Ray, 137
Marcello, Alessandro, 190, 200
Marcello, Benedetto, viii, 191-2,
    200
Marconi, Giglielmo, 328
Marey, Louis, 218
Marinetti, Filippo, 230, 259, 264
Markevitch, Igor, 321, 325
Martenot, Maurice, 335
Martinu, Bohuslav, 321, 325
Mayan civilization, 69
McCarthy, Sen. Joseph, 197
McLuhan, H. Marshall, ix, 5,
    125
McPhee, Colin, 232, 281
Meisel, Edmund, 291
Melba, Dame Nellie, 312, 314,
    331
metronome, 275-8
Messiaen, Olivier, 328, 335, 336,
    339
Michelangelo, 142
microphone, 37, 120, 332-3
Midas, King, 129-30
Miller, Dayton S., 313, 323, 324
Mon, Franz, 261
Monteverdi, Claudio, 106, 107-
    10, 114, 116, 117, 161, 168,
    189, 227, 255, 269, 320
Morrison, Herbert, 57-8, 68
Moskowski, Moritz, 244, 249
Mozart, Leopold, 169, 170, 174,
    200
Mozart, Wolfgang Amadeus,
    101, 116, 117, 154, 170-2, 173,
    184, 192, 198, 207-8, 229-30,
    257, 271, 272-4, 282, 296, 311,
    325, 345-6
Mozart effect, ix, 266
Munch, Edvard, 10
Murphy, Eddie, 24
musique concrète, 338
Mussorgsky, Modest, 320, 325
Muybridge, Eadweard, 218

Nancarrow, Conlon, 248-9, 250
nasal cavity, 46
Nayfack, Nicholas, 337, 339
Neumann, Johann von, 262
Newman, Barnett, 233
Niemecz, Friar Primitivus, 244
Nightingale, Florence, 59-61, 62,
    67, 309
Nijinsky, Vaslav, 282
Nimbus Records, 313-14
NYPD Blue, 171-2
Nyman, Michael, 281

Ockeghem, Johannes, 227
Oedipus Rex, 58
Olson, Harry, 334
omnidirectional hearing, 33,
    120-1
ondes martenot, 231, 335
ossicles, 35

Pal, George, 327
Palladio, Andrea, 104, 162
Palma, Jacopo da, 129-31
Parsons, Denys, 84-5, 99
Partch, Harry, 232, 236
Pavarotti, Luciano, 193
Penderecki, Krzysztof, 233
Perkins, Anthony, 284
Pérotin, 97-8, 99, 281
Petipa, Marius, 216
Picart, Bernard, 181
Picasso, Pablo, 230, 260, 278,

282, 347-8
Pinter, Harold, 94
Pleyela, 246
Pohlman, Johannes, 184
Pokrovsky, Dmitri, 247-8
Pratella, Balilla, 230
prelude, 267-8
prepared piano, 232-3, 236
Presley, Elvis, 240, 304
Priuli, Giovanni, 113, 117, 321
psychological time, 276-8
Public Enemy, The, 297, 301
punctuation marks, 90-1
Purcell, Henry, 160, 174
Pythagoras, School of, 85, 92, 178, 329

Queneau, Raymond, 261, 264

RCA (Radio Corporation of America), 216-17, 320
Ravel, Maurice, 294, 299, 300, 302, 319, 320, 323, 324
Read, Herbert, 142-3
Reich, Steve, 281
Reis, Claire, 231, 235
Rockmore, Clara, 331, 338
Roddenberry, Gene, 330
Rodger, Ian, 94, 99
Rodin, Auguste, 142
Rooley, Anthony, 154
Rossini, Gioachino, 290, 300, 311
Rósza, Miklos, 298
Rothko, Mark, 233
Russolo, Luigi, 264
Ryle, Gilbert, 330

St. Cecilia, 126
Sabine, Wallace, 187
Sacher, Paul, 321
Saint-Saëns, Camille, 102, 116, 290, 300
Saryglar, Alexei, 263
Satie, Erik, 230, 235, 245, 288, 315, 324, 346

scales, 53-4, 83
Scelsi, Giacinto, 233, 236
Schaeffer, Pierre, 338
Schiller, Johann Friedrich Christoph von, 198
Schmeling, Ernst, 212
Schoenberg, Arnold, 281, 282, 299, 301, 324, 328, 330, 338
Schubert, Franz, 101, 208-12, 218, 282, 290, 295, 315, 344
Schumann, Robert, 194-5, 290
Schütz, Heinrich, 113
Schwitters, Kurt, 261, 263
Scott, Léon, 305, 309
Scriabin, Alexander, 294, 300, 320, 328
Sennett, Mack, 286
Severini, Gino, 230
Shakespeare, William, 141, 149, 195, 308
Shaw, George Bernard, 259, 295
Simpsons, The, 283-4
Sinatra, Frank, 322
Singin' in the Rain, 229, 235, 285, 297
Sony, 173
Sophocles, 58
sound, speed of, 28, 186
Spielberg, Stephen, 337
stairway, acoustics of, 86
Stalling, Carl, 289, 295, 299, 300, 301
stave, musical, 90-1
Stein, Gertrude, 259-60, 262, 263, 281
Steiner, Max, 298
Sterne, Laurence, 258, 263
Stockhausen, Karlheinz, 233, 236, 241, 255, 263, 281, 323, 325, 336, 338
Stockham, Thomas, 313, 324
Stokowski, Leopold, 216-17, 219, 320
Stradivari, Antonio, 41
Strauss, Johann II, 213-16, 219, 269, 323, 345

Strauss, Richard, 105, 117, 230, 235, 291-2, 298, 300
Stravinsky, Igor, 63, 227, 246-8, 250, 282, 291, 295, 299, 300, 301, 315, 321, 323, 324, 344
Strayhorn, Billy, 322
sushi, 175
synesthesia, 328
synthesizer, 333-5

tablature, 143-5
Tainter, Charles Sumner, 306
Tallis, Thomas, 321, 325
Taylor, Deems, 218
Tchaikovsky, Peter Ilyich, 216, 219, 227
Termin, Leon, 329-30
Thalberg, Sigismond, 184
theremin, 329-30, 331, 338
trautonium, 332
Trautwein, Friedrich, 332
tuning fork, 182
Turner, Joseph Mallord William, 199
Tuva, music of, 70-3
2001: A Space Odyssey, 106, 213-16, 219, 234-5, 337, 345
Tyndall, John, 182-3
Tzara, Tristan, 261-2

Van Eyck, Hubert, 125-8, 140
Van Eyck, Jan, 125-8, 140
Varèse, Edgar, 231, 235, 247, 337
Verdi, Giuseppe, 314, 324, 334, 339
Vermeer van Delft, Jan, 138-40, 145, 146
Vitruvian figure, 88
Vitruvius (Marcus Vitruvius Pollio), 86-9, 99
Vivaldi, Antonio, 115-16, 117, 162, 169, 171, 174, 192, 321
vocal cords. 45-6
Voigt, Augustus, 287-8, 299
Voltaire, 169
Vos, Marten de, 137-9

Wagner, Richard, 230, 235, 281, 290, 296, 299, 300, 311
Warner Brothers, 229, 235, 285, 289, 295, 297, 300, 301
wavelength, 34
Waxman, Franz, 298
Weber, Max, 173
Webern, Anton, 315
Wedgwood, Josiah, 172-3
Weill, Kurt, 325
Welles, Orson, 330, 336
Wells, H.G., 310, 327, 330, 336
Whistler, James McNeill, 281
white noise, 86, 225
Whiteman, Paul, 246
Willaert, Adriaan, 227
Williams, John, 219, 331
Wolff, Christian, 179
Wyatt, Sir Thomas, 147-9, 154

Yamaha, 173

Zeno of Elea, 271

About the author

New Zealand-born composer and writer on music Robin Maconie studied piano with Christina Geel and read English literature and criticism and contemporary music at Victoria University of Wellington under Roger Savage and Frederick Page. After graduating he studied analysis at the Paris Conservatoire under famed composer Olivier Messiaen 1963-64, and the following year traveled to Cologne, Germany to study composition and electronic music under Karlheinz Stockhausen, Herbert Eimert, Bernd-Alois Zimmermann, Henri Pousseur, and others. He has held teaching appointments at the universities of Auckland, Sussex, Surrey, Oxford, and The City University, London, and is currently Professor of Media and Performing Arts at the Savannah College of Art and Design.

As a music correspondent for *The Daily Telegraph, The Times Educational Supplement* and *The Times Literary Supplement*, he gained a reputation for clarity and directness in defense of new and unfamiliar music. He was an editorial assistant to John Mansfield Thomson in the formative years of the journal *Early Music* and freelanced as an arranger-producer of pop music in London during the seventies. His landmark monograph *The Works of Karlheinz Stockhausen*, published in 1976 and extensively revised in 1990, remains the best independent assessment of one of the twentieth century's most controversial composers. *Stockhausen on Music*, his 1989 edition of lectures and interviews with the composer, was reissued in 2000 to critical acclaim.

Robin Maconie has been married twice and has one daughter. He lives in Wilmington Island, Savannah. His hobby, when he is not writing about music, is composing stories in verse for children, the first of which, *Alice and Her Fabulous Teeth*, was published in 2000.